Lecture Notes in Physics

Edited by J. Ehlers, München, K. Hepp, Zürich,
H. A. Weidenmüller, Heidelberg, and J. Zittartz, Köln
Managing Editor: W. Beiglböck, Heidelberg

57

Physics of Highly Excited States in Solids

Proceedings of the 1975 Oji Seminar at
Tomakomai, Japan, September 9–13, 1975.
Edited by M. Ueta and Y. Nishina

Springer-Verlag
Berlin Heidelberg GmbH 1976

Editors

Prof. Masayasu Ueta
Physics Department
Tohoku University
Aobayama
Sendai 980/Japan

Prof. Yuichiro Nishina
Research Institute for
Iron, Steel & Other Metals
Tohoku University
Katahira 2–Chome
Sendai 980/Japan

ISBN 978-3-540-07991-0   ISBN 978-3-540-37975-1 (eBook)
DOI 10.1007/978-3-540-37975-1

FOREWORD

The Oji Seminar on "Physics of Highly Excited States in Solids" was held from Sept. 9 to 13, 1975, in Tomakomai, Hokkaido, Japan. The Seminar was held under the auspices of the Japan Society for the Promotion of Science by the fund donated by the Fujihara Foundation of Science upon contributions from the Paper Companies of Oji, Jujo, Honshu and others in commemoration of the centennial of the production of western style papers by the old Oji Group in Japan.

It was attended by 65 participants from 9 countries of America, Asia, Australia and Europe. The present volume contains the invited lectures and original/review papers contributed by 36 research individuals and their co-workers. The contents of the volume are grouped under the following headings:

> Excitonic Molecules
> Excitonic Condensation
> Phase Transition
> Electron-Hole Drops
> Light Scattering
> Stimulated Photoluminescence
> Nonlinear Optics.

Many interesting events of the Seminar took place in the active discussions following each presentation and throughout the free hours after scheduled sessions as well, when the personal contacts could be blended with beverages of various origins. The informality of this seminar, however, made it difficult to keep the written forms of communications which brought up many important points of the above-mentioned subjects to the attention of the audiences and even contributors themselves.

Each contribution was refereed by two members of the participants whose enduring assistance was deeply acknowledged by the Organizing Committee.

July 1976

Masayasu Ueta, Editor
Yuichiro Nishina, Co-editor

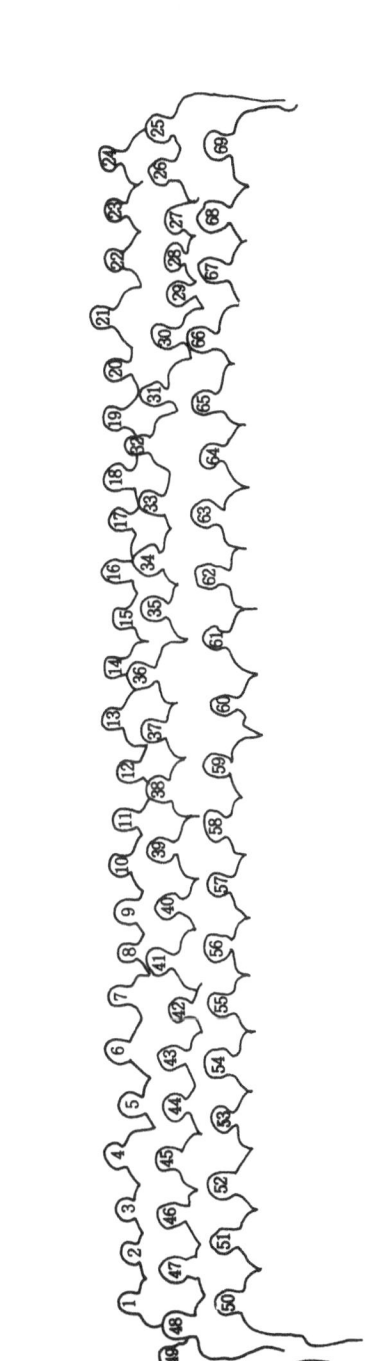

1. U. S. Tandon
2. M. Inoue
3. S. Sugano
4. A. Mysyrowicz
5. N. Kuroda
6. Y. Toyozawa
7. T. Nagashima
8. A. Nakamura
9. M. Kobayashi
10. S. Miyamoto
11. H. Saito
12. H. Fukuyama
13. Y. Oka
14. H. Haug
15. Mrs. Haug
16. Y. Segawa
17. T. Moriya
18. S. Suga

19. H. Kuroda
20. N. Nagasawa
21. U. Rößler
22. T. Ito
23. M. Umeno
24. G. O. Müller
25. T. Goto
26. M. Kagari
27. Mrs. Ito
28. Miss E. Watanabe
29. Miss T. Sanada
30. M. Morimoto
31. J. B. Grun
32. J. M. Hvam
33. H. Port
34. A. Schenzle
35. M. Glicksman
36. T. M. Rice

37. J. Treusch
38. W. Czaja
39. C. Horie
40. Y. Nagaoka
41. E. Hanamura
42. H. Kamimura
43. K. Kobayashi
44. P. Vashishta
45. K. Morigaki
46. O. Akimoto
47. Y. Kuramoto
48. K. Shirasuna
49. K. Betzler
50. T. Kushida
51. Y. Nishina
52. T. Watanabe
53. Mrs. Grun

54. S. Shionoya
55. A. Baldereschi
56. A. Morita
57. R. Conradt
58. C. D. Jeffries
59. M. Ueta
60. Y. Ichikawa
61. K. Okano
62. R. Kubo
63. C. Benoit à la Guillaume
64. I. Broser
65. W. A. Runciman
66. J. C. Hensel
67. H. Hasegawa
68. E. Otsuka
69. S. Narita

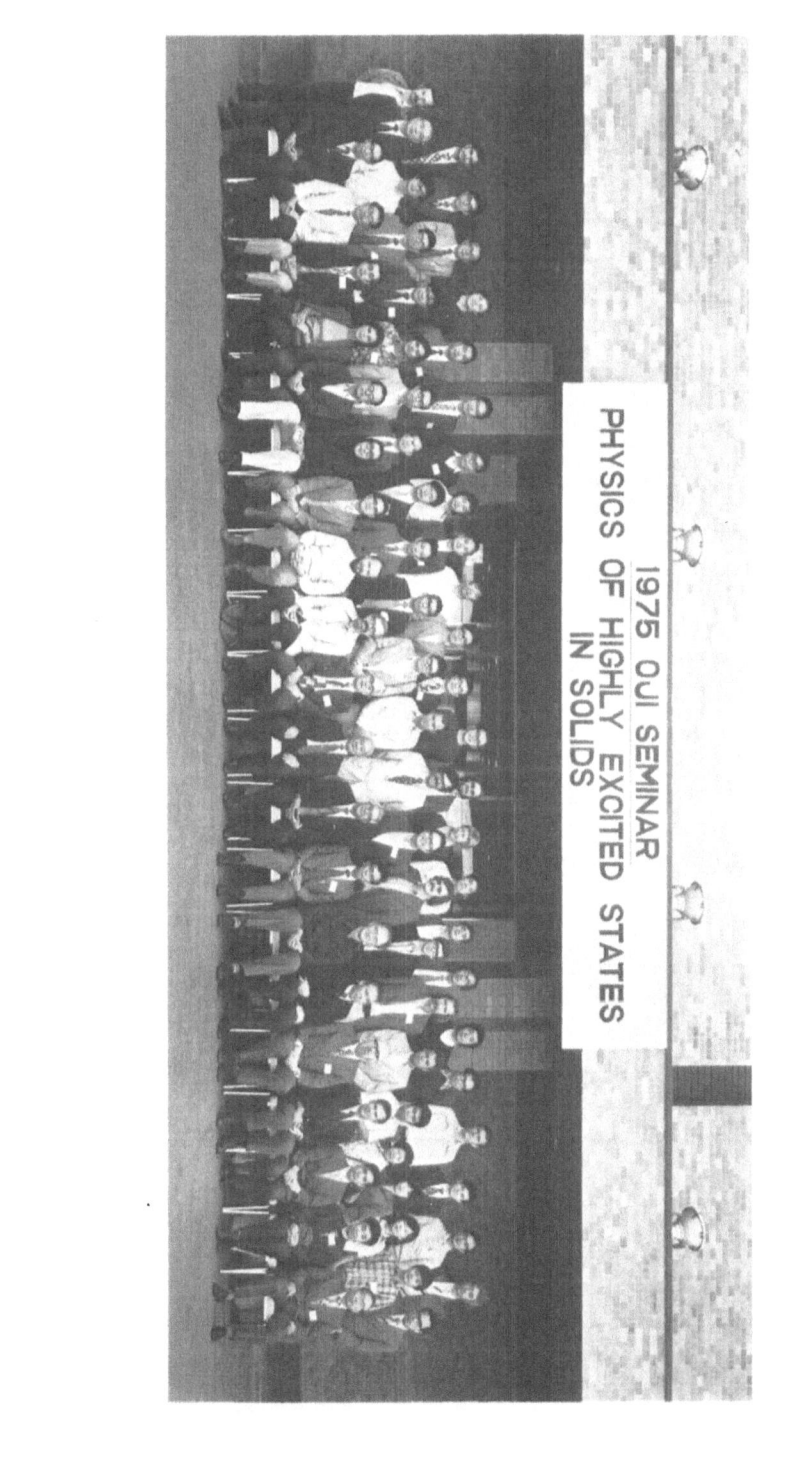

1975 OJI SEMINAR
PHYSICS OF HIGHLY EXCITED STATES IN SOLIDS

CONTENTS

EXCITONIC MOLECULES

EXCITONIC CONDENSATION

PHASE TRANSITION

ELECTRON-HOLE DROPS

LIGHT SCATTERING

STIMULATED PHOTOLUMINESCENCE

NONLINEAR OPTICS

TWO-PHOTON GENERATION OF EXCITONIC MOLECULES IN CuCl AND CuBr

Masayasu Ueta and Nobukata Nagasawa

Department of Physics,
Faculty of Science,
Tohoku University,
Sendai, Japan

ABSTRACT

Radiative recombination of excitonic molecules in CuCl and CuBr is reviewed. The emission consists of M_L and M_T bands in CuCl and of M_L, M_T and M_f bands in CuBr. These bands correspond to the recoil of an exciton into the longitudinal, transverse and triplet states, respectively. When excitonic molecules are generated indirectly by the excitation of crystals into the band-to-band region, the line shape of the bands is explained by considering that excitonic molecules are in Maxwell-Boltzmann distribution. Excitonic molecules are found to be generated directly by the giant two-photon absorption. In this case, extremely sharp emission lines appear at the high energy edges of each M band, which are attributed to the Bose condensation of excitonic molecules. The two-photon resonance Raman scattering is discussed in connection with the emission from the Bose condensed state.

I. INTRODUCTION

1.1 Excitonic Molecule

There are two main interaction products between excitons in highly excited semiconductors; an excitonic molecule and electron-hole metallic phase. The former is found in cuprous halides and probably in CdS and CdSe and the latter in Ge and Si. The Wannier exciton is analogous to the hydrogen atom so that excitonic molecules have been expected to be formed since 1958 by Lampert[1] and Moskalenko.[2] From the theoretical work[3] dealing the binding energy of the excitonic molecule, it was considered to be unstable in crystals having the electron and hole effective mass ratio, M_e/M_h, ranging between 0.2 and 0.4. Therefore, experimental work to find the excitonic molecule was directed to crystals such as CuCl having $M_e/M_h < 0.2$ and Ge or Si in which $M_e/M_h > 0.7$.

In 1966, Haynes[4] found a superlinear emission band at 1.08 eV in Si and considered it as the radiative annihilation of an excitonic molecule with leaving a free electron and hole pair behind.   In Ge, a similar emission band was later found at 0.708 ev.[5] These emission bands are now attributed to the electron-hole drops, other than the excitonic molecules.

In CuCl crystals, a new emission band was found by ruby laser excitation and it was assigned as due to the excitonic molecule by Grun *et al.*[6] and by Goto *et al.*[7], because the emission intensity increased in proportion to the square of the excitation density and the energy separation of the emission band from the free exciton band, ~40 meV, was reasonable for the binding energy of the excitonic molecule expected from the theoretical work.   Here an excitonic molecule was assumed to be annihilated radiatively with an exciton left behind.

In highly excited states, a number of elementary interactions between exciton-exciton and exciton-free charge carriers lead to super-linear emissions, the photon energies of which depend on the interaction mechanisms.   Therefore, the observation of the square dependence of emission intensity upon excitation power and the energy position of the emission band are not enough to decide uniquely the elementary processes responsible for the emissions.

The first reliable proof of the existence of excitonic molecules was given by Souma *et al.*[8] in zone refined CuCl through the line shape analysis of a new superlinear emission band, called M band, which appeared when the crystal was illuminated by a giant ruby laser at 4.2 K.   A similar emission band attributed to excitonic molecules has been found also in CuBr.[9]   After theoretical work concerning the binding energy of an excitonic molecule by Akimoto and Hanamura[10] which showed that the excitonic molecule could be formed in crystals, having any values of M_e/M_h, new emission bands similar to the asymmetric M band in CuCl and CuBr, were found by Shionoya's group successively in

CdS,[11] CdSe[12] and ZnO,[13] and they were attributed to the radiative annihilation of excitonic molecules. On the other hand, a somewhat different assignment has been made by another group that they are attributed to the stimulation of acoustic side bands of bound excitons.[14] However, the experiment on the stress effect in CdS, carried out by Segawa and Namba,[15] seems to give a confirmation of the molecule. The complex band structure, anisotropy of electron and hole masses, some difficulty of purification and small binding energies make the exciton molecule emission complex in II-VI compounds. Different exciton complexes found in different groups might have their emission bands in nearly the same energy positions.

1.2   Bose Condensation of Excitons and Excitonic Molecules

The boson-like nature of excitons and excitonic molecules has been the basis for the discussions[16] on how their statistics manifests themselves in the phase change at low temperature. The first experimental report on a narrow emission line showing the Bose condensation of excitons has been done by Akopyan et $al.$[17] in CdSe excited by the second harmonics of a neodymium laser at 4.2K. The conclusion of the Bose condensation was based on the consideration that excitons in CdSe are repulsive to each other and satisfy the necessary conditions for the Bose condensation. However as described above, excitons are attractive to form excitonic molecules. Thus, the Bose condensation of single excitons in CdSe becomes to be unexpected. In fact, the narrow line reported by Akopyan et $al.$ has not been observed by other researchers. Czaja and Schwerdtfeger[18] are the second reporter of the Bose condensation of indirect excitons in AgBr at 1.6 K. They found a narrow emission line near K=0 in the TA phonon-assisted exciton band. Together with the very low threshold temperature for the appearance of the emission line, the Bose condensation has been concluded.[19]

On the Bose condensation of excitonic molecules, Kuroda et $al.$[20]

reported an emission line on the high energy edge of the emission band of the excitonic molecule in CdSe, and attributed it to the Bose condensation of the excitonic molecule at K=0. However, Johnston and Shaklee[21] have claimed on the conclusion of Kuroda *et al.*, and assigned the narrow emission to a bound exciton complex involving a neutral donor. Thus, the Bose condensation of excitonic molecules has not been clarified yet. In the experiments above mentioned, crystals were excited into their band-to-band region and excitonic molecules are formed secondarily through the interactions between excitons generated by the recombination of hot electrons and holes. Therefore, the temperature of the excitonic molecules will not be low enough to condense as confirmed by Souma *et al.*[8]

Hanamura[22] has recently shown theoretically that excitonic molecules can be generated directly by the giant two-photon excitation with using photons, of which energy is given by

$$h\nu = E_{ex} - \frac{1}{2} E_m^b \tag{1}$$

where E_{ex} stands for the exciton energy and E_m^b the binding energy of an excitonic molecule. Furthermore, the Bose condensed molecules are expected to be created coherently with using a laser excitation. Hanamura has further shown that the absorption coefficient for the two-photon generation of excitonic molecule given by (1) depends on the photon density, and it amounts to the order of ~10^5/cm with using the photon density, of 10^{15} photons/cm^2. Gale and Mycyrowicz[23] have confirmed the two-photon generation of excitonic molecules in CuCl by observing that the absorption coefficient for the two-photon absorption increases rapidly, when the incident photon energy approaches that given by (1) and the extrapolated absorption coefficient at the peak will be the same as that for the one-photon absorption in the exciton band, ~10^5/cm, with photon density of 10^{17}/cm^2.

We could confirm also the efficient generation of the excitonic

molecules by the two-photon excitation in CuCl as well as in CuBr.
Moreover, the emission band has been found to show extremely sharp
lines at the high energy edges of the M_L and M_T bands, which is considered to show the Bose condensation of the excitonic molecule at
$K \sim 0$.

In this paper, we present experimental results in cuprous halides
carried out in our laboratory on the emission bands originating from
excitonic molecules in their thermal equilibrium and Bose condensed
states.

II. EXPERIMENTAL RESULTS AND DISCUSSIONS

2.1 Emission of Excitonic Molecules of Maxwell Distribution in CuCl
and CuBr Crystals near 4.2 K

Cuprous halides have the zincblende crystal structure, and an
exciton consists of a Γ_6 electron and a Γ_7 or Γ_8 hole. The exciton
absorption bands[24,25] in CuCl and CuBr crystals, called Z_3 and $Z_{1,2}$,
are well separated from the continuous band due to the band-to-band
transition, and thus the binding energy of the exciton is rather large,
190meV for CuCl and ~120meV for CuBr.

The M band due to the excitonic molecule was found[8] to have an
asymmetric line shape as mentioned before, and this was explained to
reflect the Maxwell-Boltzmann distribution of the kinetic energy of
excitonic molecule. The line shape was expressed by

$$I(E) \propto E^{1/2} \exp(-E/kT), \tag{2}$$

where E is measured towards the low photon energy side. The emission
is concluded to arise from the transitions between energy levels of
excitonic molecule and single exciton as shown in Fig.1. Namely, an
excitonic molecule is concluded to be annihilated radiatively with
leaving an exciton behind. From the energy separation of the peak
from the high energy edge, which is equal to 1/2kT, the temperature of

the excitonic molecule was shown to
be 26 K which was higher than the
lattice temperature in Souma's experimental condition; the crystal
was illuminated by a Giant Ruby
laser with ~50 MW/cm^2. By assuming
the above conclusion of the M-
emission mechanism correct, the
binding energy of the molecule is
determined from the energy separation between the free exciton band
and the high energy edge of the M
band to be 34 meV in CuCl[8] and
26 meV in CuBr.[9]

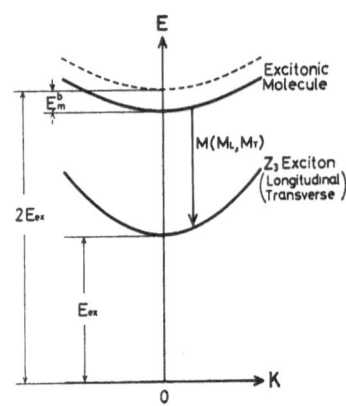

Fig.1. Schematic representation
of energy levels of ex-
citon and excitonic
molecule.

The remaining exciton can be in either longitudinal or transverse
states. With reflecting this fact, the M band was found by Koda's
group[26] to consist of two bands, called M_L and M_T bands. The energy
separation between them was in fact equal to that of the L-T splitting
of the exciton, being 5.4 meV.[27] The longitudinal state is singlet
and the transverse state is doubly degenerate, so that the M_T intensity
must be twice as much as the M_L intensity. However, as shown in Fig.
2, the intensity ratio is a function of excitation density, and with
increasing excitation density, the M_L band becomes much stronger than
the M_T band. Grun *et al.*[28] have shown the intensity ratio of 2 is
observed in a wide range of excitation density. For the dependence of
the M_L and M_T band-intensities upon excitation density, an opinion has
been proposed that the two M bands come from the different rotational
levels of the molecule.[29]

In CuBr, the M band was previously found to consist of two bands,
M_1 and M_2, and they were previously assigned as the recoil of an exciton
into the transverse and triplet states,[9] respectively. However in a

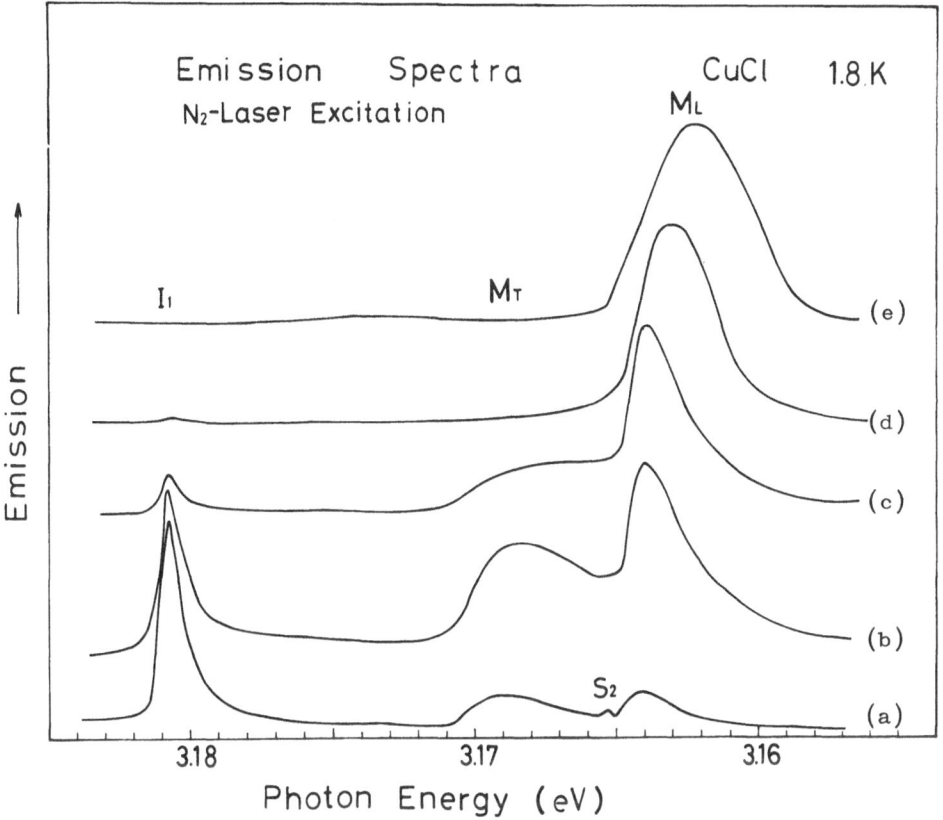

Fig.2. The variation of emission spectrum of excitonic molecule with the power of
exciting N_2-laser in CuCl. Excitation intensity increases from (a) toward
(e).

recent work, they have been clearly assigned to the M_L and M_T as in

CuCl, because the energy separation between two bands coincides also

with the L-T splitting of the exciton, 11.2 meV, which is determined

from the Kramers-Kronig analysis of a reflection spectrum.   Therefore,

the two bands in CuCl are considered to be due to the recoil of an ex-

citon into the L and T states, rather than to be attributed to the

rotational structure.

2.2 Line Shape of the M Band

Figure 3 shows a typical M emission spectrum of CuCl[30] excited

into the band-to-band region with using a N_2 laser.   The spectrum is

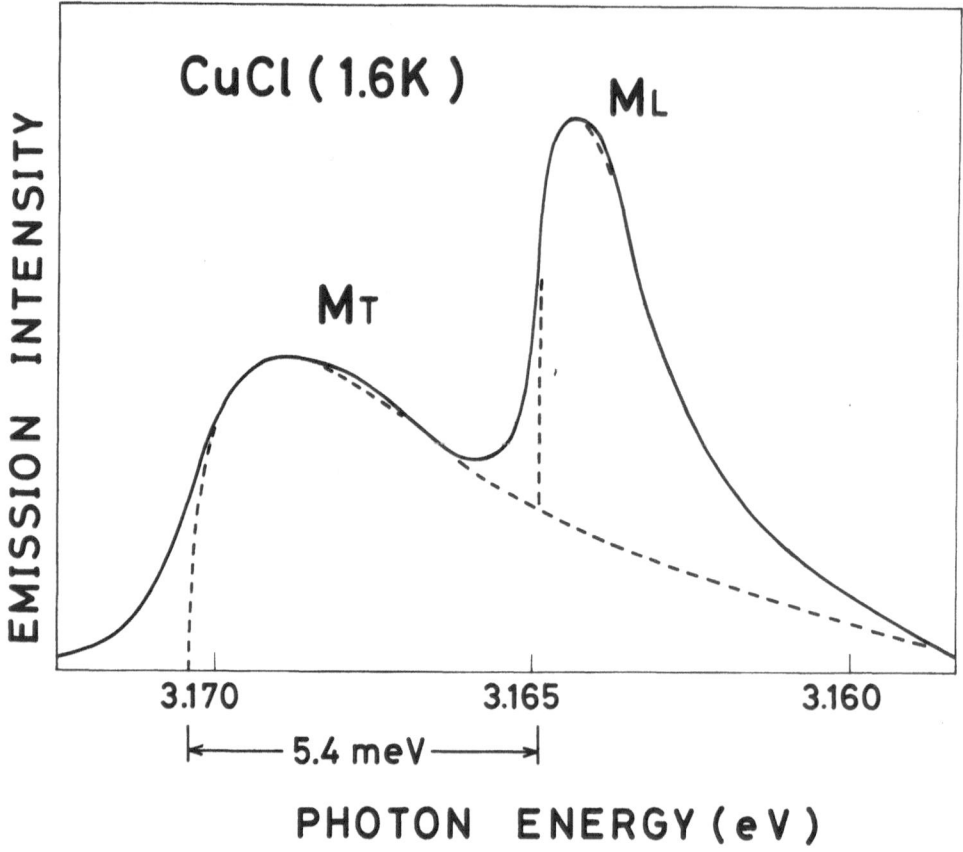

Fig.3. Line shape of the M_L and M_T emission bands of excitonic molecule in CuCl in the case of the excitation into the band-to-
band region with a N_2 laser.

the same as (b) in Fig.2 showing two bands, M_L and M_T, in the case of

relatively weak laser excitation. The line shapes are found to be

fitted to the formula (2). The high energy edges of the both bands,

corresponding to E=0, are determined to be,

$$3.1647 \text{ eV } (M_L) \text{ and } 3.1700 \text{ eV } (M_T) \text{ in CuCl.} \qquad (3)$$

The energy difference of these edges, 5.3 meV, is in agreement

with the L-T splitting of the exciton, 5.4 meV.[27] In CuBr crystal,

the M band was found to have another band, M_f, in addition to the M_L

and M_T bands and it was assigned as due to the recoil of an exciton

into the triplet state.   By a similar procedure the high energy edges
were determined to be,

$$2.930 \text{ eV } (M_L) \text{ and } 2.941 \text{ eV } (M_T) \text{ in CuBr.} \qquad (4)$$

The energy difference of these edges, 11 meV, is also in agreement
with the L-T splitting of the exciton.   For the M_f band, the determination was not clear because of the superposition of an unknown
emission band.   The high energy edges of the bands correspond to the
transitions from the state K=0 of the excitonic molecule as described
before.

With increasing the N_2 laser power, the M_L band becomes much
stronger than the M_T in both cases of CuCl and CuBr, and has a symmetric
line shape with broader band width.   The line shape was expected to be
Lorentzian which is due to the collisions between excitonic molecules.[31]
If the gain saturation of the emission, which was found by Shaklee et
al.,[32] was taken into account, the M_L band shape would be more complicated than the simple Lorentzian.

2.3 Excitation Spectra of the M_L and M_T Bands

With using a frequency tunable dye laser, excitation spectra of
the M_L band in CuCl and CuBr were measured[33,34]   near the high energy
edges of the bands as shown by upper curves in Fig.4.   Sharp excitation
peaks are observed at 3.187 eV and at 2.954 eV in CuCl and CuBr, respectively.   The photon energies of these peaks are almost exactly the
same as those expected from the formula (1) by adopting known values:

$$E_x = 3.204 \text{ eV and } E_m^b = 34 \text{ meV for CuCl and}$$

$$E_x = 2.967 \text{ eV and } E_m^b = 26 \text{ meV for CuBr.}$$

Thus, the excitonic molecule was found to be generated effectively by
the two-photon absorption.   The excitation spectrum for the M_T band in
CuCl was found to show a sharp peak at exactly the same photon energy

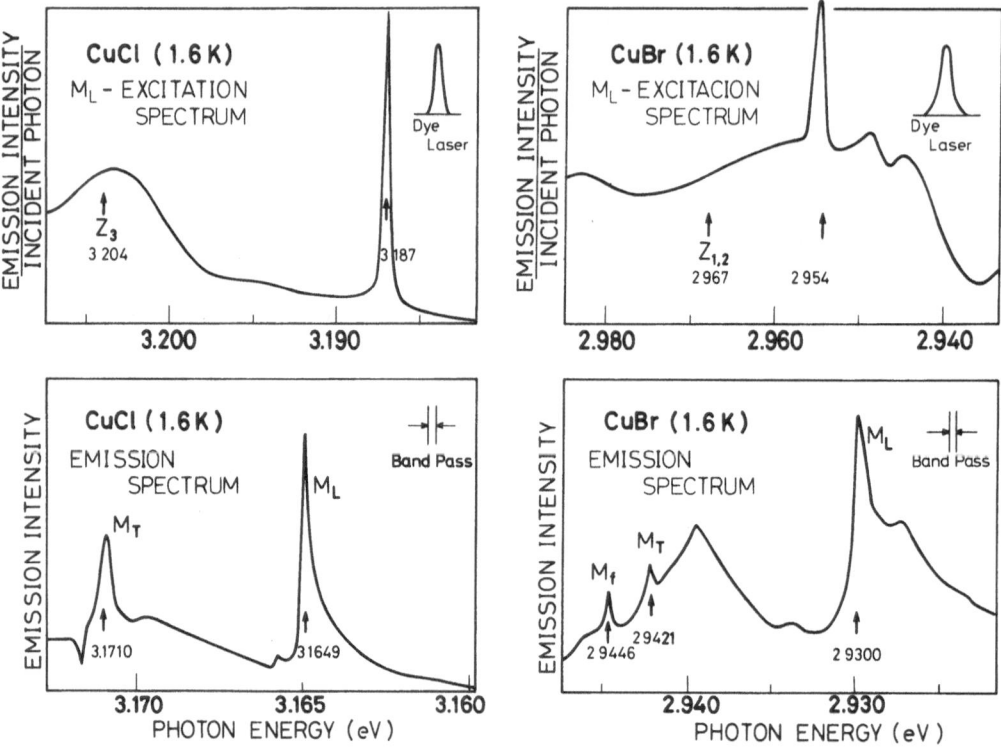

Fig.4. Upper curves: Excitation spectra of the M_L emission due to
excitonic molecule in CuCl and CuBr, indicating two-photon
excitation peaks.
Lower curves: Emission spectra of CuCl and CuBr excited into
their two-photon excitation peaks in the upper curves, showing
a sharp line at the high energy edge of the M_L band.

as in the M_L band, namely at 3.187 eV as shown by Fig.5 in ref. (30).

Gale and Mycyrowicz[35] measured firstly the excitation spectra for

M_L and M_T bands in CuCl, and found the same peak at 3.187 eV for the M_L

bands and a different peak at 3.199 eV for the M_T band. In our case,

the initial states responsible for both M_L and M_T bands can be reached

with using the same photon energy, 3.187 eV, but in the case of Gale

and Mycyrowicz the situation is somewhat different. In their case, the

initial state for the M_T band should be different from that for the M_L

band. Other emission bands due to bound excitons appeared in their M

band region, so that the identification of the M_T band seemed to be

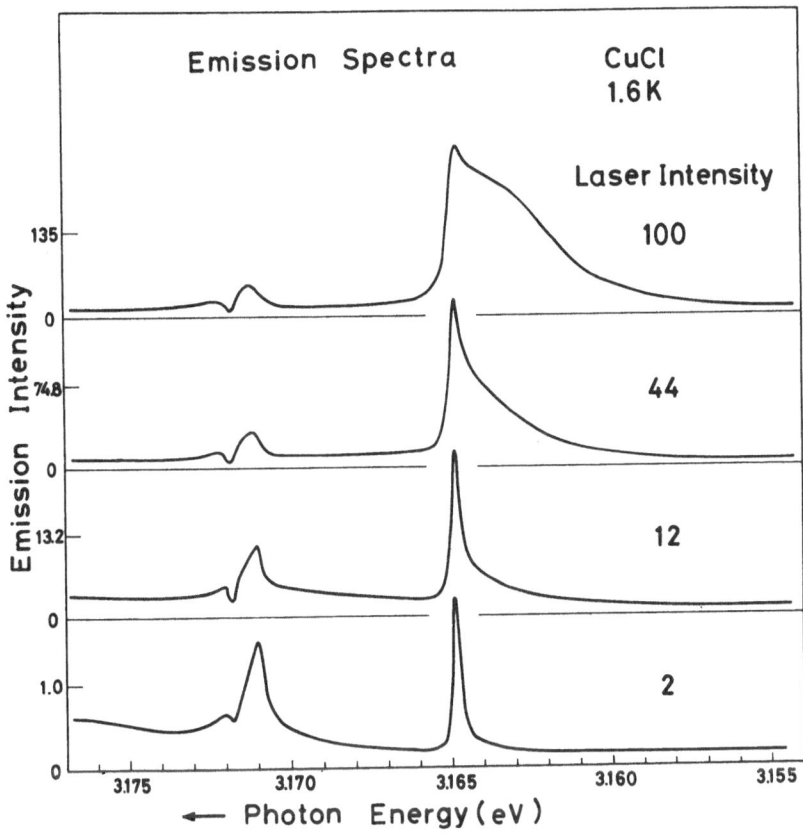

Fig.5. Variation of emission intensities of the sharp line
and its side band of the Bose condensed excitonic
molecules, with respect to the excitation power.

mistaken.

2.4 Emission Spectra of Excitonic Molecules Directly Generated and

the Bose Condensation of Excitonic Molecules[30,34]

When CuCl and CuBr crystals are excited into the sharp peaks in
the excitation spectra, very sharp emission bands are observed in each
M_L, M_T and M_f band regions as shown by lower curves in Fig.4. The
photon energies of these emission bands are:

3.1649 eV(M_L) and 3.1710 eV(M_T)   for CuCl,

(5)

2.9300 eV(M_L), 2.9421 eV(M_T) and 2.9446 eV(M_f) for CuBr.

Comparing (5) with (3) and (4), one can see that the sharp emissions of CuCl and CuBr in the M_L band region coincide in energy exactly with those of the high energy edge of the M_L band of Maxwell distribution, but the one at M_T band region is located at an about 1meV higher energy position of the edge of the M_T band. The same applies to the sharp emission line in the M_f band because its energy separation from that in the M_T band, 2.5 meV, is almost equal to the energy difference between the transverse and triplet exciton bands, 2.7 meV, determined from the reflection spectrum. A dip observed at the high energy side of the sharp emission band of M_T is due to a bound exciton.[36]

With increasing excitation density, a broad side band comes out at the low energy side of the sharp emission line in CuCl as shown in Fig. 5. With a further increase of excitation intensity, the relative intensity of the sharp emission against the side band decreases and the peak of the side band shifts to the low energy side. With the highest excitation density of our laser (~ 50 kW/cm^2), the peak separation from the sharp emission line amounts to ~ 3 meV as seen in curve a) of Fig. 6. The shape of the side band can not be expressed by the formula (2) as shown by a dotted curve. For comparison, the emission spectrum excited into the Z_3 exciton band was shown by curve b). In this case, the line shape is well expressed by (2). Thus, the emission spectrum of the excitonic molecule, directly generated by the giant two-photon excitation, consists of a sharp line and a side band which originates from the non-thermalized excitonic molecules. From the very sharpness of the line together with its energy corresponding to the transition from the K=0 state of the excitonic molecule, the Bose condensation of the molecules at K~0 has been concluded. By the simple theory,[37] the relative concentration of Bose condensed particles N_o with respect to the total one, N, is expressed by,

$$\frac{N_o}{N} = 1 - \frac{8}{3\sqrt{\pi}} a^{3/2} \left(\frac{N}{V}\right)^{1/2} ,$$

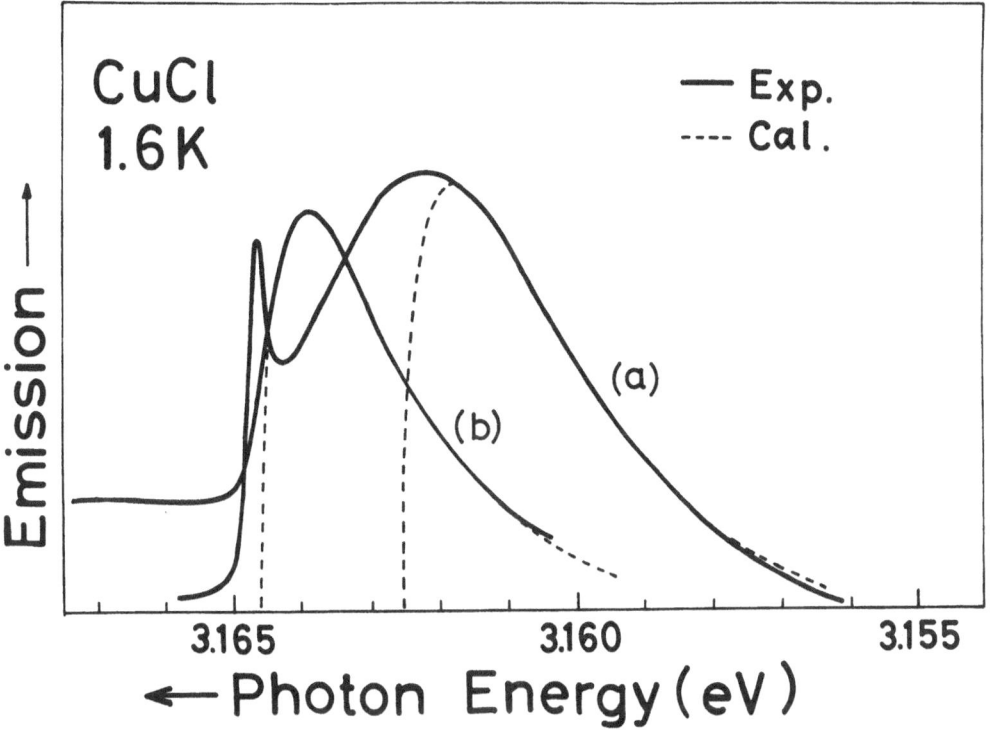

Fig.6. Comparison of M_L emission bands in different generation processes of excitonic molecules in CuCl; a) the direct formation through the two-photon excitation, and b) indirect generation through the band-to-band excitation.

where V stands for the crystal volume and a is the scattering amplitude expressing the interaction between particles. If we consider that the intensities of the sharp emission line and its side band are proportional to N_o and $N-N_o$, respectively, the decrease of sharp emission line with respect to the side band with increasing laser intensity is explained qualitatively by the theory.

Qualitative temperature dependence of the emission spectra is shown in Fig. 7. With increasing temperature, the sharp peak decreases its intensity and disappears at ~25 K. However, if one makes the excitation density increase at this stage, the sharp peak becomes to appear again. This fact shows that there is a certain critical temperature for the appearance of the sharp emission line and it depends

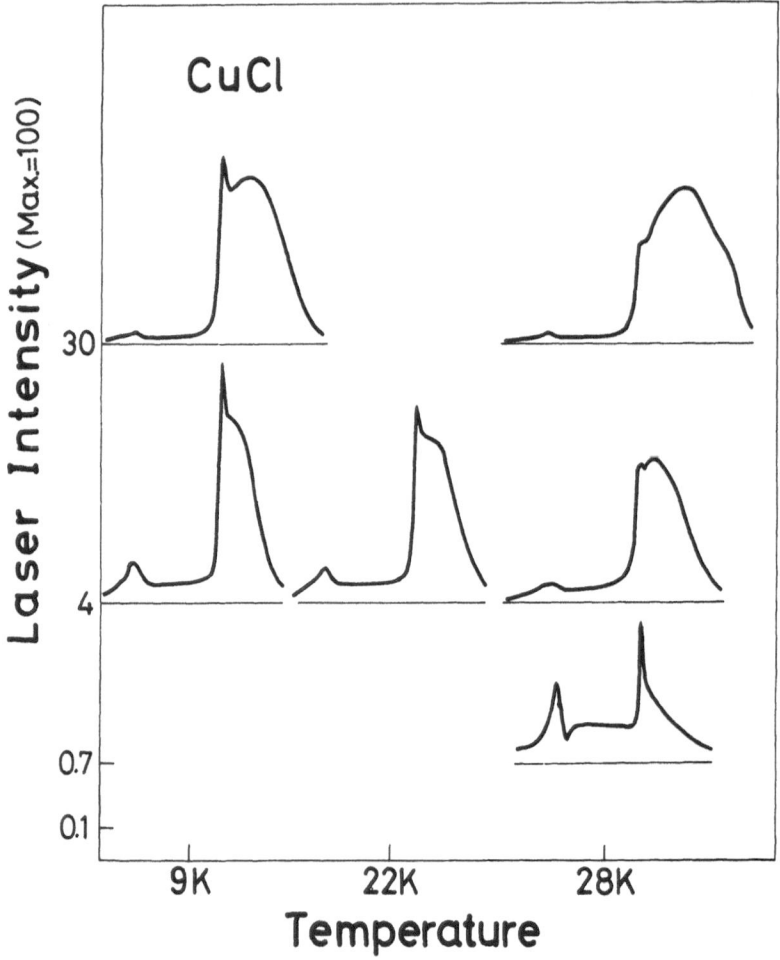

Fig.7. Variation of the M_L emission with respect to the temperature and excitation power changes.

on excitation density. These facts are also explained by the Bose condensation which shows that the critical temperature for the condensation increases in proportion to the particle concentration with

$$T_c \propto \left(\frac{N}{V}\right)^{2/3} .$$

Further, the concentration N_o is shown to decrease with the increase of T in a range $0 < T < T_c$ by,

$$\frac{N_o}{N} = \left\{ 1 - \left(\frac{T}{T_c}\right)^{3/2} \right\} ,$$

if the excitonic molecule is treated as an ideal boson gas. Thus, the excitation density and temperature dependence of the intensities of the sharp line and side band are qualitatively explained by the Bose condensation. Therefore, it is considered that the majority of the excitonic molecules, generated by the two-photon absorption, condense at the state K~0 and decay radiatively from there. The side band is considered by Hanamura[16] to arise from the collective motion of condensed molecules.

By considering the polariton nature of the transverse exciton, it is found that the photon energy of the emission from the K=0 state of the excitonic molecule to the transverse exciton ought to deviate much from the high energy edge of the M_T band. However, from the $K = 2K_o$ state, the observed deviation of 1 meV is reasonable, where K_o is the wave number of a photon given by (1). Thus the excitonic molecules are considered to condense at $K = 2K_o$.

To distinguish the emission line from the Bose-condensed state from the one due to bound exciton complexes, the study of the Zeeman splitting is most important. At the present stage of experiment, the magnetic field effect was studied for only CuCl.[30] No splitting and no shift of the 3.1649 eV emission line in the M_L band were found in a magnetic field of 60 kG. The splitting of the 3.1700 eV band in the M_T band region was not clear because of a little broader width of it and of the superposition of a bound exciton line, responsible for the dip above mentioned, which split into four components. No splitting of the 3.1649 eV emission is in agreement with the assignment that the emission is due to the transition from the Bose condensed state of the excitonic molecule to the longitudinal exciton state and is not due to bound exciton complexes.

2.5 Two-Photon Absorption Spectrum for the Generation of Excitonic
Molecules

In connection with the relaxation of the excitonic molecules to
the Bose-condensed state, and also with the resonant Raman scattering,
which is described later, we have to know the detailed line shape of
the two-pnoton absorption spectrum for the generation of excitonic
molecules. As mentioned before, the absorption cross-section depends
on the power of the exciting laser light. We could obtain the profile
of the absorption spectrum as shown in Fig. 8, with using a dye laser
of a moderate intensity having the photon density of $\sim 10^{15}/cm^2$. The
used dye laser has the 0.25 meV half-width which is narrowest in our
case. The absorption peak is located at 3.1870 eV as a matter of

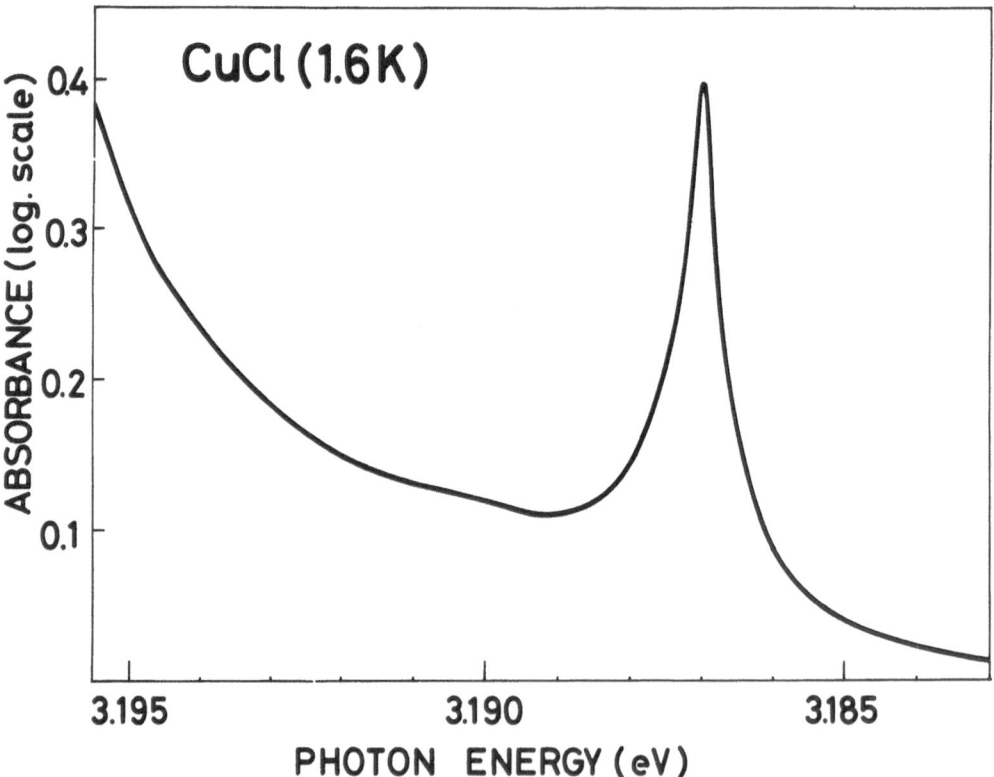

Fig.8. Two-photon absorption spectrum for the generation of excitonic molecule in
CuCl, measured with a dye laser of 0.25 meV half width.

course and its absorption coefficient amounts to ~10^3/cm.

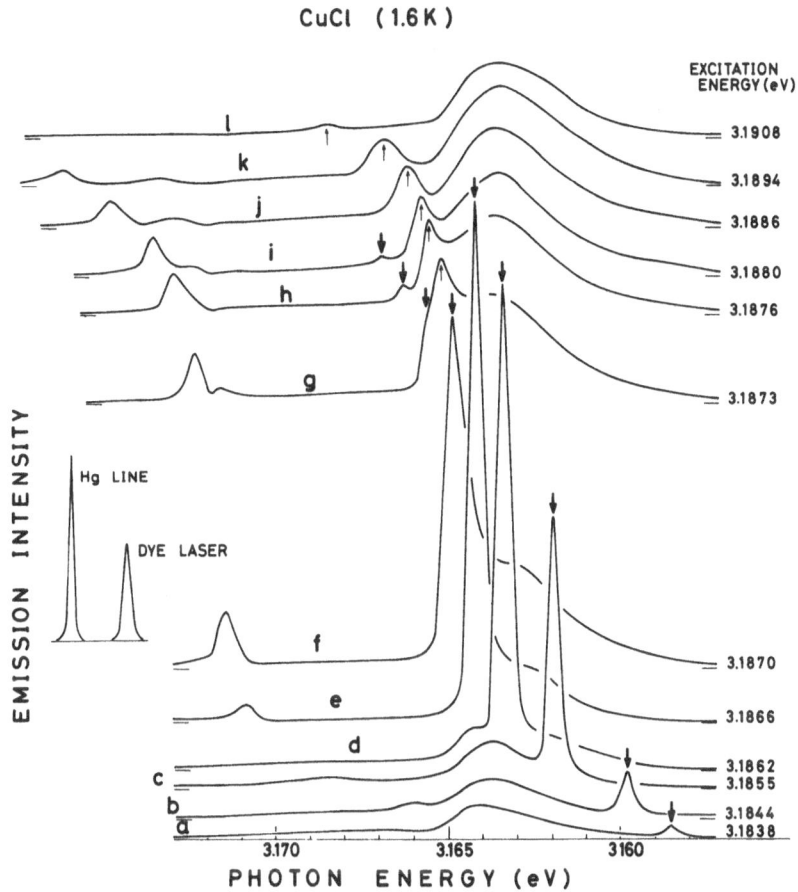

Fig.9. Two-photon resonance Raman spectrum of CuCl with using a dye-laser of 0.25 meV half width.

2.6 Two-Photon Resonance Raman Scattering

In regard to a question that the present sharp emission line, which we considered as being due to the Bose-condensed excitonic mole-cules, might be due to the two-photon resonance Raman scattering, emission spectra were studied in CuCl with shifting the energy of the dye laser. In the present case of the Raman process, the excitonic molecule and exciton are the intermediate and final states, respec-

tively.   Figure 9 shows the Raman spectra measured with using a dye
laser of 0.25 meV half-width.   The photon energy of the excitation
laser was shifted in a range from 3.184 eV to 3.191 eV which covers the
absorption spectrum in Fig. 8.   When the energy of the dye laser

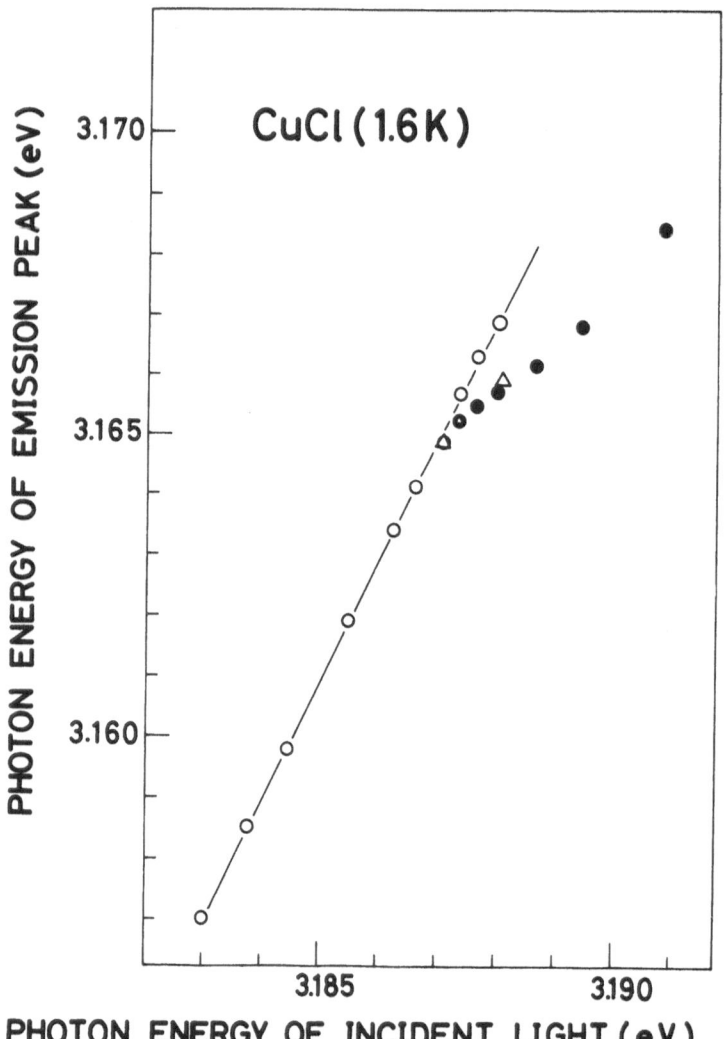

Fig.10. Shift of the Raman and emission peaks in Fig.9 versus
        the shift of the dye-laser energy.   Data marked by
        Δ was obtained with using two photons, $\hbar\omega_1 + \hbar\omega_2$ having
        a small energy difference.

approaches 3.1870 eV, sharp peaks, marked by thick arrows, are much
enhanced, and at the off-resonance in the high energy side from
3.1870 eV, the peak intensity decreases very rapidly. The energy of
the sharp peaks was plotted by open circles with respect to that of
the exciting dye laser in Fig. 10. With the shift of the exciting
light, $\Delta\nu$, the peak is found to shift by $2\Delta\nu$, indicating that these
peaks are the two-photon Raman bands.

The highest intensity of the Raman band occurs at 3.1642 ± 0.0001
eV with the excitation at 3.1866 ± 0.0001 eV. Namely, it does not occur
with the excitation at the two-photon absorption peak, but does so at
a little lower energy position by 0.4 meV.

As seen in curves from g) to k), another bands marked by thin
arrows are observed when the excitation energy is larger than 3.1870 eV.
The shift of these bands deviates from the straight line in Fig. 10 as
shown by closed circles. A broad band, of which the peak is located
at ~3.1635 eV, is found in all curves, and becomes to predominate with
the increase of the excitation energy. The line shape of this band
seems to be expressed by the formula (2), so that it is considered to
originate from the one-photon excitation in the exciton absorption
tail which overlaps with the two-photon absorption spectrum as seen in
Fig. 8. The Raman bands have ~0.45 meV half-width, which is a little
broader than that of the dye laser. This fact is also valid when the
dye laser band width is broadened to 0.45 meV as shown previously.[34]

Small bands which appeared near the 3.172 eV (M_T band) region
were previously attributed to the emission from the Bose condensed
excitonic molecules with recoiling transverse excitons. These peaks
are also found to shift by $2\Delta\nu$ with respect to that, $\Delta\nu$, of the
excitation laser energy. Thus, the bands are attributed to the two-
photon resonance Raman scattering having the transverse exciton as
the final state.

When the band width of the dye laser is broadened to 2 meV in such a way that it covers the whole region of the excitation spectrum for the generation of excitonic molecules as shown in Fig. 11 a), the emission spectra show a sharp peak at 3.165 eV and a little broader one at 3.171 eV in common, and two-photon resonance Raman peaks which shift by $2\Delta\nu$ were not observed as seen in figure b) as reported in ref. (34). Curves $^{1-5)}$ are the spectra with the excitation of the laser at respectively numbered regions in a). It is emphasized that the band width of the 3.165 eV emission is narrower than 0.15 meV, that is nearly the same as the band pass of the detecting monochromator. Further, the peak positions of two bands at 3.165 eV and at 3.171 eV do not change with the shift of exciting laser light.

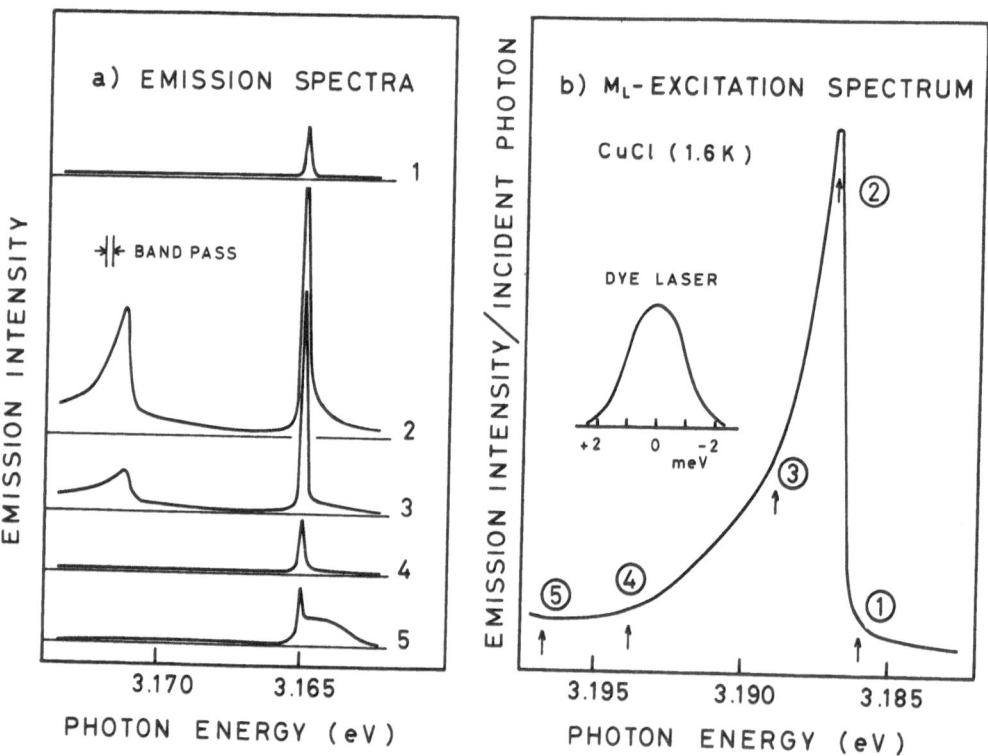

Fig.11. a): Two-photon excitation spectrum for the generation of excitonic molecule, measured with a dye-laser of 2 meV half width.
b): Emission spectra excited at the two-photon absorption region in a) with a dye laser of 2 meV half width.

A problem becomes very stimulating whether or not the distinction is possible between the resonance Raman scattering due to the one-step process and the luminescence due to the two-step process of the absorption and subsequent emission of photons. An approach to this problem has been proposed by Kubo *et al.*[38] By the theory, the scattered spectrum consists of Raman and luminescence parts and the luminescence part remains when the band width of the excitation dye laser is much wider than the modulation of excitonic molecule as discussed by Hanamura[39] in connection with our results shown in Fig. 11.

The extreme narrowness of the 3.165 eV emission band and no shift of this band as well as the one at 3.171 eV with respect to the shift of the excitation light energy, shown in Fig. 11, are not in contradiction to the consideration of the Bose condensation of excitonic molecules.

When the crystal is strongly excited by the two-photon process with using a wide band dye laser of 3.187 eV energy, the generated molecules give rise to the Bose condensation. On the other hand, the weaker excitation with narrow band laser brings about the Raman scattering, and the sharp emission from the condensed state is not observed. From these facts, the critical density of the excitonic molecules is shown to be necessary for the Bose condensation in the present case as well. If we could use a more intense dye laser, both the two-photon resonance Raman and the emission from the Bose condensed state might be expected to be observed simultaneously even in the case of making the laser band width narrow. The emission bands marked by thin arrows in Fig. 9, which are enhanced only when the crystal is excited above 3.187 eV, might be the luminescence from the unrelaxed states of excitonic molecules. It is probable that such an emission and also the one from the Bose condensed state are considered to be hidden in the Raman peak at 3.165 eV.

For making the Bose condensation of excitonic molecules and distinction between Raman and luminescence clearer, the studies with using an intense dye laser of tunable band width and time resolved spectroscopy of the sharp line and its side band are indispensable. The precise determination of the laser intensity on the sample is also very important.

REFERENCES

1) M. A. Lampert: Phys. Rev. Letters $\underline{1}$ (1958) 450.

2) S. A. Moskalenko: Optika Spectrosk. $\underline{5}$ (1958) 147.

3) R. R. Sharma: Phys. Rev. $\underline{170}$ (1968) 770.

   R. K. Wehner: Solid State Commun. $\underline{7}$ (1969) 2.

4) J. R. Haynes: Phys. Rev. Letters $\underline{4}$ (1960) 361.

5) C. Benoît à la Guillaume, F. Salvan and M. Voos: J. Luminescence $\underline{1,2}$ (1970) 315.

6) J. B. Grun, S. Nikitine, A. Bivas and R. Levy: J. Luminescence $\underline{1,2}$ (1970) 241.

7) T. Goto, H. Souma and M. Ueta: J. Luminescence $\underline{1,2}$ (1970) 231.

8) H. Souma, T. Goto, T. Ohta and M. Ueta: J. Phys. Soc. Japan $\underline{29}$ (1970) 697.

9) H. Souma, H. Koike, K. Suzuki and M. Ueta: J. Phys. Soc. Japan $\underline{31}$ (1971) 1285.

10) O. Akimoto and E. Hanamura: Solid State Commun. $\underline{10}$ (1972) 253.

   J. Phys. Soc. Japan $\underline{33}$ (1972) 1537.

11) S. Shionoya, H. Saito, E. Hanamura and O. Akimoto: Solid State Commun. $\underline{12}$ (1973) 223.

12) H. Saito and S. Shionoya: *Luminescence of Crystals, Molecules and Solutions,* ed. F. Williams (Plenum, N.Y. 1973) p.104.

13) S. Miyamoto and S. Shionoya: J. Luminescence $\underline{12/13}$ (1976) 563.

14) A. F. Dite, V. I. Revenko, V. B. Timofeev and P. E. Altukhov: JETP Lett. $\underline{18}$ (1973) 579.

15) Y. Segawa and S. Namba: Solid State Commun. $\underline{17}$ (1975) 489.

16) S. A. Moskalenko: Soviet Physics–Solid State $\underline{4}$ (1962) 199;

J. M. Blatt, K. W. Boer and W. Brandt: Phys. Rev. $\underline{126}$ (1962) 1961;

R. C. Casella: J. Phys. Chem. Solids $\underline{24}$ (1963) 19;

17) L. Kh. Akopyan, E. F. Gross and B. S. Razbirin: JETP Letters $\underline{12}$ (1970) 251.

18) W. Czaja and C. F. Schweidtfeger: Solid State Commun. $\underline{15}$ (1974) 87 and *Proc. of Oji Seminar* (this issue).

19) Some umbiguity was cast by Von der Osten that the emission line has connection with a bound exciton: Private communication from H. Haug.

20) H. Kuroda, S. Shionoya, H. Saito and E. Hanamura: Solid State Commun. $\underline{12}$ (1973) 533.

21) W. D. Johnston Jr. and K. L. Shaklee: Solid State Commun. $\underline{15}$ (1974) 73.

22) E. Hanamura: Solid State Commun. $\underline{12}$ (1973) 951.

23) G. M. Gale and A. Mycyrowicz: Phys. Rev. Letters $\underline{32}$ (1974) 727.

24) S. Nikitine: *Progress in Semiconductors* $\underline{6}$ (1962) 235.

25) M. Cardona: Phys. Rev. $\underline{129}$ (1963) 69.

26) S. Suga and T. Koda: Phys. Status solidi (b) $\underline{61}$ (1974) 291.

27) W. Staude: Phys. Status solidi (b) $\underline{43}$ (1971) 367.

28) J. B. Grun: J. Luminescence $\underline{12/13}$ (1976) 581.

29) S. Nikitine: *Excitons at High Density:* ed. H. Haken and S. Nikitine, Springer Tracts in Modern Physics (1973) p.18.

30) N. Nagasawa, N. Nakata, Y. Doi and M. Ueta: J. Phys. Soc. Japan $\underline{39}$ (1975) 987.

31) E. Hanamura: *Luminescence of Crystals, Molecules and Solutions,* ed. F. Williams (Plenum, N. Y. 1973) p.121.

32) K. L. Shaklee: *Excitons at High Density:* ed. H. Haken and S. Nikitine, Springer Tracts in Modern Physics (1973) p.221.

33) N. Nagasawa, N. Nakata, Y. Doi and M. Ueta: J. Phys. Soc. Japan $\underline{38}$ (1975) 593.

N. Nakata, N. Nagasawa, Y. Doi and M. Ueta: J. Phys. Soc. Japan
<u>38</u> (1975) 903.

34) N. Nagasawa, S. Koizumi, T. Mita and M. Ueta: J. Luminescence
<u>12/13</u> (1976) 587.

35) G. M. Gale and A. Mycyrowicz: *Proc. XIIth Intern. Conf. Semiconductors,*
Stuttgart, 1974, ed. M. H. Pilkuhn (Teubner, Stuttgart, 1974) p.133.

36) T. Anzai, T. Goto and M. Ueta: J. Phys. Soc. Japan <u>38</u> (1975) 774.

37) N. N. Bogolyubov: J. Phys. USSR <u>11</u> (1947) 23.

38) R. Kubo, T. Takagawara and E. Hanamura: *Proc. Oji Seminar* (this
issue).

39) E. Hanamura: *Proc. Oji Seminar* (this issue).

EXCITONIC MOLECULE AND ITS BOSE CONDENSATION

Eiichi Hanamura and Masahiro Inoue

The Institute for Solid State Physics
The University of Tokyo,
Roppongi, Minato-ku,
Tokyo

ABSTRACT

The three optical responses of excitonic molecule; giant two-photon absorption, luminescence and optical conversion from single exciton are theoretically discussed. Then the luminescence spectrum characteristic of Bose-condensed excitonic molecules is presented by taking account of the final state interaction between the excitonic molecule and the single exciton at the finite system temperature. Lastly is answered a question whether the third order optical process of the giant two-photon absorption due to the excitonic molecule and its subsequent emission leaving the single exciton behind is considered as two separable processes of absorption and emission or as an inseparable process like two-photon Raman scattering.

I. INTRODUCTION

The excitonic molecule[1] is the bound state of two single excitons and its internal motion is visualized in analogy to a hydrogen molecule, the positive holes playing the role of the protons. However, this excitonic molecule has two prominent characters in contrast with the hydrogen molecule. The first character of the excitonic molecule gas is its large quantum effect as an assembly of Bose-particles, which comes from much lighter translational mass of the excitonic molecule. The Bose-condensation of excitonic molecules may be expected from this character.[2] The second prominent character comes from the fact that the excitonic molecule gas constitutes the open system in close contact with radiation field. By making the best use of this character, firstly the Bose-condensed system of the excitonic molecules can be created directly through giant two-photon excitation.[3] Secondly, evidence for Bose-condensation of excitonic molecules can be found in the luminescence spectrum due to radiative decay of the excitonic molecules.[4] Thirdly, the Bose-condensed excitonic molecules oscillate with a common phase with respect to the ground state, and the whole system behaves as a macroscopic quadrupole-moment. Therefore it can interact coherently with the coherent radiation field through two-photon transitions.[5] This point is in a striking contrast with the quantum fluid near the ground state. As an example, self-induced transparency associated with the two-photon transition due to the excitonic molecule may take place.[5]

Since optical methods are used not only to generate the Bose-condensed system of excitonic molecules but also to experimentally confirm the Bose-condensation, the complete knowledge of electronic structure[6] and optical properties[7] of excitonic molecule is required for studying its Bose-condensation. This is discussed in Sec. II. The emission spectrum characteristic of Bose-condensed excitonic molecules[8] is discussed in Sec. III. As a reflection of the second character, there arises a question whether the third order optical process of two-photon absorption by the excitonic molecule and its subsequent emission leaving behind a single exciton, can be considered as two separable processes of absorption and emission or as an inseparable process of Raman scattering.[9] This is answered in Sec. IV. Section V is devoted to the conclusion.

II. OPTICAL PROPERTIES OF EXCITONIC MOLECULES[6,7]

We have three optical processes relevant to the excitonic molecule; giant two-photon absorption to make the excitonic molecule, luminescence due to radiative decay of the excitonic molecule into a single exciton and optical conversion of an exciton into the excitonic molecule by one photon absorption; the reverse process of the second. Selection rules and several related topics for these optical processes are discussed for CuCl, CuBr and CdS.

2-1. Excitonic Molecule in CuCl.

Electronic structure of CuCl crystal is characterized by Γ_7 valence band and Γ_6 conduction band. Two kinds of excitons with symmetry Γ_2 and Γ_5 are formed by superposing excitations of a valence electron into the conduction band. Γ_2 exciton is optically inactive due to pure triplet spin structure and Γ_5 exciton is optically active. State of the excitonic molecule is also described by superposition of excitations of two valence electrons into the conduction band. In semiconductors, the exchange interaction within the same band are important for the formation of the excitonic molecule. Therefore, two-electron state and two-hole state with zero angular momentum, respectively, are a good approximation to compose an excitonic molecule, and the interband electron-hole exchange interaction is taken into account as the perturbation. It is noted that the intra-band hole-hole exchange energy is estimated to be 30 meV and the inter-band electron-hole exchange energy is 3 meV. On the other hand, the lowest state of the excitonic molecule in CuCl is described as a linear combination of two optically active Γ_5 and two-optically inactive Γ_2 exciton states:

$$| \Gamma_1^{mol} \rangle\rangle = \frac{1}{\sqrt{V}} e^{i\mathbf{K}\cdot\mathbf{R}} {}_0 \Psi^{++}(r,r',R) | 0,0 \rangle\rangle , \qquad (2\text{-}1)$$

where

$$| 0,0 \rangle\rangle = \frac{1}{2} | J_t = 0; (J_{exc}=0)^2 \rangle + \frac{\sqrt{3}}{2} | J_t = 0; (J_{exc}= 1)^2 \rangle$$

and Ψ^{++} describes the internal motion of two holes R and two pairs of an electron and a hole r and r'. This electronic structure in semiconductors gives a hint to the idea of exciting the excitonic molecule directly by two-photon excitation through the optically active part and observing its existence in the luminescence due to its radiative decay into a composing exciton. The next lowest state of the excitonic molecule in CuCl is made in terms of the envelope function which is odd to the exchange of two holes. This sacrifices the hole-hole intra-band exchange energy. This state, however, is optically forbidden for the following three processes; the two-photon excitation, the luminescence due to the radiative decay into any exciton state and the optical conversion of a single exciton of any kind into this state of the excitonic molecule. The selection rules for these three optical processes are listed in Fig. 1. The second lowest state of the excitonic molecule in

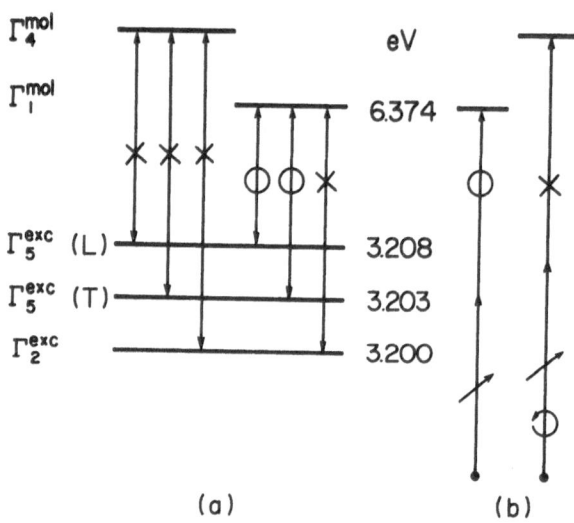

(a)   (b)

Fig.1. Energy diagram of exciton and excitonic
molecule in CuCl.
(a) shows the selection rule for the
luminescence of excitonic molecule and
the optical conversion of an exciton
into an excitonic molecule and (b) shows
that of the giant two-photon absorption
due to excitonic molecule. $\bigcirc$ and $\times$ mean
the allowed and forbidden transitions,
respectively.

CuCl has Γ_4 character for the Bloch function part and this is multiplied by the envelope function which is odd with respect to exchange of two holes. The Bloch function part has the finite matrix element for the second order electron-radiation interaction as realized from the fact that $\Gamma_5 * \Gamma_5$ $(=\Gamma_1 + \Gamma_3 + \Gamma_4 + \Gamma_5)$ contains Γ_4 state. However it vanishes when the integration in the coordinate of hole-hole separation is performed for the final molecular state due to its odd character. The lumiminescence from this molecular state and the optical conversion to this molecular state from any exciton state are also forbidden due to the same reason.

2-2. Excitonic Molecule in CuBr

In the crystal of CuBr, the bottom of the conduction band has the same symmetry Γ_6 as CuCl but the top of the valence band is composed of four-fold degenerate Γ_8 state. Two kinds of excitons with total angular momentum $J_{exc} = 1$ (Γ_5) and $J_{exc} = 2$ ($\Gamma_3 + \Gamma_4$) are formed from super-position of excitations of a valence electron into the conduction band. An exciton of $J_{exc} = 2$ ($\Gamma_3 + \Gamma_4$) has the pure triplet electron-hole spin structure and is optically inactive. In this material also, the electron-electron and the hole-hole intraband exchange energies which are taken into account in the effective mass calculation are much larger than the electron-hole interband exchange energy. Therefore as the basis functions of the excitonic molecule, the zero angular momentum state $| 0,0 \rangle_e$ of two electrons is accepted because it minimizes the electron-electron intraband exchange energy, and the $J_h = 0$ and $J_h = 2$ states $| J_h, m_h \rangle_h$ are chosen due to the same reason from four states $J_h = 0,1,2$ and 3 which are made from two $j = 3/2$ holes. The basis functions for the low lying states of an excitonic molecule with the total angular momentum $J_t = 0$ and 2, are given by

$$| J_t, m_t \rangle\rangle = | 0,0 \rangle_e \cdot | J_h, m_h \rangle_h \quad (J_t = 0,2 \text{ and } J_h = 0,2). \qquad (2-2)$$

The state of the excitonic molecule is described in terms of these basis functions and the envelope function of the internal motion Ψ as follows;

$$| \Gamma^{mol}(J_t, m_t) \rangle\rangle = \frac{1}{\sqrt{V}} e^{i\mathbf{K} \cdot \mathbf{R}_0} \Psi^{++}(r, r', R) | J_t, m_t \rangle\rangle . \qquad (2-3)$$

The molecular state $| 0,0 \rangle\rangle$ is decomposed into two parts which are made from basis functions of two Γ_5 ($J_{exc} = 1$) excitons and two $\Gamma_3 + \Gamma_4$ ($J_{exc} = 2$) excitons as follows;

$$| 0,0 \rangle \rangle = \sqrt{\tfrac{3}{8}} \; | 0,0 \colon (J_{exc}=1)^2 \rangle \; + \sqrt{\tfrac{5}{8}} \; | 0,0 \colon (J_{exc}=2)^2 \rangle \quad . \qquad (2\text{-}4)$$

On the other hand, the molecular state of $| 2,m_t \rangle \rangle$ is divided into three parts of $(J_{exc} = 1)^2$, $(J_{exc} = 1) \times (J_{exc} = 2)$ and $(J_{exc} = 2)^2$. Because of this cross term, the $J_{exc} = 2$ excitonic molecule can radiate by annihilating a $J_{exc} = 1$ exciton radiatively and leaving a spin triplet exciton $J_{exc} = 2$. The components of $J_{exc} = 1$ in $| 2, m_t \rangle \rangle$ are equal to each other for all $m_t = \pm2$, ±1, 0 as well as equal to that of $| 0,0 \rangle \rangle$ and therefore these states have the same value of the electron-hole exchange energy $3/4 \; \Delta_{exc}^{exch}$. Here the internal motion of an exciton is assumed not to be modified in the excitonic molecule and Δ_{exc}^{exch} is the energy splitting of $J_{exc} = 1$ and 2 excitons due to the electron-hole exchange interaction. The interband matrix elements of hole-hole Coulomb interaction[6,7] are expressed by three parameters Δ, Δ' and Δ'' as the perturbations on the solutions obtained by the effective mass theory. When these interactions are taken into account, the energy spectrum of an excitonic molecule in CuBr is given as follows;[6,7]

$$E_{mol}(\Gamma_1) = \Delta + 2\Delta' + \Delta''$$
$$E_{mol}(\Gamma_3) = \Delta + \Delta'' \qquad\qquad (2\text{-}5)$$
$$E_{mol}(\Gamma_5) = \Delta - \Delta'' \quad ,$$

where the energy reference is taken at $2\varepsilon_{1s} - E_{mol}^b + 3/4\Delta_{exc}^{exch}$.

The selection rules of these three optical processes relevant to the excitonic molecule in CuBr are listed in Fig.2. Giant two-photon absorption due to the excitonic molecule shows dependence of its intensity on the polarization of exciting light as shown in Table 1. Here the direction of the light propagation is taken in the z-axis of the crystal and the polarization of the light makes an angle ϕ against the y-axis in the x-y plane. It is noted that the three levels show different angular dependence in the giant two-photon absorption for linearly polarized light. On the other hand, Γ_1 excitonic molecule is

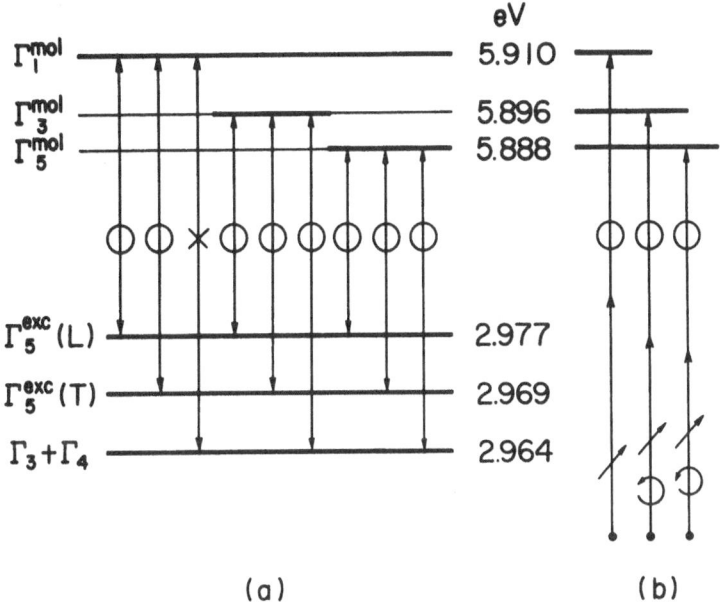

$$\Gamma_1^{mol} \quad\text{———} \quad 5.910$$
$$\Gamma_3^{mol} \quad\text{———} \quad 5.896$$
$$\Gamma_5^{mol} \quad\text{———} \quad 5.888$$

$$\Gamma_5^{exc}(L) \quad 2.977$$
$$\Gamma_5^{exc}(T) \quad 2.969$$
$$\Gamma_3+\Gamma_4 \quad 2.964$$

(a)   (b)

Fig.2. Energy diagram of exciton and excitonic molecule
in CuBr. (a) shows the selection rule for the
luminescence and the optical conversion of an
exciton into an excitonic molecule and (b) shows
that of the giant two-photon transition of an
excitonic molecule.

Molecule	$E_{mol}/2$ in eV	relative absorption intensity	
		linear pol.	circular pol.
Γ_1	2.955	4	0
Γ_3	2.948	$6\sin^2 2\phi$	1
Γ_5	2.944	$6\cos^2 2\phi + 1$	1

Table 1. Giant two-photon absorption due to excitonic molecule in CuBr.
The direction of the light propagation is taken in the z-axis
of the crystal and the polarization of the light makes an angle
ϕ against the y-axis in the x-y plane.

forbidden and Γ_3 and Γ_5 are allowed with equal intensity in the giant

two-photon absorption for circularly polarized lights.

2-3. Excitonic Molecule in CdS

The bottom of the conduction band and the top of the valence band are described by Γ_7 and Γ_9 states, respectively. The lowest excitonic and molecular states are constructed from properly symmetrized electron functions Γ_7 and hole functions Γ_9. Optically active Γ_5 exciton and inactive Γ_6 exciton of pure spin triplet are formed, and the ground state of the excitonic molecule is described in terms of the exciton basis functions as follows,[6,7]

$$|\Gamma_1^{mol}\gg = \frac{1}{\sqrt{V}}\ e^{i\mathbf{K}\cdot\mathbf{R}_0}\ \Psi^{++}(r,r',R)\ |\Gamma_1\gg \qquad (2\text{-}6)$$

with

$$|\Gamma_1\gg = \frac{1}{2}\ [\,|\Gamma_{5x}\rangle\ |\Gamma_{5x}\rangle\ -\ |\Gamma_{5y}\rangle\ |\Gamma_{5y}\rangle$$

$$+\ |\Gamma_{6-}\rangle\ |\Gamma_{6-}\rangle\ -\ |\Gamma_{6+}\rangle\ |\Gamma_{6+}\rangle\,] \qquad (2\text{-}7)$$

where $(\ |\Gamma_{5x}\rangle\ ,\ |\Gamma_{5y}\rangle\ ) = \frac{1}{\sqrt{2}}[\frac{+}{-}|P_1^1\rangle - |P_1^{-1}\rangle\,]$ are optically active to the light with polarization in x- and y-directions, respectively, and $|\Gamma_{6\pm}\rangle = \frac{1}{\sqrt{2}}[\,|P_2^2\rangle \pm |P_2^{-2}\rangle\,]$ are optically inactive excitons with pure triplet spin structure. The selection rules are listed in Fig. 3.

When the uniaxial field (σ) of the stress is applied perpendicular to the c-axis and the x-direction is chosen in this field, the Γ_5 exciton is split into two components Γ_{5x} $(E /\!/ \sigma)$ and Γ_{5y} $(E \perp \sigma)$ with polarization in the x- and y-axis, respectively. On the other hand, the emission line due to decay of the excitonic molecule splits also into two lines under the uniaxial stress perpendicular to the c-axis, but the polarization characteristics are reversed.[10] This can be understood in terms of eq. (2-6) as follows: When one Γ_{5x} (or Γ_{5y}) exciton is radiatively annihilated in the radiative decay of the excitonic molecule, the other Γ_{5x} (or Γ_{5y}) exciton should remain according to eq. (2-6). Therefore the $E/\!/\sigma$ and $E\perp\sigma$ polarized lights

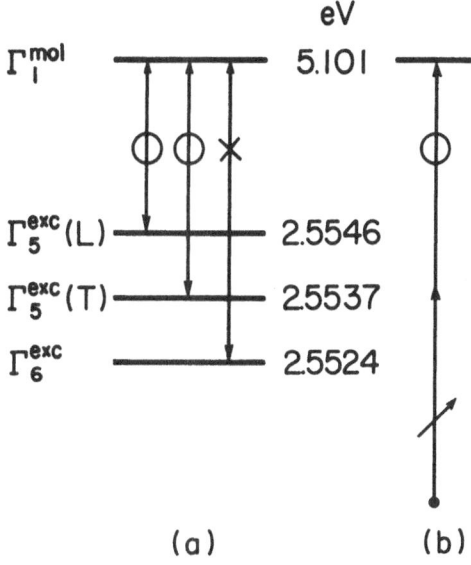

eV

Γ_1^{mol} ———— 5.101 ————

$\Gamma_5^{exc}(L)$ ———— 2.5546

$\Gamma_5^{exc}(T)$ ———— 2.5537

Γ_6^{exc} ———— 2.5524

(a)   (b)

Fig.3. Energy diagram of exciton and excitonic
        molecule in CdS. (a) shows the selec-
        tion rule for the luminescence and the
        optical conversion of an exciton into
        an excitonic molecule and (b) shows
        that of giant two-photon absorption due
        to an excitonic molecule.

should be observed at $h\nu_{\parallel}(\Gamma_{5x}) = E_{mol} - E_{exc}(\Gamma_{5x})$ and $h\nu_{\perp}(\Gamma_{5y}) = E_{mol}$
$- E_{exc}(\Gamma_{5y})$, respectively. Therefore, the reversed characteristics of
the polarization between the reflection spectrum due to excitons and
the emission spectrum of the M-line can be explained. On the other hand,
the bound exciton should not split as observed in the same reference.[10]
As a result, this experiment gave a strong support of the M-line to the
radiative decay of the excitonic molecule in CdS and ZnO.

III. EMISSION SPECTRUM FROM BOSE-CONDENSED EXCITONIC MOLECULES[4,8]

        The excitonic molecules can be accumulated at high density by the
giant two-photon excitation discussed in the preceding section.   These
excitonic molecules are expected to show the Bose-condensed state from
the large quantum effect as Bose particles. On the other hand, the

excitonic molecules with any value of the translational momentum are radiative through the process of annihilating one electron-hole pair radiatively and leaving the other pair as a free exciton. As a result, detailed information about the state of the excitonic molecule gas, *e.g.*, evidence of its Bose-condensation, can be obtained very clearly from the emission spectrum.

In this section, the characteristic emission spectrum is calculated for the Bose-condensed system of the excitonic molecules at finite temperature. Here we take into account the effect on the emission spectrum due to the final state interaction between the excitonic molecule which has been radiatively annihilated and the single exciton created in this emission process.

The emission spectrum in the linear response is given, apart from the unimportant factor, by

$$I(\omega) = 2\pi \sum_{i,f} |\langle f|p|i\rangle|^2 e^{-\beta E_i} \delta(\omega + E_f - E_i) , \qquad (3-1)$$

where $p = g \sum_K a_K^+ b_K$ is the dipole moment operator corresponding to the radiative annihilation of an excitonic molecule (b_K) with leaving a single exciton (a_K^+), and $\hbar\omega$ is the energy of the emitted photon. Here $I(\omega)$ is expressed in terms of the response function $\chi(i\omega_n)$ as follows;

$$I(\omega) = 2g^2 n(\bar{\omega}) \, \text{Im} \left[\chi(i\omega_n) |_{\bar{\omega}+i\delta} \right] , \qquad (3-2)$$

and $\chi(i\omega_n)$ is defined by $\chi(i\omega_n) \equiv \frac{1}{2} \int_{-\beta}^{\beta} d\tau e^{i\omega_n \tau} \chi(\tau)$, and

$$\chi(\tau) = \sum_{K,K'} \langle T_\tau a_K^+(\tau) b_K(\tau) b_{K'}^+ a_{K'} \rangle ,$$

where $\omega_n = 2\pi n/\beta$ and $\beta = 1/kT$, $\bar{\omega} = \omega + \mu_{exc} - \mu_{mol}$ (the photon energy is measured from the chemical potential difference $(\mu_{mol} - \mu_{exc})$ of the excitonic molecule and the exciton) and $n(\bar{\omega})$ is the boson distribution function. The contribution to the response function from the Bose-condensed excitonic molecules as well as from the molecules with the finite momentum are drawn in Fig. 4. Firstly, the final state

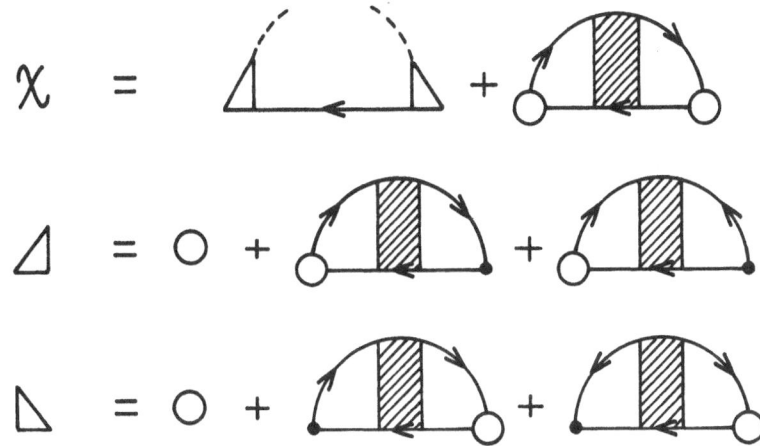

Fig.4. Response function corresponding to the emission of Bose-condensed excitonic molecule (the first term) and of the excitonic molecule in the finite momentum state (the second term). The white circle describes the electron-radiation interaction and the triangle is the electron-radiation interaction modified by the final state interaction. The shaded part represents the repeated scattering between the exciton and the excitonic molecule.

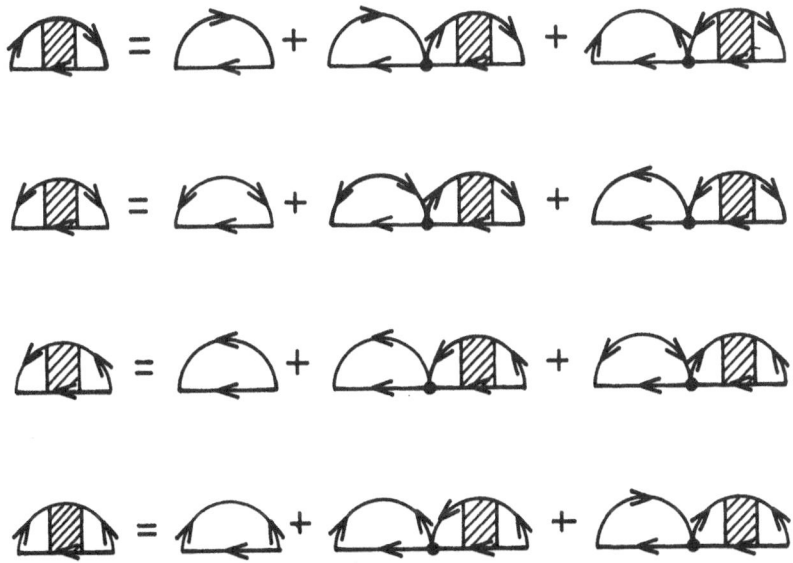

Fig.5. The final state interaction between the excitonic molecule and the exciton is taken into account by the coupled Dyson equations.

interaction between the single exciton and the excitonic molecule is taken into account (1) through the process of exciting the condensed excitonic molecule into the finite momentum state and its reverse process in the response function and (2) by the repeated scattering between the exciton and the excitonic molecule in the finite momentum state. The first process is evaluated by the vertex correction of the emission process drawn in Fig. 4 and the second effect is calculated by the coupled Dyson equations for four kinds of polarizabilities drawn in Fig. 5. Secondly, the interaction of the optically created exciton with the medium of excitonic molecules is also taken into account (3) by replacing the bare interaction V working between the exciton and one of the excitonic molecules by the renormalized t-matrix V* described in Fig. 6 in the expression of the response function $\chi(z)$ and in solving the coupled Dyson equations in Fig. 5 and (4) by the self-energy of the exciton due to interactions with the condensed excitonic molecules described in Fig. 7. The medium of the excitonic molecules is assumed to be described in the framework of the Bogoliubov approximation in terms of the t-matrix W between two excitonic molecules in the crystal vacuum. Then the response function $\chi(z)$ is described in terms of V* and W as well as four kinds of polarizabilities $P_i(z)$ (i = 1, 2, 3 and 4) where

$$P_i(z) = \frac{1}{\beta} \sum_n G_i(z) G_{exc}(z-i\,\omega_n)$$

and $G_1(z') = G_2(z') = \dfrac{u_p^2}{z'-\omega_p} - \dfrac{v_p^2}{z'+\omega_p}$ and $G_3(z') = G_4(z') = -u_p v_p$

$\times \{ \dfrac{1}{z'-\omega_p} - \dfrac{1}{z'+\omega_p} \}$ are the Green functions of the excitonic molecule

and G_{exc} is one of the single exciton with the self-energy described by Fig. 7. Here $u_p^2 = \dfrac{p^2/2M + n_0 W + \omega_p}{2\omega_p}$, $v_p^2 = \dfrac{p^2/2M + n_0 W - \omega_p}{2\omega_p}$, ω_p^2

$= \dfrac{p^2}{M} \times n_0 W + \dfrac{p^4}{4M^2}$.

The emission spectrum of eq. (3-2) is evaluated for several values

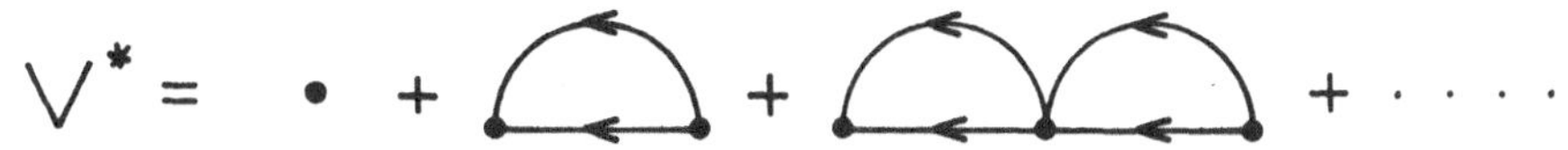

$$V^* = \quad \bullet \quad + \quad \text{(diagram)} \quad + \quad \text{(diagram)} \quad + \quad \cdots \cdots$$

Fig.6. The interaction of the single exciton with the medium of Bose-condensed
system of excitonic molecules is taken into account by replacing V by the t-
matrix V* due to the repeated scattering with one of the excitonic molecules.

$$\Sigma = \quad V + \text{(diagram)} + \text{(diagram)} + \text{(diagram)} + \text{(diagram)} + \text{(diagram)}$$

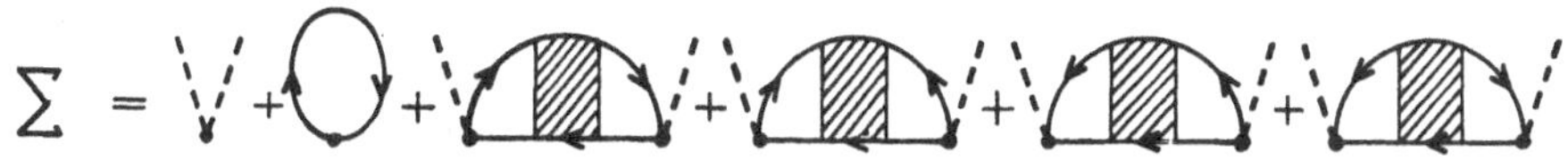

Fig.7. The self-energy of the single exciton due to the interaction with the
condensed excitonic molecules.

of molecular concentrations and system temperatures in which the

Bogoliubov approximation is justified. Here the photon energy $\hbar\omega$ is

normalized by $\hbar^2/Mf_0^2$, where f_0 is the scattering length which is

related to W by $W = 4\pi\hbar^2 f_0/M$, and V*/W is taken as a constant parameter.

The emission spectra are drawn in Fig. 8 for several values of V*/W and

of concentrations of the excitonic molecules.

Some of the interesting results are listed:

(1) The sharp emission line due to the radiative decay of Bose-condensed

excitonic molecules is somewhat broadened due to scattering of the ex-

citon in the final state with the medium of the Bose-condensed excitonic

molecules. The width of this sharp line depends on the concentration of

the excitonic molecules and the system temperature as shown in Fig. 8.

Note the order change of the abscissa of the photon energy in Fig. 8.

(2) The side band appears in the lower energy side due to the two

origins; (a) the radiative decay of a Bose-condensed excitonic molecule

accompanied with excitation of the individual and collective motions of

the excitonic molecules and (b) the radiative decay of the excitonic

molecules with the finite momentum. Both contributions are described

by the second process of Fig. 4 but a part of this side band intensity

is transferred into the sharp line through the final state interaction

described by the vertex correction in the first process of Fig. 4.

(3) The emission singularity appears at $\hbar\omega = (2-\sqrt{3})n_0 W$ due to the van

Hove singularity of the joint state density of the excitonic molecule

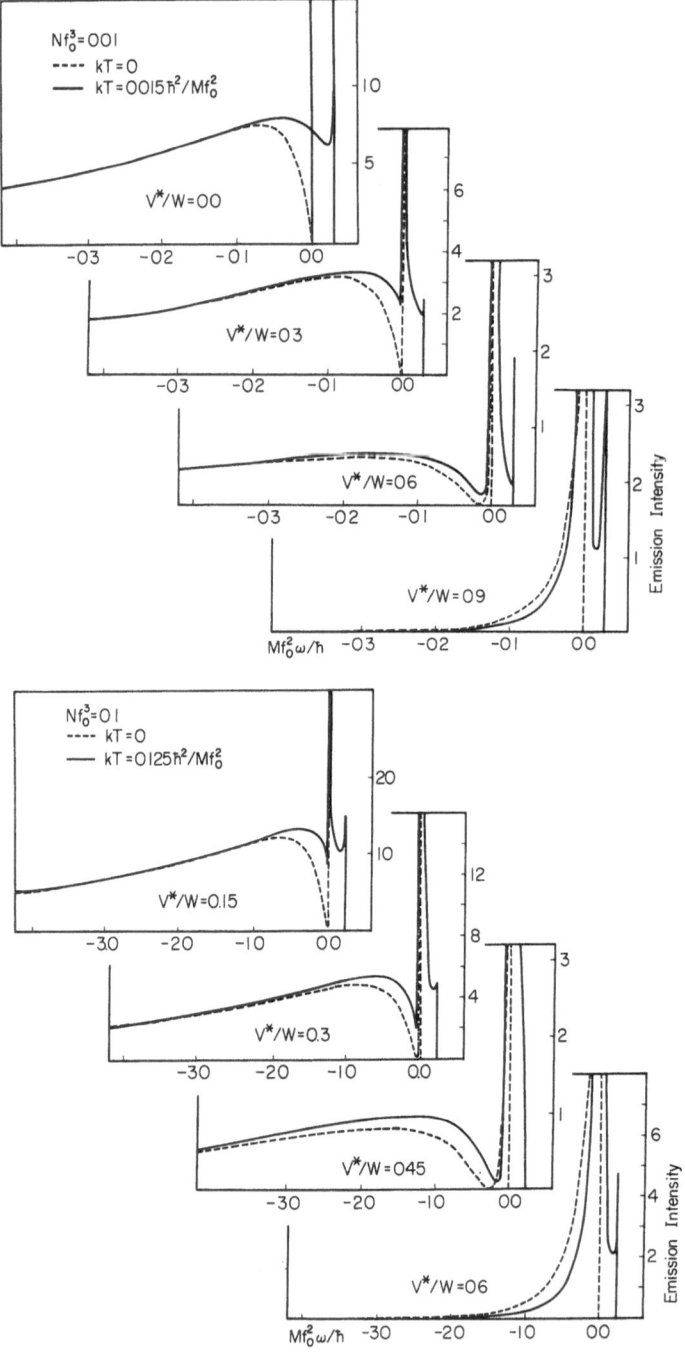

Fig.8. The emission spectra characteristic of the Bose-condensed excitonic molecules are shown for various values of the system temperature and the concentration with parameters V*/W.

and the single exciton. However it will be smeared or weakened when

the finite value of the momentum of the emitted light is taken into account.

(4) At the lower concentration of the total excitonic molecules than the case of Fig. 8, the ratio of the sharp line intensity to the total area is very close to unity and it decreases with the increase of the total concentration and the rise of the system temperature as shown in Fig. 8.

(5) As the parameter V^*/W is varied, the profile of the emission spectrum is changed very much around $V^*/W = 0.5$. The curve in Fig. 8 can fit the observed emission spectra[11] for a wide range of concentrations with respect to the position and the profile of the sharp line and the band width and the profile of the side band besides the lower energy tail. The theoretical curve has a rather longer tail. The side band peak is located at the photon energy shifted by an order of magnitude of $n_0 W$ below the sharp peak. n_0 is approximately proportional to the total concentration n of the excitonic molecules and therefore the side band is observable only in the case of high density excitation of the excitonic molecules under the present experimental resolution in the emission spectrum. When the exciton mass $M/2$ is 0.4 m + 4.2 m and the scattering length f_0 is estimated to be $4a_0$ (a_0 is the exciton Bohr radius and $7\mathring{A}$ for CuCl), the peak separation of the side band from the sharp line is evaluated to be 1.6 meV for $nf_0^3 = 0.1$ (n = 5×10^{18}/cc) and 0.2 meV for $nf_0^3 = 0.01$ (n = 5×10^{17}/cc).

IV. LUMINESCENCE OR RAMAN SCATTERING?

By using the second character of excitonic molecules, the Bose-condensed system of excitonic molecules can be directly created by the giant two-photon absorption and the evidence of Bose-condensation can be observed in the characteristic emission spectrum of excitonic molecules. On the other hand, here arises a question whether the third order optical process of two-photon absorption due to the excitonic molecule and its subsequent emission, leaving a single exciton behind, can be considered as two separable processes of absorption and emission or an inseparable process like Raman scattering.

To answer this question, the transition rate of this third order optical process is evaluated in terms of the damping theory for the interaction of the excitonic molecule in the intermediate state and the single exciton in the final state with their surrounding reservoir. The initial state is composed of the ground state and the radiation field (n_1, n_2) for the incoming light of frequency ω_1 and the outgoing light of frequency ω_2. The intermediate state of an excitonic molecule is connected with the ground state through two-photon transition of frequency ω_1 by V_1^+ and with the final state of single exciton through one-photon transition of frequency ω_2 by V_2^+. Starting from the initial state $i = (g, n_1, n_2)$ at $t = 0$, the probability $P(t)$ of finding the final state $f = (m, n_1-2, n_2+1)$ at time t is given by

$$P(t) = \int_0^t dt_1 \int_0^{t_1} dt_2 \int_0^{t_2} dt_3 \int_0^t dt_1' \int_0^{t_1'} dt_2' \int_0^{t_2'} dt_3' \; Tr_R$$

$$\cdot \langle f | e^{-i(t-t_1')H} V_2^+ e^{-i(t_1'-t_2')H} V_1 e^{-i(t_2'-t_3')H} V_1 e^{-it_3'H}$$

$$\cdot \rho_0(H) e^{it_3 H} V_2 e^{i(t_2-t_3)H} V_1^+ e^{i(t_1-t_2)H} V_1^+ e^{i(t-t_1)H} | f \rangle \quad .$$

$$(4-1)$$

Here Tr_R describes taking the trace over the coordinates of the surrounding reservoir and $\rho_0(H)$ is the distribution function of the total system. The process of two-photon absorption due to the excitonic molecule is assumed to be completed instantaneously because ω_1 is enough off-resonant to an exciton level. Then we have three kinds of time orderings for the contribution to $P(t)$ as shown in Fig. 9. The first contribution is expressed as follows;

$$P_i(T) = 2Re \int_0^t dt_1 \int_0^{t_1} dt_2 \int_0^{t_2} dt_3 \int_0^{t_3} dt_1' \int_0^{t_1'} dt_2' \int_0^{t_2'} dt_3'$$

$$\cdot Tr_R \langle f | \{\{ e^{-i(t_1-t_2)H} [[e^{-i(t_2-t_3)H} [e^{-i(t_3-t_1')H} (V_2^+)_{c'c}$$

$$\cdot \{e^{-i(t_1'-t_2')H}(V_1)_{b'b}(e^{-i(t_2'-t_3')H}(V_1)_{a'a}\rho_0\ e^{i(t_2'-t_3')H})_0$$

$$\cdot\ e^{i(t_1'-t_2')H}\}_n\ e^{i(t_3-t_1')H}]\ (V_1^+)_{d'd}\ e^{i(t_2-t_3)H}]]\ (V_1^+)_{e'e}$$

$$\cdot\ e^{i(t_1-t_2)H}\}\}(V_2)_{f'f}\ |\ f\rangle\ e^{i\omega_1(t_2+t_3-t_2'-t_3')}e^{-i\omega_2(t_1-t_1')}$$

$$\tag{4-2}$$

$$= 2t\ \text{Re} \iiint_0^\infty\ d\tau_3 d\tau_2 d\tau_1\ \Sigma \ll ff'\ |\ \hat{U}(\tau_3)\ |\ \ell e \gg$$

$$\cdot \ll md'\ |\ \hat{U}(\tau_2)\ |\ c'n \gg\ll cn\ |\ \hat{U}(\tau_1)\ |\ b'0 \gg \rho_{aa}\ e^{-i\omega_2\tau_3}\ e^{i2\omega_1\tau_1}$$

$$e^{i(2\omega_1-\omega_2)\tau_2}\ (V_2^+)_{c'c}(W_1)_{b'a}(W_1^!)_{d'c}(V_2)_{f'f}\ ,\tag{4-3}$$

where $(W_1)_{b'a} \equiv i(V_1)_{b'b}(V_1)_{a'a}/(\omega_{a'a}-\omega_1)$ is the two-photon transition matrix element via the intermediate exciton state and $(W_1)_{b'a}$ and $(W_1^+)_{d'c}$ are obtained after integrations with respect to $\tau' = t_2' - t_3'$

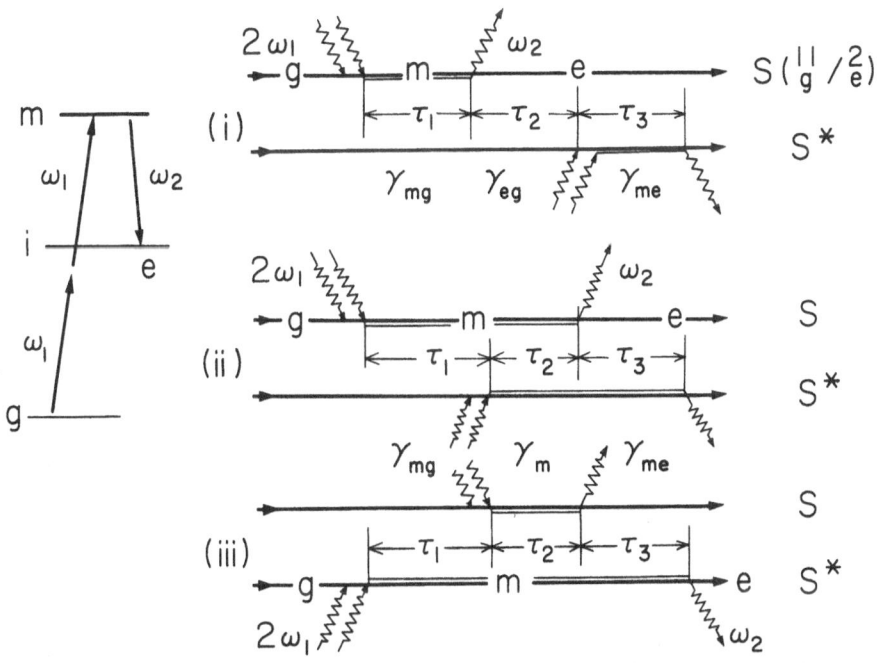

Fig. 9. The third order optical process is composed of three types of contributions which depend on time orderings of the interaction with the radiation field.

and $\tau = t_2 - t_3$. Here the damping theory is used to take into account the effect of the surrounding reservoir as follows;

$$\ll ab \mid \hat{U}(\tau) \mid cd \gg \equiv \{ \langle a \mid e^{-iH\tau} \mid c \rangle \langle d \mid e^{iH\tau} \mid b \rangle \}_{av}$$

$$= \delta_{ac}\delta_{db} e^{-i\omega_{ab}\tau - \Phi_{ab}\tau} , \qquad (4\text{-}4)$$

where $\quad \Phi_{ab} = \Phi_{ba}^* = \gamma_{ab} + i\Delta_{ab}$

$$\gamma_{ab} = \gamma_{ab}^c + \frac{1}{2}(\Gamma_a + \Gamma_b)$$

$$\Phi_{aa} = \gamma_a = \gamma_a^c + \Gamma_a . \qquad (4\text{-}5)$$

$(1/\Gamma_a)$ is the life-time of level a, γ_a^c is the inelastic collision frequency of level a and γ_{ab}^c is composed of the elastic as well as inelastic collisions of levels a and b. By a similar way, the second and third contributions of diagrams (ii) and (iii) are evaluated as follows;

$$P_{ii}(t) = 2t \text{ Re} \iint\limits_0^\infty \int d\tau_3 d\tau_2 d\tau_1 \sum \rho_{aa} \ll ff' \mid \hat{U}(\tau_3) \mid c'0 \gg$$

$$\cdot \ll c0 \mid \hat{U}(\tau_2) \mid \ell e \gg \ll md' \mid \hat{U}(\tau_1) \mid b'n \gg$$

$$\cdot (V_2^+)_{c'c}(W_1)_{b'a}(W_1^+)_{d'e}(V_2)_{f'f} e^{2i\omega_1\tau_1} e^{-i\omega_2\tau_3} \qquad (4\text{-}6)$$

$$P_{iii}(t) = 2t \text{ Re} \iint\limits_0^\infty \int d\tau_3 d\tau_2 d\tau_1 \sum \ll ff' \mid \hat{U}(\tau_3) \mid c'0 \gg$$

$$\cdot \ll c0 \mid \hat{U}(\tau_2) \mid b'n \gg \ll am \mid \hat{U}(\tau_1) \mid \ell e \gg$$

$$\cdot \rho_{d'd'} (V_2^+)_{b'b}(W_1)_{b'a}(W_1^+)_{d'e}(V_2)_{f'f} e^{-2i\omega_1\tau_1 - i\omega_2\tau_3} . \qquad (4\text{-}7)$$

The expression of eq.(4-4) is inserted into eqs.(4-3), (4-6) and (4-7) and time integrals are performed. Then the transition rate representing the output of ω_2-photon for an influx of ω_2 photon is described finally as follows;

$$F(2\omega_1, \omega_2) = \lim_{t\to\infty} P(t)/t$$

$$= A \frac{\gamma_{mg}}{(2\omega_1 - \omega_{mg})^2 + \gamma_{mg}^2} \cdot \frac{\gamma_{me}}{(\omega_2 - \omega_{me})^2 + \gamma_{me}^2} \left[\frac{2}{\gamma_m}\right.$$

$$+ \frac{\gamma_{eg}\{1 + (2\omega_1 - \omega_{mg})(\omega_2 - \omega_{me})/\gamma_{mg}\gamma_{me}\}}{(2\omega_1 - \omega_2 - \omega_{eg})^2 + \gamma_{eg}^2}$$

$$- \frac{(2\omega_1 - \omega_2 - \omega_{eg})\{(2\omega_1 - \omega_{mg})/\gamma_{mg} - (\omega_2 - \omega_{me})/\gamma_{me}\}}{(2\omega_1 - \omega_2 - \omega_{eg})^2 + \gamma_{eg}^2} \left.\right] \qquad (4\text{-}8)$$

where $A \equiv n_1(n_1-1)(n_2+1) \mid (W_1)_{mg} \mid^2 \mid (V_2)_{me} \mid^2$.

In the first process, the difference of phase modulations between the molecule state and the ground state, between the final exciton state and the ground state and between the final exciton state and the molecule state work in the time intervals τ_1, τ_2 and τ_3, respectively. This gives the second and third terms in the bracketed passage of eq. (4-8). On the other hand, in the second and third processes, the excitonic molecule is really created in the time interval τ_2 and only the radiative damping works in this time interval. Both phase modulation and radiative damping work in other time intervals. These two processes give the first term in the bracketed passage in eq. (4-8). The first process contains the Raman scattering term and the second and third processes give the luminescence terms. The first term is called luminescence type because the frequency ω_2 of the emission peak is independent of the frequency ω_1 of the incoming laser light and the second and third terms are of Raman type in the sense that ω_2 is correlated with ω_1 as shown in this expression. Let us discuss the following two limiting cases. In the first case, only the molecular state is strongly modulated, which is described by the inequalities:

(a) $\gamma_{mg} \simeq \gamma_{me} = \bar{\gamma} \gg 2\gamma_{eg} > \gamma_m$. The i-th contribution, *i.e.*, the second and third terms in the bracketed passage are simplified into the following two terms: $\pi\delta(2\omega_1-\omega_2-\omega_{eg})\{(2\omega_1-\omega_{mg})^2 + \bar{\gamma}^2\}$ and $-\bar{\gamma}$, respectively. When the spectral width of the incoming laser light is much wider than $\bar{\gamma}$, the emitted spectrum is given by integrating $F(2\omega_1, \omega_2)$ with respect to ω_1, which is denoted as $F(\omega_2)$. The i-th contribution is cancelled out each other and the only luminescence part remains finite: $F(\omega_2) = \pi A(2\bar{\gamma}/\gamma_m)f(\omega_2-\omega_{me}, \bar{\gamma})$, where $f(x,y) = x/(x^2+y^2)$. In the second case, the phase only of final exciton state is strongly modulated, which is described by the following inequalities: (b) $\gamma_{eg} \simeq \gamma_{me} \equiv \bar{\gamma} \gg 2\gamma_{mg} > \gamma_m$. When the spectral width of the incoming laser light is much larger than $\bar{\gamma}$, two components of emitted lights are given by the following equations: $F(\omega_2) = \pi Af(\omega_2 - \omega_{me}, \bar{\gamma})[\bar{\gamma}f(\omega_2- \omega_{me}, \bar{\gamma}) + 2\bar{\gamma}/\gamma_m]$. Therefore the peaks of two emission components are coincident with each other and the integrated intensity has the ratio 1 to $4\bar{\gamma}/\gamma_m$. As a result, the luminescence component is dominating as far as the radiative damping γ_m is much smaller than the phase modulation of the exciton in the final state. The emission peak was observed to be independent of the peak frequency of the incoming laser light as far as the spectrum of the laser light covers the excitation frequency of the excitonic molecule.

When the monochromatic laser light is used, the more information is obtained. Then the Raman and luminescence components show the different ω_1 and ω_2 dependences and the detailed analysis of experimental data on the basis of eq. (4-8) gives us information of the relaxation times and the radiative life-time of the excitonic molecule as well as their change due to Bose-condensation. In Fig. 10, some examples of emission spectra are shown for various values of ω_1 with parameters of $\bar{\gamma}/\gamma_{eg}$ and $\bar{\gamma}/\gamma_m$. With increasing the value of $\bar{\gamma}/\gamma_m$, the luminescence component is shown to be overcoming the Raman component, although the latter is still observable due to the sharpness of the Raman spectrum.

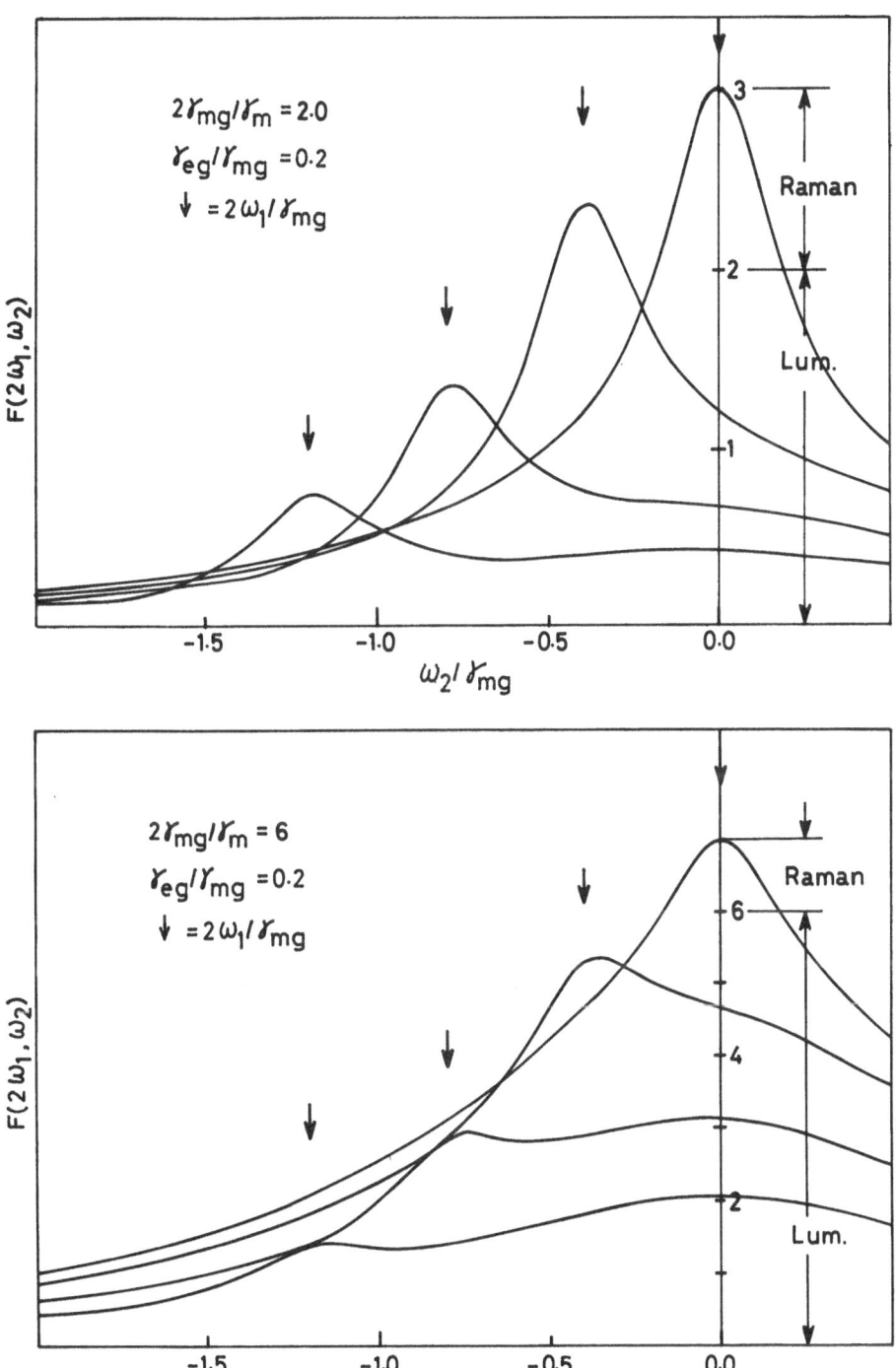

Fig.10. (See next page.)

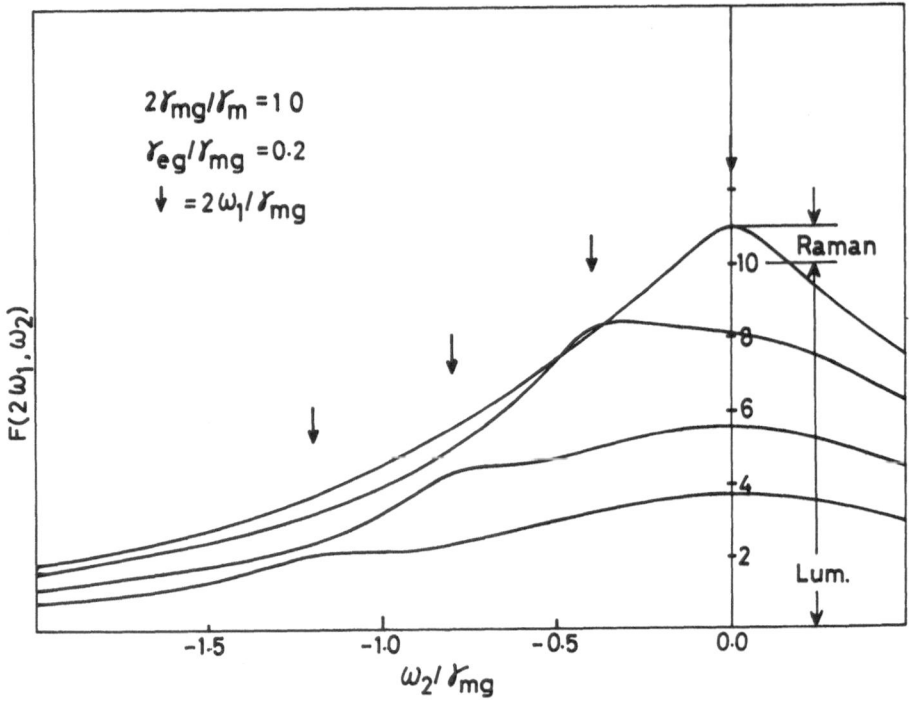

Fig.10. The emission spectra are composed of the luminescence and Raman contributions. As γ/γ_m increases, the luminescence part overcomes the Raman part.

V. CONCLUSION

The giant two-photon absorption due to excitonic molecules have been observed in CuCl,[11] CuBr[11] and CdS and the existence of the excitonic molecule was confirmed in these materials. Furthermore the study of the Bose-condensation of excitonic molecules was made possible by the direct excitation of excitonic molecules through the giant two-photon excitation. As discussed in Luminescence Conference[12] and introduced by Ueta,[13] we may conclude that the evidence of the Bose-condensation of excitonic molecules was found in the following two experimental facts: First, the sharp emission line and its characteristic side band were observed in CuCl and CuBr under the giant two-photon excitation of excitonic molecules in agreement with theoretical expectations with respects to the position and the profile of the sharp line and the separation of the side band peak from the sharp line and the profile of the side band, and second, the relative intensity of the sharp line to the total spectral area was observed to decrease with the rise of molecular concentration and the system temperature in agreement at least qualitatively with theoretical expectation.

Finally the following three points will become important in the near future to study the Bose-condensation of the excitonic molecules in more detail. First, the analysis of Raman and luminescence components will give us the information on the relaxation process of the excitonic molecules and the exciton and its change due to the Bose-condensation. Second, the time analized emission spectrum of the excitonic molecules will be interesting under the giant two-photon excitation by the pico-second laser pulse. This will make clear the dynamical mechanism of the Bose-condensation and its breaking process due to the relaxation and damping process. The third point is the most interesting. The coherent property of Bose-condensed excitonic molecules is most clearly reflected on the coherent interaction with the coherent radiation field.[5] Therefore the observation of the coherent optical process re-levant to the Bose-condensation such as the self-induced transparency of two-photon transition due to excitonic molecules will bring about more direct and clear evidence for the Bose-condensation of excitonic molecules.[5]

REFERENCES

1) E. Hanamura: *Optical Properties of Solids*, ed. by B. O. Seraphin (North-Holland, Amsterdam, 1976) p.81.

2) E. Hanamura and M. Inoue: *Proc. Intern. Conf. on Phys. of Semiconductors, Warszawa 1972*, ed. M. Miasek *et al.* (Elsevier, N. Y., 1972) p.711.

3) E. Hanamura: Solid State Comm. **12** (1973) 951.

4) E. Hanamura: *Luminescence of Crystals, Molecules and Solutions, Leningrad 1972* ed. F. Williams (Plenum Press, N. Y., 1973) p.121.

5) E. Hanamura: *Proc. Intern. Conf. on Phys. of Semiconductors, Stuttgart 1974.* ed. M. H. Pilkuhn (Teubner, Stuttgart, 1974) p.137.

6) E. Hanamura: J. Phys. Soc. Japan <u>39</u> (1975) 1506.

7) E. Hanamura: J. Phys. Soc. Japan <u>39</u> (1975) 1516.

8) M. Inoue and E. Hanamura: to be submitted to J. Phys. Soc. Japan.

9) R. Kubo, T. Takagawara and E. Hanamura: *Proc. of Oji Seminar* (this issue).

10) Y. Segawa and S. Namba: Solid State Commun. <u>17</u> (1975) 489.

11) N. Nagasawa, N. Nakata, K. Doi and M. Ueta: J. Phys. Soc. Japan <u>38</u> (1975) 593.

12) E. Hanamura: J. of Luminescence, <u>12/13</u> (1976) 119.

13) M. Ueta and N. Nagasawa: *Proc. of Oji Seminar* (this issue).

BIEXCITON LUMINESCENCE IN CuCl AND CuBr

J. B. Grun, R. Levy, E. Ostertag, H. Vu Duy Phach and H. Port*

Laboratoire de Spectroscopie et d'Optique du Corps Solide
(associé au C.N.R.S. n° 232)
Université Louis Pasteur
5, rue de l'Université
67000 Strasbourg, France

ABSTRACT

When CuCl and CuBr are excited in their band to band absorption,
biexcitons are generated from carriers and excitons. They are characterized by a Boltzmann distribution function with a temperature higher
than that of the sample. They decay radiatively from their ground
state to exciton-polariton states. Broad emission bands are observed.

When biexcitons are directly created by a resonant two-photon
absorption in their ground state, they may form a Bose-Einstein condensate. When they recombine radiatively, sharp emission bands are
indeed observed on the high energy edge of the broad emission bands.

Additional emission lines have been observed on the low energy
side of the free exciton emission in CuCl and CuBr, under high excitation intensities at low temperatures. They have been previously attributed to the radiative decay of biexcitons from their ground state
to the ground states of excitons.[1,2] As will be shown, different
emission spectra are obtained when the biexcitons are generated from
the population of carriers and excitons, or created directly.

I.   GENERATION OF BIEXCITONS FROM THE POPULATION
OF CARRIERS AND EXCITONS

When the crystals are excited in their band to band absorption by
an intense ultraviolet light, (1 MW Lambda Physik Nitrogen laser, 3.677
eV) a large number of free carriers are created. They thermalized
rapidly and form excitons.[3] Biexcitons are then generated from this
large pool of excitons.

---

*Present address: Physikalisches Institut, Teil 3, Universität Stuttgart, Germany

The spectrum obtained for CuCl at 4.2 K under an excitation intensity of 8×10^{21} UV photons/cm^2 sec is drawn in solid line in Fig. 1.

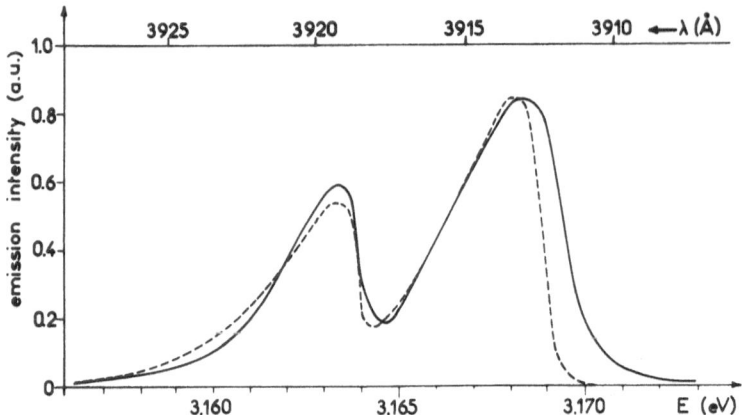

Fig.1. Experimental luminescence spectrum of CuCl (solid line) under strong excitation (8×10^{21} photons/cm^2 sec) in its band to band absorption at 4.2 K. Theoretical curve (dashed line) with T = 18 K and E_M^b = 28 meV.

Two broad emission bands called B_T (3.168 eV) and B_L (3.164 eV) are observed. Their line shape has been calculated assuming transitions from the biexciton ground state Γ_1[4] to the longitudinal Γ_5 exciton and the corresponding upper and lower polaritons.[5] The transition towards the lower Γ_2 exciton level is forbidden.

The biexcitons are assumed to be thermalized in their ground state and characterized by a Boltzmann distribution function with a temperature T.[*] Their binding energy E_M^b is defined relatively to the energy of the lowest Γ_2 exciton state. These two quantities have been used as adjustable parameters in order to fit best theoretical and experimental spectra.

The calculated curve, corresponding to the experimental spectrum given in Fig. 1, is drawn in dashed line in the same figure. The values

---

[*] Souma et al.[6] had previously shown that the B_L band could be characterized by a Boltzmann function.

of the parameters are T = 18 K and E_M^b = 28 meV.

An increasing discrepancy exists between the calculated curve and the experimental spectrum at higher excitation intensities. In the experimental spectrum, the B_T line broadens more and more, while the B_L line remains almost unchanged. The discrepancy is certainly due to simplifying assumptions made in our calculations. However, the experimental intensities ratio of B_T and B_L remains equal to two for all excitation intensities, corresponding to the statistical creation of one longitudinal exciton for two transverse excitons in the radiative decay of biexcitons.

When we excite high quality single crystals of CuBr in their band to band absorption, three broad emission bands are observed. Their peaks are approximatively at 2.945 eV for the B_t band at 2.940 eV and 2.929 eV for the B_T and B_L bands[7] respectively. The spectrum obtained at an excitation intensity of 6×10^{22} UV photons/cm^2 sec at 1.6 K is drawn in Fig. 2.

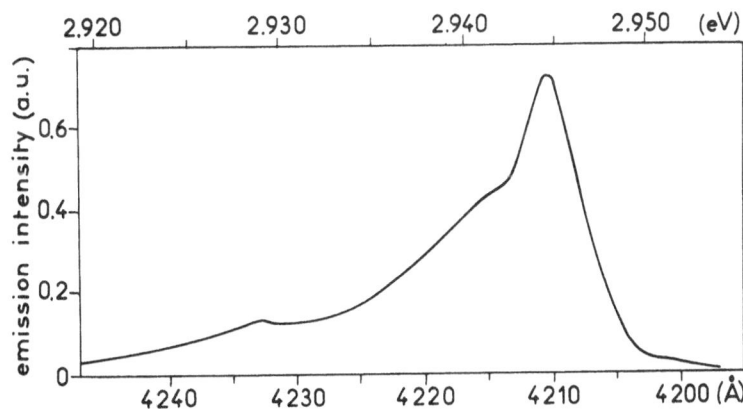

Fig.2. Experimental luminescence spectrum of CuBr under strong excitation (6×10^{22} photons/cm^2 sec) in its band to band absorption at 1.7 K.

We know from Comte's theoretical investigation[8] that there are three possible symmetries Γ_1, Γ_3 and Γ_5 for the most stable states of the biexciton corresponding to an envelope function which is symmetric under electron and hole permutations. These three states are not split by the electron-hole exchange interaction. Therefore, it we assume that these states remain degenerate, the observed emission can be simply explained by the radiative decay of the biexcitons.

The B_t line may be due to the allowed transitions from the biexciton ground state to the degenerate Γ_3, Γ_4 lower exciton state in CuBr. The binding energy of the biexciton in its ground state, defined relatively to the lower Γ_3, Γ_4 exciton state, is shown to be equal to 17 meV. The two other emission bands B_T and B_L can be related to the radiative decay of biexcitons from their ground state to the longitudinal Γ_5 exciton state and the corresponding lower polariton state, as in CuCl.

II.   DIRECT GENERATION OF BIEXCITONS

The direct creation of biexcitons by a resonant two-photon absorption has been predicted theoretically by Hanamura[9] and first observed by Gale and Mysyrowicz in CuCl.[10] All biexcitons are then generated with the same translational momentum $2 \vec{K}_o$, $\vec{K}_o$ being the wave-vector of the exciting photons.

In Fig. 3, are drawn the emission spectra of a cleaved single crystal of CuCl, excited at 1.7 K by a tunable dye laser [2×10^{-3} M/1 Butyl PBD in toluene pumped by the Lambda Physik Nitrogen laser] emitting at 3890 Å (3.187 eV) with a spectral width of 2.5 meV. Two photons simultaneously absorbed directly create a biexciton in its Γ_1 ground state[11] as it can be deduced from the known biexciton binding energy and as it has been observed by Ueta and his group in the two-photon excitation spectrum.[12]

We have observed two narrow emission peaks as in ref.(12) showing an instrument limited linewidth of 0.6 meV on the high energy edge of the two broad emission bands B_T and B_L at 3909.4 Å (3.1713 eV) and

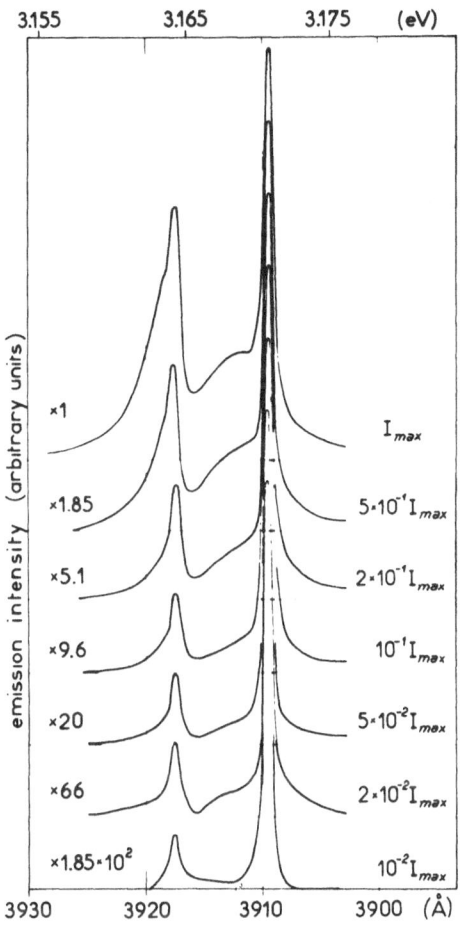

Fig.3. Experimental luminescence spectra of
CuCl under two-photon excitation at
3.187 eV. The normalization factor
for each spectrum is given on the
left hand side of the figure.

3917.3 Å (3.165 eV) respectively. The first line is several times more
intense than the second one.

The line intensities show a pronounced dependence on excitation
intensity. At low excitation intensities, only the sharp lines are
present. When the excitation intensity increases, the broad emission
bands B_T and B_L appear and increase more rapidly than the sharp ones.
They vary as the quadratic power of the excitation intensity as expected

from a two-photon excitation, while the sharp lines vary linearly in
this range of excitation, after a quadratic increase at low intensities.

These sharp emission lines are also very sensitive to the temperature. They decrease and disappear very rapidly when the temperature
increases $(T \simeq 20$ to 30 K$)$.

The sharp lines have been attributed by Ueta *et al.*[12] to the
radiative decay of biexcitons condensed in the $\vec{K} = 0$ or $2\,\vec{K}_o$ state as
predicted by Hanamura.[9] These sharp lines cannot be related to any
stimulated process, their study in function of the excitation intensity
is not showing any exponential growth. They can neither be identified
with known impurity emission lines.

Figure 4 shows the emission spectra of a cleaved single crystal of

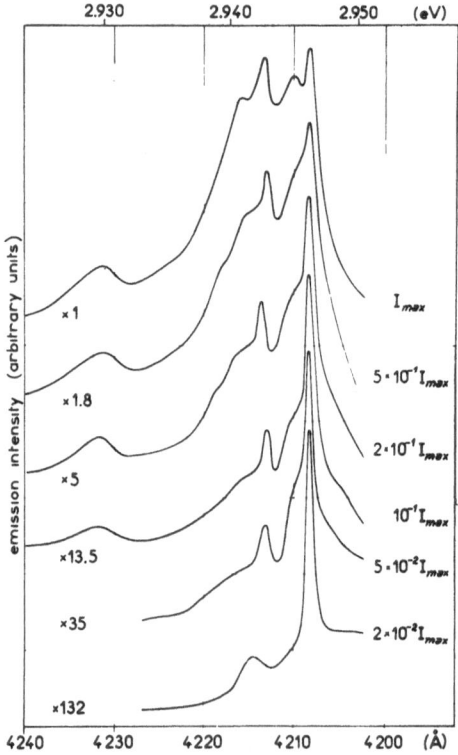

Fig.4. Experimental luminescence spectra of CuBr under two-
photon excitation at 2.955 eV. The normalization
factor for each spectrum is given on the left hand
side of the figure.

CuBr at 1.7 K. In this case, the tunable dye laser [10^{-3} M/1 POPOP in Toluene pumped by the nitrogen laser] emits a 4196 Å (2.955 eV) light with a spectral width of 1.25 meV. This laser generates biexcitons in their ground state, by resonant two-photon absorption, as can be deduced from the value of their binding energy.

As in CuCl, two narrow lines are observed on the high energy edge of the B_t and B_T bands respectively at 4208.3 Å (2.946 eV) and 4213.5 Å (2.9424 eV). Ueta and his group have only observed a sharp line on the high energy edge of the third B_L line, which was not found in our experiments.[13]*

The observation of these lines in CuBr at the spectral position expected from the radiative decay of biexcitons condensed in the $\vec{K} = 0$ or 2 $\vec{K}_o$ state is another argument in favour of Hanamura's interpretation.[9]

III. CONCLUSION

We have shown that biexcitons decay radiatively from their ground state to exciton-polariton states. When biexcitons are generated from carriers and excitons, broad emission bands are observed, due to the large population of biexcitons in their ground state corresponding to temperature higher than that of the sample. When they are directly generated, sharp emission lines are observed on the high energy edge of the broad bands. They can be tentatively attributed to a Bose-Einstein condensation of biexcitons.

ACKNOWLEDGEMENT

We would like to thank Professor. H. Haken and Professor S. Nikitine for many stimulating discussions. We are also indebted to Professor Hanamura for many fruitful suggestions.

---

* Nagasawa *et al.*[14] have reported the observation of three sharp lines on the high energy edges of the B_L, B_T and B_t.

REFERENCES

1) A. Mysyrowicz, J. B. Grun, R. Levy, A. Bivas and S. Nikittine: Phys. Lett., 26A (1968) 615.

   J. B. Grun, S. Nikitine, A. Bivas and R. Levy: J. Luminescence, 1-2 (1970) 241.

2) S. A. Moskalenko: *Proc. of the XIIth European Congress on Molecular Spectroscopy, Strasbourg* (1975), ed. M. Grossmann, S. G. Elkomoss and J. Ringeissen (Elsevier, Amsterdam, 1976) p.45.

3) Y. Toyozawa: Suppl. Prog. theor. Phys., 12 (1959) 111.

4) F. Bassani, J. J. Forney and A. Quattropani: Phys. Stat. sol.(b), 65 (1974) 591, and Helv. phys. Acta, 47 (1974) 201.

5) E. Ostertag, R. Levy, J. B. Grun: Phys. Stat. sol., 69 (1975) 629.

6) H. Souma, T. Goto, T. Ohta and M. Ueta: J. Phys. Soc. Japan, 29 (1970) 697.

7) R. Levy, A. Bivas, C. Comte and E. Ostertag: *Proc. of the XIIth European Congress on Molecular Spectroscopy, Strasbourg* (1975), ed. M. Grossmann, S. G. Elkomoss and J. Ringeissen (Elsevier, Amsterdam, 1976) p.97.

8) C. Comte: Opt. Comm., 14 (1975) 79.

9) E. Hanamura: Solid State Commun., 12 (1973) 951.

10) G. M. Gale and A. Mysyrowicz: Phys. Rev. Lett., 32 (1974) 727.

11) E. Doni, R. Girlanda and G. Pastori Parravicini: to be published in Solid State Commun.

12) N. Nagasawa, N. Nakata, Y. Doi and M. Ueta: J. Phys. Soc. Japan, 38 (1975) 593.

13) N. Nakata, N. Nagasawa, Y. Doi and M. Ueta: J. Phys. Soc. Japan, 38 (1975) 903.

14) N. Nagasawa, S. Koizumi, T. Mita and M. Ueta: *Proc. Intern. Conf. Luminescence. Tokyo,* 1975, ed. S. Shionoya, S. Nagakura and S. Sugano (North Holland, Amsterdam, 1976) p.587.

BIEXCITONS IN CuCl

A. Mysyrowicz[*]

Laboratoire d'Optique Quantique du C.N.R.S.
Ecole Polytechnique
91120 Palaiseau (France)

ABSTRACT

The experimental evidence for the existence of the biexciton in
CuCl is discussed. It relies on the analysis of the luminescence of
strongly excited crystals, on the two-photon absorption and excitation
spectrum. In the second part, measurement of the biexciton lifetime at
4.2°K is reported.

The study of strongly excited crystals has become a subject of increasing interest during the last few years. If a high density of
electron-hole pairs is generated, new collective excitation states
appear in the system. The simplest such collective excitation is represented by the biexciton, an excitonic molecule composed of two
electron-hole pairs bound by direct Coulomb and exchange interaction.
This quasiparticle has been reported in a large number of crystals. In
this paper, some of the experimental evidence for the existence of
biexcitons in CuCl is discussed.

The first kind of evidence relies on the appearance of a new emission, located on the low energy side of the free exciton luminescence
in pure CuCl crystals subject to an intense irradiation with UV laser
light. At 4.2°K, this new emission consists of four lines, which are
observed in a large variety of samples, provided the excitation rate
is important enough (see Fig. 1). The origin of the lines B and C at
25567 cm^{-1} and 25522 cm^{-1} has been attributed to the presence of biexcitons in the crystal.[1)] The main argument for this interpretation relies
on the particular intensity dependence of these lines upon excitation
intensity. In a large range of excitation, the lines show a quadratic

* Present address: Groupe de Physique des Solides de l'E.N.S., Tour 23 2, Place
Jussieu, 75221 Paris Cedex 05. FRANCE.

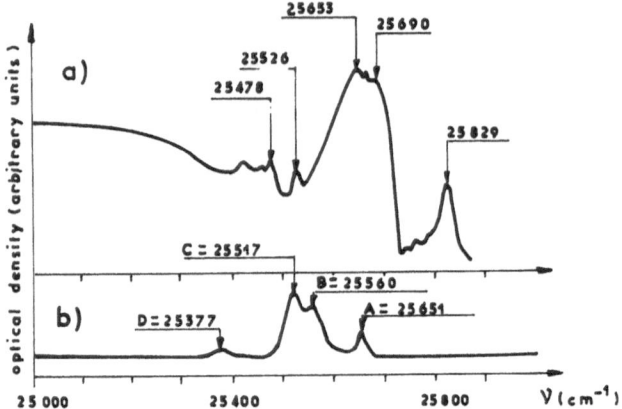

Fig.1. Microdensitometer trace of CuCl luminescence
at 4.2 °K when excited
a) by a low intensity light source
b) by the high intensity light from a UV
laser ($\hbar\omega$ = 288000 cm⁻¹).

dependence until they become linear, at higher input intensities. The
free exciton decay line, in contrast, remains linear in the whole excitation range. Such a behaviour may be understood if one assumes that
the new luminescence results from the radiative decay of biexcitons
into photons and free excitons. The biexcitons are formed because of
the high density of hot free carriers generated by incident light absorption so that the quadratic dependence is observed since two photons
are required to produce a single molecule. The linear law at higher
input intensities is due to the fact that a free exciton is generated
in the decay process. A detailed analysis of the kinetics has been
given by Knox *et al.*[2]

The position of the lines yields a value for the binding energy of
the molecule which is in good agreement with the calculations.[3,4] The
large value E_B = 275 cm⁻¹ (comparable to the exciton binding energy in
many II-VI compounds) is due to the important admixture of d-like
orbitals in the upper valence band (80%),[5] resulting in a very small
electron-to-hole effective mass ratio (0.02) for the quasi-particle,
which has thus a strong similarity with the hydrogen molecule.

A confirmation for the above interpretation has been given by Ueta
et al. from an analysis of the luminescence line shape.[6] It was found

that the band-to-band two-photon excited luminescence line shape could be fitted with a model in which the initial biexciton population had a Maxwell velocity distribution, as expected for a gas of free particles in interaction with the lattice. The effective temperature of the biexciton gas was found to be significantly higher than the lattice temperature.

More recently, Koda and Suga have analysed the luminescence of strongly excited mixed CuCl-CuBr crystals.[7] From the fact that the splitting between lines B and C varied as a function of composition in the same way as the splitting between the longitudinal and transverse Γ_5 exciton, they assigned the two lines to a transition from the same biexciton level to the transverse and longitudinal branch of the exciton-like polariton.

It must be noted that this set of results, although convincing, does not constitutes a definite proof for the existence of biexcitons in CuCl. This can be best seen from the example of Si and Ge, where the existence of excitonic molecules was first inferred from a similar kind of experimental evidence,[8,9] until the concept of electron-hole drops was proposed[10] and further experiments showed that the observed emission was related to the presence of metallic drops in these materials.[11]

A different type of evidence may be obtained from the study of the absorption spectrum of the biexciton. Such an approach has been proposed first by Hanamura.[12] It provides a powerful method of investigation because here one measures directly the biexciton resonance energy, in contrast with the previous method which relies on the analysis of a decay process. As noted by Hanamura, the two-photon transition rate from the ground state to the molecular state exceeds by several orders of magnitude the usual two-photon band-to-band transition rate in the same spectral region and takes exceptionally large values of the order 10^4 cm^{-1} MW^{-1}. There are two reasons for this giant nonlinear absorption coefficient; first, the proximity of the exciton

state from the single photon energy, at $E_{exc} = \hbar\omega + E_B/2$ which acts as

a quasi-resonant intermediate state; secondly, the particular (two-

electron) type of transition further enhances the non-linear response

of the medium.

A direct measurement of the two-photon absorption coefficient has

been performed, in a limited spectral range, around $2\hbar\omega = E_M = 2E_{exc} - E_B$

in CuCl by Gale and Mysyrowicz.[13] This can be done by measuring the

transmitted intensity of a parallel propagating intense monochromatic

light beam through the sample as a function of input intensity. In

presence of linear loss α and non-linear loss δI_o, the transmission law

reads:

$$I/I_o = \exp(-\alpha \ell)/[1 + \alpha^{-1} \delta I_o \{1 - \exp(-\alpha \ell)\}]. \tag{1}$$

A plot of I/I_o *versus* I_o, the incident intensity, directly yields α and

δ, the two-photon transition cross-section. In agreement with the

theoretical predictions, a strong and sudden increase of δ was measured

in the expected region from values $\leq 10^{-49}$ cm^4 sec typical of band-to-

band transitions in solids, up to $\delta = 10^{-44}$ cm^4 sec at 214 cm^{-1} below

the Γ_5 exciton line. Unfortunately, this kind of measurement could not

be performed at higher frequencies, close to the exciton resonance,

because of the rapidly increasing linear and non-linear losses, which

prevented any significant amount of light to be transmitted through the

sample, so that no independent measurement of the biexciton internal

energy E_M could be obtained by this method. A complete two-photon ab-

sorption spectrum, extending over an entire line profile, would be very

useful. The interpretation of such a spectrum will suffer from two

complications; first, one must consider the influence of a resonant

enhancement in the intermediate state, whenever the incident photon

energy coincides with a real level of the crystal. Such levels exist

close to the exciton, they are due to donor impurities. Even if the number of centers is small (10^{17}), it can increase the two-photon

transition rate considerably and thus distort the true biexciton ab-

sorption spectrum. Secondly, a structure in reflectivity has been ob-

served around $\hbar\omega = 2E_{exc}$ by Cardona[14] at room and nitrogen temperature.

It arises from an anomaly in the joint density-of-states due to the

second conduction band. In a non-centrosymmetric crystal like CuCl,

this structure can also be observed in the two-photon absorption

spectrum. Its contribution to the measured spectrum must be evaluated;

this may require more linear absorption or reflection data, especially

at liquid helium temperature.

Another method of investigation, related to the previous one, con-

sists in measuring the biexciton excitation spectrum. Whenever the

incident photon energy of the intense light source equals a half the

biexciton energy, one expects an increase of the characteristic mole-

cular luminescence, because a large number of molecules is directly

created by giant two-photon absorption. Hence, by monitoring the out-

put intensity as a function of input frequency (with constant input

intensity) one can get informations upon the biexciton resonance energy.

Such a method of investigation has the advantage of a great sensitivity

but requires caution in the interpretation. The excitation spectrum

does not reproduce directly the two-photon absorption spectrum, but is

affected by the competition between linear and non-linear attenuation

of the light beam in the medium. Further, the spectrum shape depends

upon the experimental conditions. In particular, the detection geometry

is crucial since the penetration depth of the beam in the sample varies

considerably with α and δI.

The total molecular luminescence excited by two-photon absorption

is of the form:

$$I_M = c \int_0^\ell I^2 (x) \, dx , \qquad (2)$$

where ℓ is the sample thickness.

For a parallel propagating beam $I(x)$ is given by eq.(1)

Integration of (2) yields:

$$I_M = c\ I_o \left\{ \frac{(1 + \varepsilon)\ \{1 - \exp(-\alpha\ell)\}}{1 + \varepsilon\ \{1 - \exp(-\alpha\ell)\}} - \frac{1}{\varepsilon}\ \ln\left[1 + \varepsilon\ \{1 - \exp(-\alpha\ell)\}\ \right] \right\} . \quad (3)$$

Thus, the output luminescence is a complicated function of ε, the ratio between two-photon and linear absorption and only in special conditions will it have a quadratic law dependence upon I_o. For example, in the case of important linear loss ($\alpha\ell > 1$) eq.(3) reduces to

$$I_M = c\ I_o \left\{ 1 - \frac{1}{\varepsilon}\ \ln(1 + \varepsilon)\ \right\} \quad (4)$$

and I_M is insensitive to variations of the two-photon absorption co-efficient, resulting in a broadening of the resonances. This behaviour reflects the fact that all incident photons are absorbed in the medium.

The molecular excitation spectrum of CuCl at 4.2°K and 77°K has been measured by Gale and Mysyrowicz.[13,15] (see Figs. 2 and 3). In this experiment, a backward geometry was used and care was taken that the luminescence from the whole sample depth was collected. In this way, the molecular excitation could be compared, through eq.(3) with the direct measurement of δ performed in the same spectral range. Excellent agreement was found between both methods. (see Fig. 2). The behaviour of lines B and C is consistent with the interpretation of these lines given earlier. They exhibit a peak in the excitation spectrum at the expected frequency $\hbar\omega = E_M/2$. (the relative intensity of these two lines, however, changes with input frequency, with a maximum for the ratio C/B at $\hbar\omega = 25815$ cm^{-1}, the energy of the for-bidden Γ_2 exciton, a fact presently not understood). As expected from eq.(3), the intensity dependence of the lines upon I_o varies between a quadratic and a linear law. The highest slope is observed at the lower excitation frequencies, becoming close to one near the excitation spectrum peak and remaining there at higher frequencies. Further the

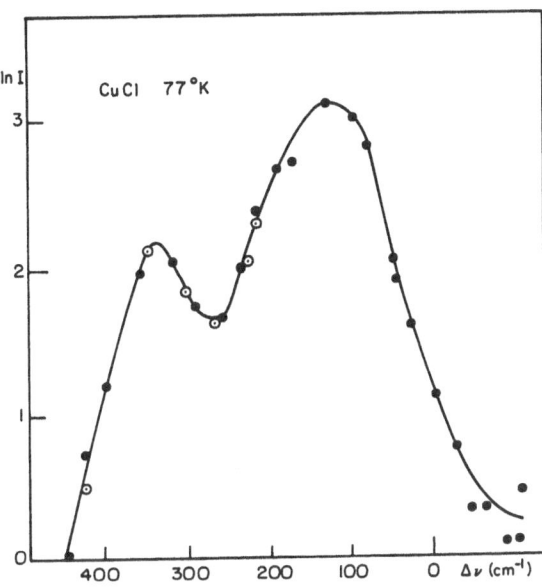

Fig.2. Two-photon excitation
spectrum of CuCl at
77 °K plotted in func-
tion of the difference
between the Γ_5 exciton
frequency and that of
the incident light.
The open circles are
obtained from the
direct measurement of δ.
The input intensity is
100 kW.

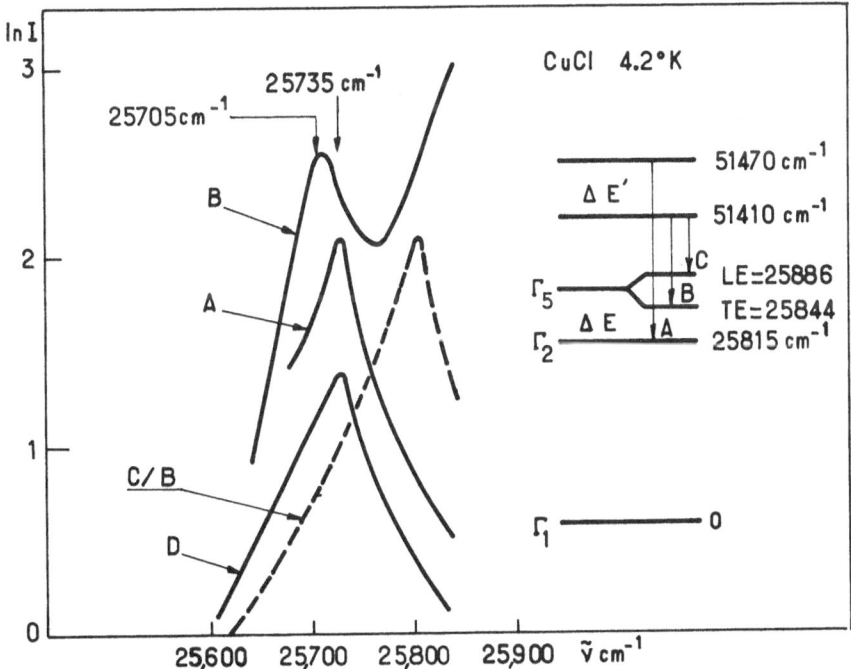

Fig.3. Two-photon excitation spectrum of the various emission lines of CuCl at
4.2 °K plotted in function of input light frequency. The input intensity
is 100 kW. Curve C/B gives the ratio of the intensities of lines C and
B.

width of the lines is smaller at the peak of the excitation spectrum
where 2 $\hbar\omega$ = E_M (see Fig. 4); this results from the fact that here
cold biexcitons (with a wave vector 2 k_o (k_o = incident photon wave
vector) are directly created. In that case the thermalisation of the

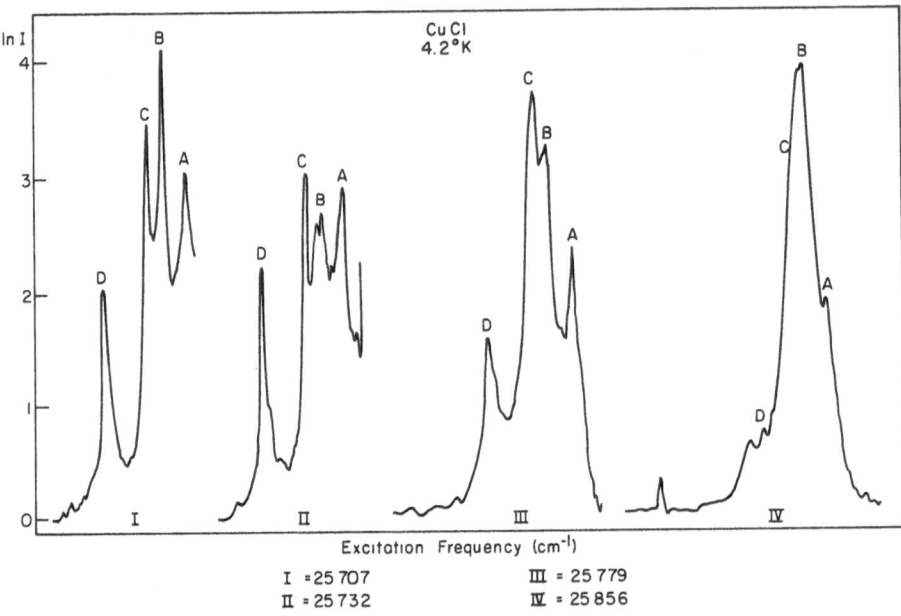

Excitation Frequency (cm⁻¹)

I = 25 707   III = 25 779
II = 25 732   IV = 25 856

Fig.4. Luminescence spectrum of CuCl at 4.2 °K, for different input frequencies.
The input intensity is approximately equal in each case = 100 kW. Curve
I corresponds to resonant two-photon excitation of biexcitons. In curve
IV, the input photon energy is resonant with the Γ_5 free exciton.

particles with respect to the lattice involves an interaction with
acoustic phonons and occurs in a time comparable with the lifetime of
the particles. Therefore, the thermalisation process is incomplete and
the decaying biexcitons have an effective temperature lower than the
lattice temperature. Therefore, the molecular emission will be narrow,
since the biexciton and exciton bands are flat, due to the large mass
of the particles (m_B = 2m_{exc} $\simeq$ 40 m_o; m_o = free electron mass). By
contrast, at higher input frequencies, an excess energy is delivered to
the system, which must be dissipated locally, leading to a higher
effective temperature for the gas of biexcitons as deduced from the

lineshape analysis.[6] Collisions with other particles present in that
case (excitons, free carriers) also contribute to the broadening of the
lines.

A different explanation of the observed narrowing at resonant two-
photon excitation, in terms of the occurrence of a Bose-Einstein condensation of biexcitons has been proposed by Ueta et $al.$ (see this
volume).

Other features observed in the excitation spectrum cannot be explained simply. The broad new emission appeared at 77°K, as well as
lines A and D at 4.2°K, shows a peak of the excitation spectrum at a
different, higher input photon energy. This has been attributed to the
presence of an excited molecular state, 60 cm^{-1} above the lowest free
biexciton state.[15] According to Bassani et $al.$[16] a state of symmetry
Γ_4, corresponding to the first rotational level of the orthomolecule of
hydrogen, is stable in a compound like CuCl. However, the two-photon
transition from the ground state Γ_1 to a state of symmetry Γ_4 with a
single light beam is forbidden in first approximation, so that the
direct creation of biexcitons in the excited state Γ_4 appears unlikely.

In the last part of this paper, the luminescence obtained with
intense, picosecond pulse excitation is described and a measurement of
the biexciton lifetime is reported.[17] The experimental set-up is shown
in Fig. 5. A pulse, of duration 25 ps is switched out from a single
transverse mode (TEM$_{oo}$) mode locked ruby laser pulse train and after
subsequent amplification to a final energy of up to 10 mJ illuminates
the sample held at liquid helium temperature. The luminescence is
detected in the backward geometry. The microdensitometer trace of a
typical emission spectrum, obtained in a single shot exposure is shown
in Fig. 6. It differs from that obtained when the same sample is
irradiated with a nanosecond UV pulse (see same figure) in that only
two of the four characteristic lines are seen, while a new emission
peak E appears at 25455 cm^{-1}.

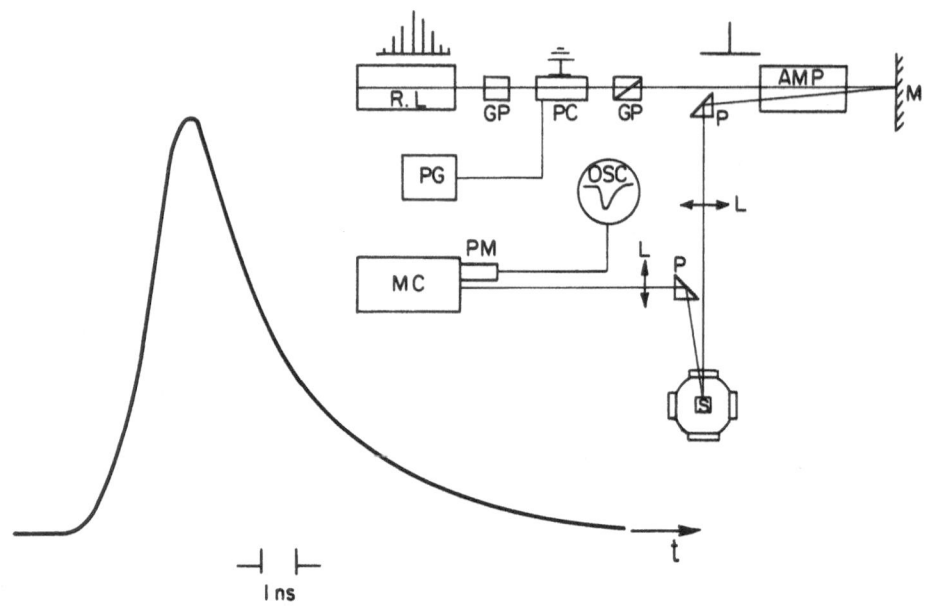

Fig.5. Time resolved luminescence at 25510 cm^{-1} from picosecond – excited CuCl at
4.2°K.
Experimental set-up (inset)

R.L – Mode-locked ruby laser
G.P – Glan-Thompson polarising prism
P.C – Pockel's cell
AMP – Ruby amplifier
M   – Mirror
P   – Prism
PG   – Pulse generator
L   – Lens
S   – CuCl sample at 4.2°K or 77°K
MC   – Monochromator
P.M – Photomultiplier.

The disappearance of lines A and B can be explained by considering
the particular conditions of excitation in this case: The generation
of one electron-hole pair requires the absorption of two red photons.
Since the two-photon absorption coefficient in this spectral region
($2\hbar\omega = 28800$ cm^{-1}) is small, of the order 0.1 cm^{-1} MW^{-1}, excitation
occurs in the volume of the sample.

After fast thermalisation of the hot free carriers, a relatively
low density of excitons (10^{16} cm^{-3}) is formed; the total, instantaneous
number of excitons is large, however, since the exciting pulse is short

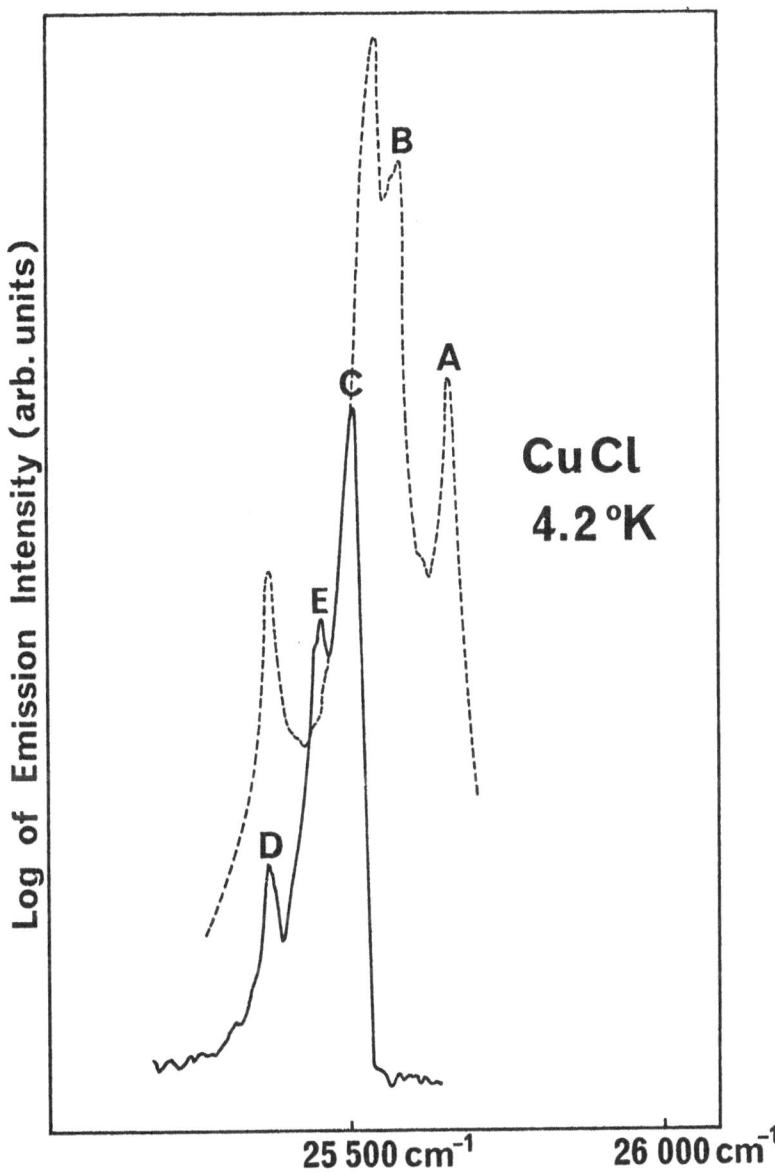

Fig.6. Solid line: Microdensitometer trace of CuCl luminescence at 4.2°K excited by single ruby picosecond pulse. Dotted line: Luminescence from the same sample excited by a ns UV pulse at 25780 cm^{-1}.

compared to the lifetime of excitons. Of these, only a small fraction will couple into molecules (10^{-3}) because of the large mean separation between the particles. In such a situation, where the decaying biexcitons are surrounded by a large number of excitons, part of the volume emission will experience reabsorption before reaching the sample surface and being detected. This reabsorption occurs because the reverse process takes place, in which photons are destroyed in the vicinity of existing excitons, to create biexcitons. This process has a very large probability, as shown by Golovin and Rashba.[18] It occurs for line B, but not C, indicating that only the lower lying transverse branch of the exciton-like polariton is substantially populated from the relaxation of the free carriers. Reabsorption of line A takes place because of the appreciable linear losses at this frequency. There are two reasons which render the molecular emission predominant, even if the number of biexcitons is small compared to that of the excitons: First, all biexcitons can participate in the decay, independent of their initial momentum while conserving both energy and momentum; by contrast, only optical excitons, with a k-vector of the order $kc = \omega$ can recombine. Secondly, the transition has a giant oscillator strength, as mentioned before.[18]

The decay time of line C has been measured[17] using a fast photomultiplier and a wide-band oscilloscope. A decay time of 3 ± 1 ns has been found at 4.2°K. This value is in agreement with the prediction of Golovin and Rashba.[18] At 77°K, no significant deviation from an instrument limited pulse could be seen, indicating a biexciton lifetime inferior to 0.5 ns at this temperature.

In conclusion, even if no single experiment at the present time proves in a totally irrefutable way the existence of the excitonic molecule in CuCl, a series of different results render its presence highly plausible.

It is a pleasure to acknowledge many fruitful discussions with Dr. G. M. Gale.

REFERENCES

1) A. Mysyrowicz, J. B. Grun, R. Levy, A. Bivas and S. Nikitine: Phys. Lett. 26A (1968) 61.

2) R. S. Knox, S. Nikitine and A. Mysyrowicz: Optics Comm. 1 (1969) 19.

3) O. Akimoto and E. Hanamura: J. Phys. Soc. Japan 33 (1972) 1531.

4) W. F. Brinkman, R. M. Rice and B. Bell: Phys. Rev. B 8 (1973) 1570.

5) K. S. Song: Journal de Physique 28 (1967) C3-43.

6) H. Souma, R. Goto, T. Ohta and M. Ueta: J. Phys. Soc. Japan 29 (1970) 697.

7) S. Suga and T. Koda: Phys. Stat. sol. B 61 (1974) 291.

8) J. R. Haynes: Phys. Rev. Letters 17 (1966) 860.

9) C. Benoît à la Guillaume, F. Salvan and M. Voos: J. Luminescence. 1 (1970) 315.

10) L. V. Keldysh: Proc. IX Intern. Conf. Phys. Semicond., Moscow, 1968, ed. S. M. Ryvkin (Nauka, Leningrad, 1968) p.1303.

11) See for example: M. Voos: Proc. XII Intern. Conf. Phys. Semicond. Stuttgart, 1974, ed. M. H. Pilkuhn (Teubner, Stuttgart, 1974) p.33.

12) E. Hanamura: Solid State Comm. 12 (1973) 951.

13) G. M. Gale and A. Mysyrowicz: Phys. Rev. Letters 32 (1974) 727.

14) M. Cardona: Phys. Rev. 129 (1963) 69.

15) G. M. Gale and A. Mysyrowicz: Proc. XII Intern. Conf. Phys. Semicond. Stuttgart, 1974, ed. M. H. Pilkuhn (Teubner, Stuttgart, 1974) p.133.

16) F. Bassani, J. J. Forney and A. Quatropani: Phys. Stat. sol. B (1975).

17) G. M. Gale, A. Mysyrowicz: Phys. Letters 54A (1975) 321.

18) A. A. Golvin and E. I. Rashba: JETP Letters 17 (1973) 478.

BOSE-EINSTEIN CONDENSATION OF FREE EXCITONS IN AgBr

W. Czaja

Laboratories RCA Ltd.,
Badenerstrasse 569
CH-8048 Zürich
Switzerland

ABSTRACT

Triplet excitons in AgBr are observed in photoluminescence at low temperatures with continuous near UV-excitation. The triplet exciton emission intensity varies strongly with temperature and excitation density. In particular this latter dependence supports the interpretation that these triplet excitons show Bose-Einstein condensation. The experiments also allow a determination of the critical temperature T_c. At low exciton densities ρ, T_c changes according to the ideal gas behaviour ($\sim \rho^{2/3}$) but differently at high exciton densities. By analysing the experimental data we arrive at some general conclusion about the observability of Bose-Einstein condensation of excitons.

I. INTRODUCTION

The AgBr samples were cut from very pure crystals originating from two different sources,[1] and were kept in the dark until they were immersed in liquid helium. Care was taken in mounting the crystals in order to minimize strains. Further experimental details can be found elsewhere.[2,3] Figure 1 shows the "TA_L"-phonon replicum of the exciton emission obtained with continuous near UV-excitation from an Ar^+-laser. The assignment of the phonon agrees with the recent inelastic neutron scattering data:[4] It is the TO_L-phonon with the symmetry of the TA_L-mode. The broader emission in Fig. 1 has a line shape determined by a Boltzmann distribution with a temperature of about 10 K, independent of experimental conditions.[2,5] This emission is due to the dipole allowed, phonon assisted, radiative recombination of singlet excitons. The sharp emission in Fig. 1 is attributed to a weakly dipole allowed, phonon assisted, radiative recombination of triplet excitons. The zero field singlet-triplet splitting is in agreement with magneto-absorption

experiments[6] and the
weak dipole character
of this transition is
due to an admixture
of Ag d-states to the
valence-band states
at the L-point.[3,7]
Thus triplet excitons
are expected to have
longer lifetimes than
singlet excitons and
consequently their
temperature is lower.

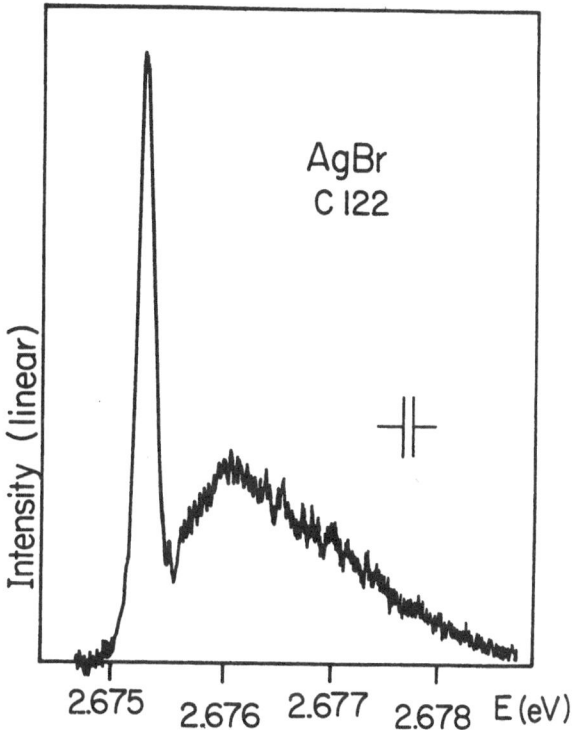

Fig.1. "TA$_L$"-phonon assisted exciton
spectrum of AgBr obtained with
laser excitation (3.40/3.52 eV)
at 1.27 K.

II. PROPERTIES OF TRIPLET EXCITON LUMINESCENCE

The intensity of the triplet exciton luminescence depends strongly
on temperature and on excitation density, compare Fig. 2.   In these
experiments[8] the laser power has been kept constant but the beam has
been focused differently.   It is mainly the excitation density dependence which supports our interpretation: The rather sharp luminescence
line of Fig. 1 is due to radiative recombination of triplet excitons
which show Bose-Einstein condensation (BEC), $i.e.$ which are in the total
momentum K = 0 state.   This model qualitatively explains our experiments.[3]   The intensity is proportional to N_o, the number of triplet
excitons in the BEC-state which is given by

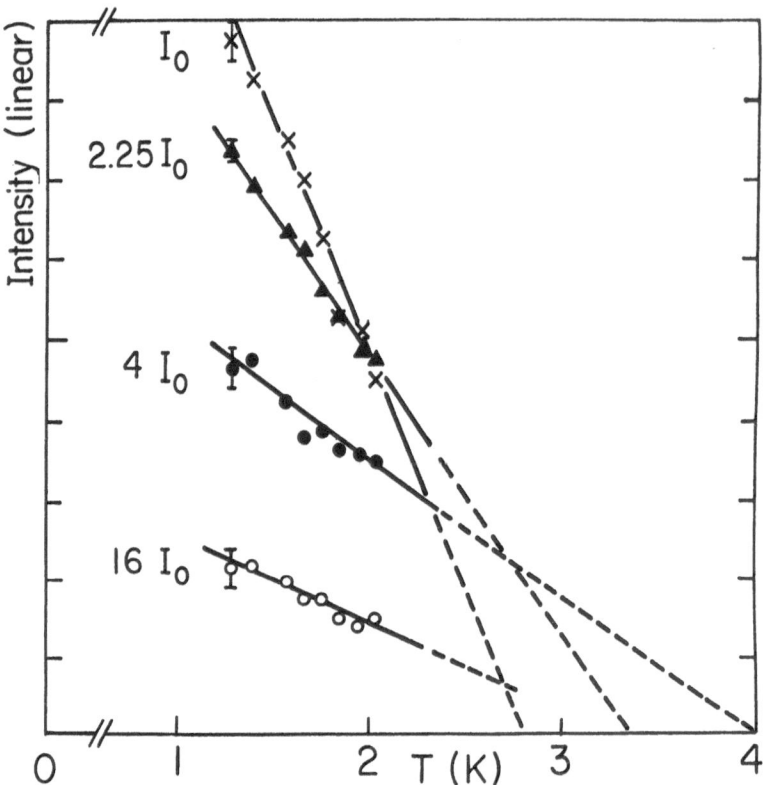

Fig.2. Intensity as a function of temperature and excitation
density (unnormalized). The numbers are relative excitation densities.

$$N_o = N(1 - (\frac{T}{T_c})^{3/2}) , \qquad (1)$$

for an ideal gas. N is the total number of excitons and T_c is the
critical temperature as obtained by a linear extrapolation of the
luminescence intensity as a function of temperature to the value zero.
The only free parameter is N. As exemplified in Fig. 3 our most
accurate data can be fitted quite well to a dependence according to eq.
(1). Also the variation with excitation density displayed in Fig. 2
is in qualitative agreement with our model: By increasing the excitation
density one increases the total exciton density ρ and consequently T_c
will increase. For an ideal gas, for instance, $T_c(\rho)$ is given by

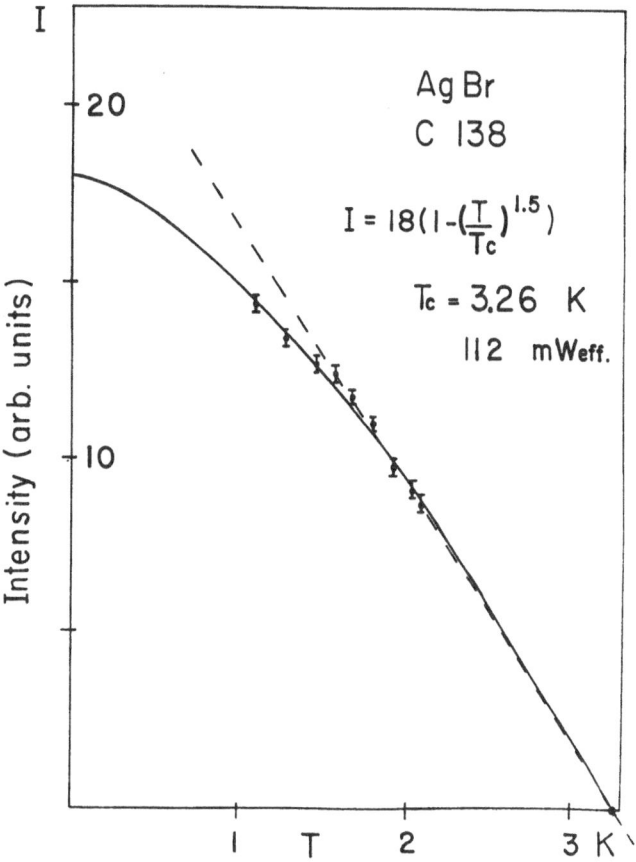

Fig.3. Experimental intensities obtained for i_{eff} = 112 mW fitted to an ideal Bose gas temperature dependence.

$$T_c = \frac{2\pi\hbar^2}{Mk} \, (\rho)^{2/3} \, \frac{1}{[\zeta(3/2)]^{2/3}} \, , \quad \zeta(3/2) = 2.612\ldots \quad (2)$$

Furthermore, one would argue that with increased density a possibly existing exciton-exciton interaction becomes increasingly effective and therefore the low temperature intensities will be decreased as is expected from the theory of a weakly interacting Bose gas, compare for instance.[3]

III.   EXCITATION DENSITY DEPENDENCE

By observing the luminescence spectrum Fig. 1 as a function of excitation density we have checked at several points that the result is independent whether the laser power is varied at constant illuminated area or whether the focus has been changed at constant laser power, provided the data are normalized with respect to the same illuminated area. We therefore believe that drastic heating effects are not present.

The exciton spectrum in Fig. 1 is inherently broadened by the width of the emitted phonon (~ 0.14 meV $\simeq 1.68$ K$^{2)}$) due to the indirect band gap in AgBr. Thus the observed intensity $I(\omega)$ is proportional to the density of states convoluted with this effective "slit width" ω_o.

$$I(\omega) \sim \frac{1}{\omega_o} \int_{\omega-\frac{1}{2}\omega_o}^{\omega+\frac{1}{2}\omega_o} d\omega' \sqrt{\omega'} \; n(\omega') \quad , \tag{3}$$

where $n(\omega')$ is the triplet exciton distribution. It is therefore impossible to perform a detailed line-shape analysis since ω_o is of the order of the bath temperature and since the high energy tail of the spectrum is masked by the much broader singlet exciton emission. However, some information about the exciton statistics can be obtained from an analysis of the observed dependence of the luminescence intensity on excitation density. To simplify the analysis we consider in Fig. 4 only the difference, ΔI, of the intensities of the sharp line of Fig. 1 observed at two different temperatures as a function of excitation power i_{eff}.[9] At very weak excitation the difference varies linear with i_{eff} and therefore also linear with ρ. This is expected if the exciton follows Boltzmann statistics. At some higher value of the excitation a drastic deviation occurs until ΔI is practically independent of the excitation density and therefore also of the total exciton density. This behaviour is expected if BEC occurs in an ideal gas, since all particles added to the system go into the condensate, whereas the thermal occupation depends only on temperature and not on density. In forming the intensity difference ΔI the density dependent terms cancel. The full horizontal line in Fig. 4 corresponds practi-

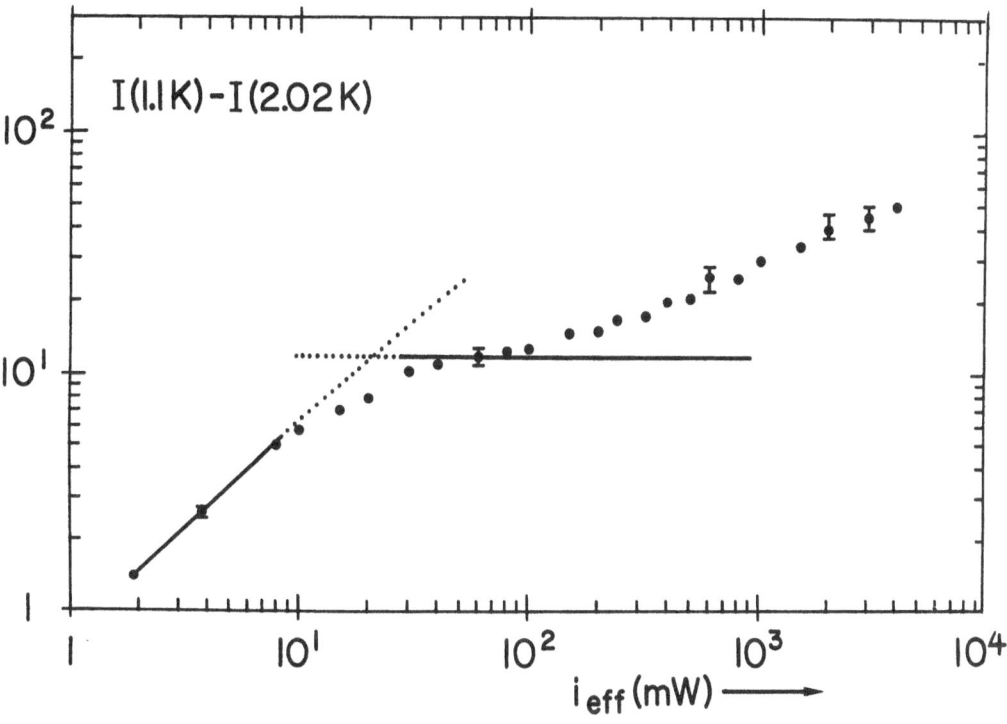

Fig.4. Intensity difference $\Delta I = I(1.1\ K) - I(2.02\ K)$ as a function
of laser power.

cally to the excitation-density range shown in Fig. 2.   It is important
to note that in agreement with the above explanation the temperature
dependence of the intensity Fig. 3 only applies to the nearly horizontal
range.   Equation (1) cannot be fitted to the temperature dependences at
very low excitation.   At very high excitations we expect to leave the
region for BEC again, as will be explained in the next section.   In
the whole range of Fig. 4 the observed line shape does not change
since it is broadened by ω_o as explained above.   This is also the
reason why no excitation or temperature thresholds have been observed
at the onset of BEC.

In the intermediate range between the linear and the almost horizontal part in Fig. 4 we have $1.1\ K \lesssim T_c \lesssim 2.02\ K$ and the upper and
lower limit allow a simple correlation between T_c and ρ.   To explain

the finer details of Fig. 4 a more refined model has to be created which among other effects has to include the exciton-exciton interaction, for which direct experimental evidence exists.[9]

IV. PHASE DIAGRAM FOR BEC

In order to obtain more quantitative information about $T_c(\rho)$ we have studied the following model:[10]   A Bose system whose particles interact only *via* two-body forces is subjected to the Bogoliubov treatment for weak interactions (the operators a_o and a_o^+ are replace by $\sqrt{N_o}$; only those interaction terms are retained in the Hamiltonian which are linear in N_o).   Furthermore it is assumed that the interaction can be described by specifying a scattering length, a.   Using the identity

$$< N_o > = N - \sum_{k}{}' < a_k^+ a_k > \quad , \qquad (4)$$

where the sum represents the thermally excited bosons, the average number $< N_o >$ of particles in the condensate can be calculated as a function of T.   The critical temperature is specified by the condition $< N_o > = 0$, whose solution is obtained numerically as a universal relation between t_c and ρa^3.   The dimensionless temperature t is defined in Fig. 5.   The region below the full curve is the two phase region for BEC (condensate and thermally excited bosons), above this curve only thermally excited bosons (normal gas) exist.   In the low density region the theory is expected to be of reasonable accuracy.   At high densities the results are only qualitatively correct.   Also shown in Fig. 5 are experimentally determined T_c for excitons from two different AgBr samples.   A scattering length $a = 2a_{ex}$ has been chosen, *i.e.* excitons are treated as hard spheres with a radius a_{ex} ($a_{ex} \approx 50\text{Å}$ exciton Bohr radius).   One experimental value ($\bullet$) is fitted to the theoretical curve.   However, the corresponding density agrees reasonably well with various other estimates.[3]   The points ($\blacktriangle$) are the values determined

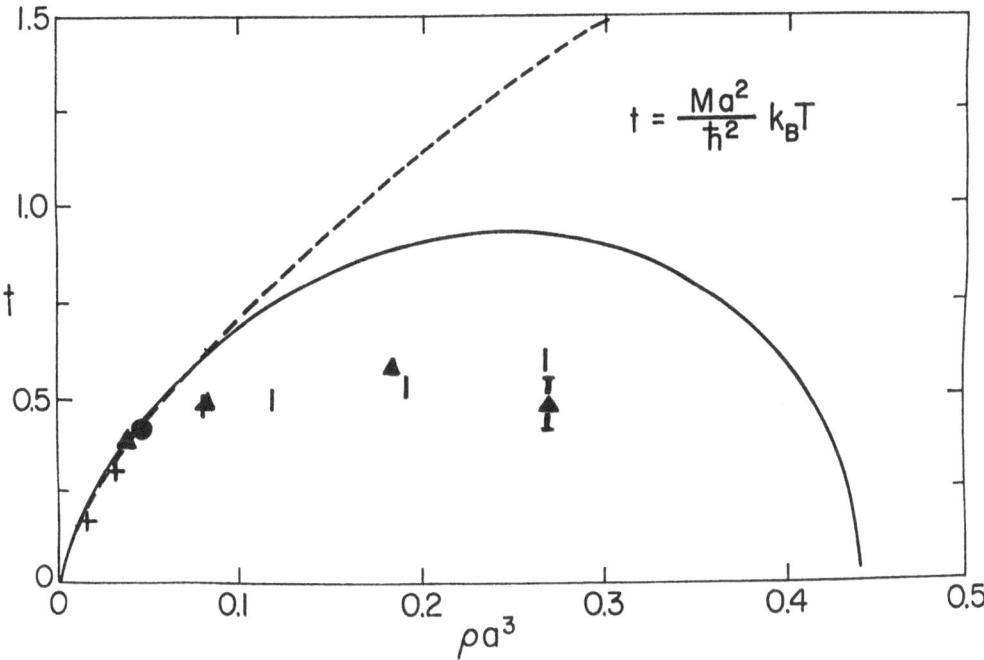

Fig. 5. Temperature vs. density (normalized units). Dashed curve: Ideal Bose gas. Solid line: Weakly interacting Bose gas. Experiments: See text.

from the measurements shown in Fig. 3. For a second sample the points (+) are determined as described in the section above, the bars (⌐) have been obtained by extrapolation (compare Fig. 3). The agreement is reasonable considering the as yet unsatisfactory experimental determination of ρ and T_c and considering the simplifications of the theoretical model.

V. CONCLUSIONS

The model calculation and the analysis of our experiments suggest the following more general statements:

1) Excitons form an interacting Bose gas which implies the existence of a maximum critical temperature (Fig. 5),

$$T_{c\ max} \simeq \frac{\hbar^2}{Mka^2} \qquad (\sim 7\ K\ for\ AgBr) \qquad (5)$$

At temperatures above $T_{c\ max}$ BEC cannot exist at all. Therefore thermalization of the excitons is important which requires long exciton lifetimes.

2) As a further consequence of the interaction there exists a maximum density above which no BEC will occur even at $T = 0$. Therefore, in order to observe BEC of excitons, the excitation should not be too high, independent of heating effects.

Since these considerations are general they will apply to the case of BEC of biexcitons as well. The following statements, however, are more related to BEC of excitons in indirect materials:

3) Due to the finite phonon width no sharp thresholds due to the onset of BEC will be observed in the emission and absorption spectra.

4) Our analysis demonstrates the importance of experimental information in addition to bare luminescence spectra, as for instance their dependences on temperature and excitation density.

5) The importance of some material parameters has been discussed in ref. 3). Another important property is the exciton-exciton interaction. An estimate[11] of the influence of the electron-phonon interaction in the rather strongly polar AgBr seems to indicate that the attractive interaction between excitons is considerably reduced.[12] However, a more detailed investigation is necessary.

The author is grateful to Dr. D. Baeriswyl for his help in preparing the material for this contribution.

REFERENCES

1) P. Junod, CIBA-GEIGY Photochemie, Fribourg, Switzerland; J. Malinovski, Institute of Photoprocesses, Bulgarian Academy of Sciences, Sofia, Bulgaria.

2) W. Czaja and C. F. Schwerdtfeger: Solid State Comm. 15 (1974) 87.

3) D. Baeriswyl and W. Czaja: RCA Review 36 (1975) 5.

4) H. Kanzaki, S. Sakuragi, S. Hoshino, G. Shirane and Y. Fujii: Solid

State Comm. <u>15</u> (1974) 1547;

W. von der Osten and B. Dorner: Solid State Comm. <u>16</u> (1975) 43.

5) W. von der Osten and J. Weber: Solid State Comm. <u>14</u> (1974) 1133.

6) M. Matsushita: J. Phys. Soc. Japan <u>35</u> (1973) 1688.

7) G. M. Mason: Phys. Rev. <u>B11</u> (1975) 5094;

J. Tejeda, N. J. Shevchik, W. Braun, A. Goldmann and M. Cardona: Phys. Rev. B (1975) to be published.

8) W. Czaja and C. F. Schwerdtfeger: *Proc. XII Int. Conf. Physics of Semicond., Stuttgart,* 1974, ed. M. H. Pilkuhn (Teubner, Stuttgart, 1974) p.142.

9) W. Czaja: *Proc. XII Congrès Européen de Spectroscopie Moléculaire, Strasbourg, France* 1975, ed. M. Grossmann, S. G. Elkomoss, and J. Ringeissen (Elsevier, Amsterdam 1976) p.155.

10) W. Czaja and D. Baeriswyl: *Proc. 14th Int. Conf. Low Temp. Physics, Helsinki* 1975 ed. M. Krusius and M. Vuorio, Vol.4 (North Holland Pub. Co., Amsterdam, 1976) p.429.

11) E. Tosatti: Phys. Rev. Letters <u>33</u> (1974) 1092.

12) Personal communication E. Tosatti: 1975. Although the mechanisms considered in ref. 11) are basically correct, the influence of the repulsive part has been overestimated.

DECAY OF TWO INDIRECT EXCITONS TO DIRECT AND
TO INDIRECT EXCITON STATE IN THALLOUS HALIDES

Jun'ichirō Nakahara and Koichi Kobayashi

Institute for Solid State Physics,
University of Tokyo
Roppongi, Minato-ku,
Tokyo, 106 Japan

ABSTRACT

Emissions from highly purified TlCl and TlBr are studied. The emission due to the radiative decay of the molecule of $X_6^+ \times R_6^-$ excitons by leaving one $X_6^+ \times R_6^-$ exciton and one M-point phonon through indirect process is found in the vicinity of the absorption edge. The emission due to the radiative decay of two $X_6^+ \times R_6^-$ excitons at the same M-point with leaving one $X_6^+ \times X_6^-$ exciton at the Γ-point through zero-phonon process is observed at the energy far lower than the edge.

I.   INTRODUCTION

In thallous halides, whose Brillouin zone is simple cubic, the maximum of the valence band is at the X-point where the symmetry of the wave function is X_6^+. There are two minina in the conduction band at the X-point and the R-point where the symmetries of the wave functions are X_6^- and R_6^- respectively. It has been shown recently that the R-point is lower than the X-point in the conduction band and the absorption edges of thallous halides at low temperatures originate in the excitation of the $X_6^+ \times R_6^-$ exciton by a phonon-forbidden indirect process with the emission of an M-point LA phonon.[1]

Let us take a system of two $X_6^+ \times R_6^-$ excitons. Since the wave vector of the $X_6^+ \times R_6^-$ exciton is at the M-point and there are three non-equivalent M-points (M_1, M_2 and M_3), the sum of the wave vectors of these two excitons is either at one of the three non-equivalent M-points ($M_1 + M_2 \rightarrow M_3$ etc.) or at the Γ-point ($M_1 + M_1 \rightarrow \Gamma$ etc.). We call them as the M-point two excitons and the Γ-point two excitons respectively.

The M-point two excitons will decay radiatively to the $X_6^+ \times R_6^-$ exciton state at a different M-point by the forbidden indirect one – phonon process when they form an excitonic molecule (molecule $M_3 \rightarrow$ exciton M_1 + phonon M_2 + photon). When they do not form a molecule, this emission is nothing but the emission due to a single exciton decay. Also they may decay to the $X_6^+ \times R_6^-$ exciton state at the same M-point by the zero phonon process (molecule $M_3 \rightarrow$ exciton M_3 + photon or exciton M_1 + exciton $M_2 \rightarrow$ exciton M_3 + photon). The emission due to the former decay must be at the energy of $E (X_6^+ \times R_6^-) - E_m^b - \hbar\omega_p$ where $E (X_6^+ \times R_6^-)$ is the energy of the $X_6^+ \times R_6^-$ exciton, E_m^b the binding energy of the exci- tonic molecule and $\hbar\omega_p$ the energy of the associated phonon. In the latter decay, the emission intensity would be quite small because two scattering, that the electron at the R-point is scattered to the X- point and the hole at the X-point is scattered to the different X- point, should occur at the same time.

The Γ-point $X_6^+ \times R_6^-$ excitons will decay radiatively to the M-point $X_6^+ \times R_6^-$ exciton by a forbidden indirect one-phonon process just same as the case of the M-point two excitons. The Γ-point two excitons may also decay radiatively to the Γ-point $X_6^+ \times X_6^-$ exciton state by a zero-phonon process because the conservation of the total momentum is satisfied. In this case, the emitted light must be at the energy of $2E (X_6^+ \times R_6^-) -$ $E (X_6^+ \times X_6^-)$ when the two excitons do not form a molecule and $2 E (X_6^+ \times R_6^-) - E_m^b - E (X_6^+ \times X_6^-)$ when they form a molecule where $E (X_6^+ \times X_6^-)$ is the energy of the direct $X_6^+ \times X_6^-$ exciton. We note that this process is particularly interesting because the $X_6^+ \times R_6^-$ exciton is converted to the $X_6^+ \times X_6^-$ exciton by the interaction between two $X_6^+ \times R_6^-$ excitons.

II. SAMPLES

Since our interest is focused on the intrinsic emissions, we used highly purified TlCl and TlBr. The powders of thallous halides were prepared from chemically purified $TlNO_3$ and hydro-halogenic acid. They

were filtered and distilled in molten state in vacuum and then were
zone-refined in respective hydrogen halide atmosphere. Their single
crystals were grown in vacuum.

The concentration of metal ion impurity was less than 10^{-7} mole
fraction. The concentration of iodine ion was less than 10^{-6} mole
fraction in TlCl and was probably the same order in TlBr. The concentration of other halogen impurity was less than 10^{-4} mole fraction
and probably much lower. The broad band emissions which are believed
to be due to impurities and lattice defects and have been found in any
thallous halide samples used in the previous works were not observed in
our TlBr and were very faint in our TlCl.

III. RESULTS AND DISCUSSION

Figures 1 and 2 are the emission spectra at 4.2 K in TlCl and TlBr
respectively which are excited by a 500 W mercury lamp. All emission
intensities are weak. In each figure, the zero-phonon energy of the
$X_6^+ \times R_6^-$ exciton is indicated as E $(X_6^+ \times R_6^-)$.[1] We note that the general
features of both the spectra in Figs. 1 and 2 are almost same.

Each spectrum is divided into two groups. The first one is the
group of several sharp emission lines observed near the energy of E
$(X_6^+ \times R_6^-)$ and they are the edge emissions. Among the edge emissions,
the emissions a,b and c are due to the radiative decays of the $X_6^+ \times R_6^-$
exciton from its three levels through the phonon-forbidden indirect
process with emitting an M-point LA phonon.[2] The emissions d and h in
TlCl and TlBr grow nonlinearly and they are one of the objects of this
study. Another object is the emissions α, β and γ in TlCl and the
emission α' in TlBr whose intensities are very weak and whose energies
are far lower than the edge emissions.
a) Decay of the molecule of the $X_6^+ \times R_6^-$ excitons to the $X_6^+ \times R_6^-$ exciton.

Figures 3 and 4 show the change of the emission spectra in TlCl
and TlBr at 4.2 K with the increase of the excitation intensity. The

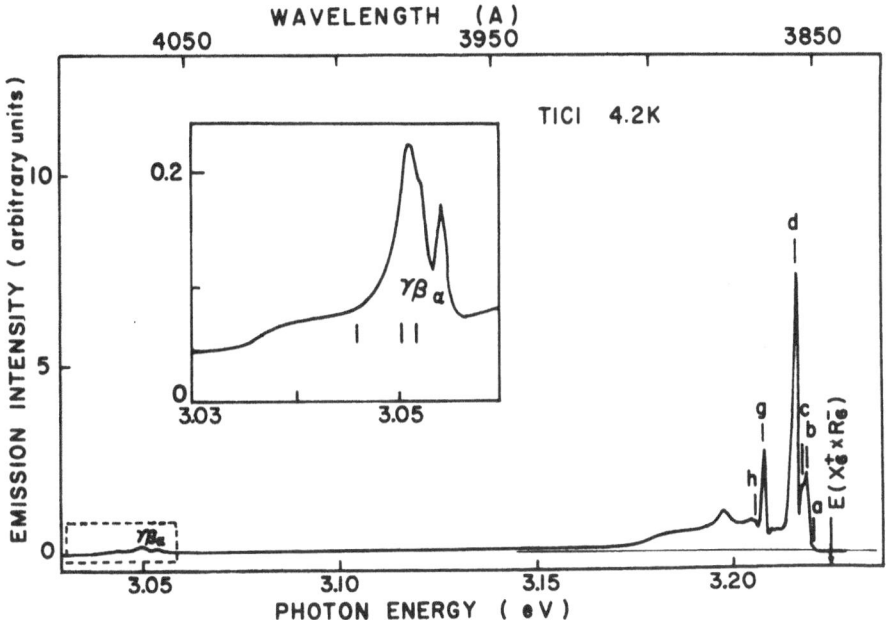

Fig.1. Emission spectrum in TlCl at 4.2 K. The arrow indicated by
E $(X_6^+ \times R_6^-)$ shows the energy of the zero-phonon $X_6^+ \times R_6^-$ exciton.
In the inset, the enlarged spectrum near the emission α, β
and γ is shown with the values of 2 E $(X_6^+ \times R_6^-)$ - E $(X_6^+ \times X_6^-)$
which are indicated by three vertical lines.

lower two curves and the upper two curves are obtained by the excitations with the mercury lamp and the nitrogen laser respectively. At
the high exitation intensity, only the emissions d and h are observed.
In TlCl, the emission d grows according to $I_0^{1.8}$ in the range of low
excitation intensity of $10^{16} \sim 10^{17}$ photons cm^{-2} sec^{-1} whereas it grows
as $I_0^{1.1}$ in the range of $10^{20} \sim 10^{23}$ photons cm^{-2} sec^{-1}. The change from
nearly I_0^2 to I_0 in the dependence on the excitation intensity suggests
that the emission d originates in the decay of an exciton through the
process involving two excitons.

The emission d in TlCl would not be the radiative decay of the M-
point exciton by the indirect process after the inelastic collision of
two M-point excitons with exciting the other M-point exciton to its

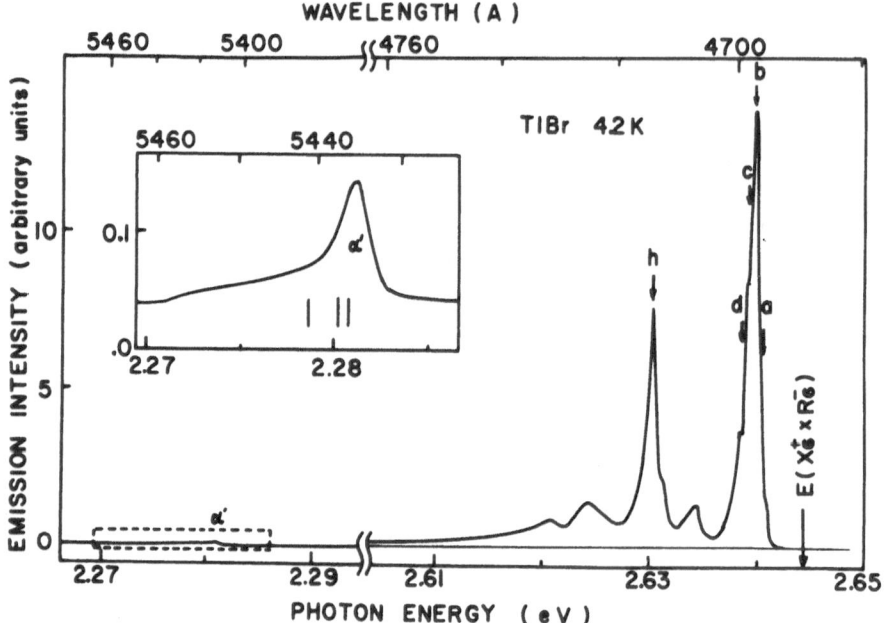

Fig.2. Emission spectrum in TlBr and 4.2 K. The arrow indicated by
E $(X_6^+ \times R_6^-)$ shows the energy of the zero-phonon $X_6^+ \times R_6^-$ exciton.
In the inset, the enlarged spectrum near the emission α' is
shown with the values of 2 E $(X_6^+ \times R_6^-)$ – E $(X_6^+ \times X_6^-)$ which are
indicated by three vertical lines.

higher envelope state. This is because the emission by this process
must be at the energy lower than the energy of the exciton emission (a,
b and c) by the amount of at least three quarters of the exciton binding
energy which is the order of 20 meV in TlCl.[2] The peak energy of the
emission d is lower than those of the emissions a, b and c by 3.8, 2.6
and 1.7 meV respectively in TlCl so that the emission d is not due to
exciton-exciton inelastic collisions.

By using the Cho's method,[3] we calculate the emission line shape
due to the radiative decay of a molecule of indirect excitons through
the phonon-forbidden indirect process with leaving one indirect exciton
and one phonon. The calculated result is in good agreement with the
experimental line shape of the emission d by taking the temperature of

Fig.3. Edge emission spectra
in TlCl at 4.2 K under
the excitations of
various intensities.
The relative magni-
tudes of the inten-
sities between the
spectra are arbitrary.
The number given in
each curve is the ex-
citation intensity
expressed by the
number of incident
photons per cm² sec.

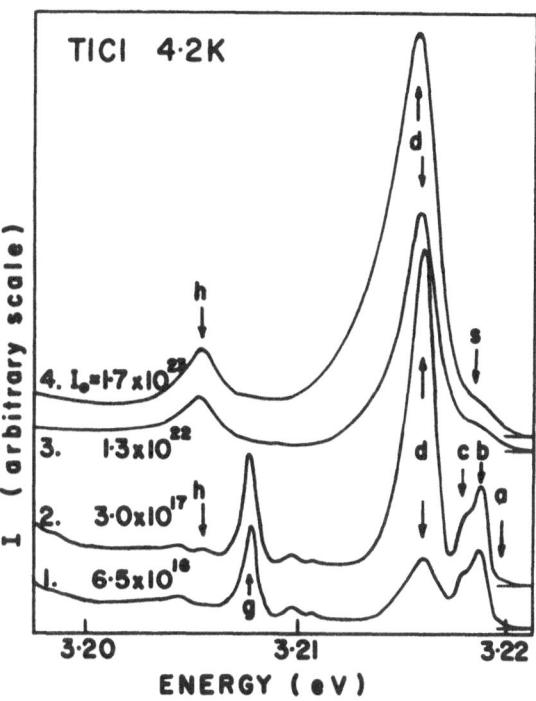

Fig.4. Edge emission spectra
in TlBr at 4.2 K under
the excitations of
vaious intensities.
The relative magni-
tudes of the inten-
sities between the
spectra are arbitrary.
The number given in
each curve is the ex-
citation intensity
expressed by the
number of incident
photons per cm² sec.

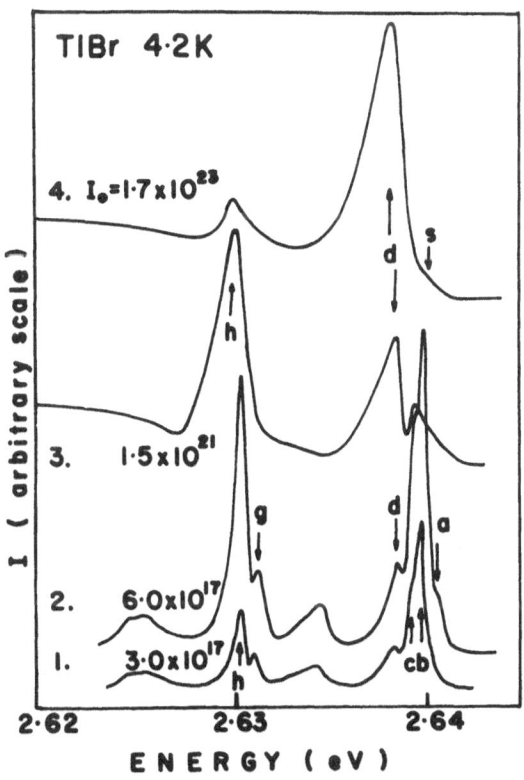

the molecule being 4.9 K in the lattice at 4.2 K. Based upon the picture that the emission d is due to the decay of the molecule of the forbidden indirect excitons, the binding energy of the molecule is estimated as 1.5 ± 0.3 meV from the emissions c and d in TlCl. The binding energy estimated from the theory by Akimoto et al.[4] is 0.9 ± 0.2 meV which is in good agreement with the experimental value.

From the above consideration, we propose that the emission d in TlCl originates in the radiative decay of the molecule of the $X_6^+ \times R_6^-$ excitons at the M- or Γ-point through a phonon-forbidden indirect process with leaving one $X_6^+ \times R_6^-$ exciton and emitting one M-point LA phonon. The emission d in TlBr behaves almost similarly to the emission d in TlCl and the binding energy of the molecule of the $X_6^+ \times R_6^-$ excitons is obtained as 0.7 ± 0.4.

The emissions h in TlCl and TlBr grow superlinearly ($I_0^{1.5}$ in TlCl and $I_0^{1.8}$ in TlBr) at low excitation intensity and do linearly at high excitation intensity. Thus the emission h must relate to the process involving two $X_6^+ \times R_6^-$ excitons. However we have no conclusive explanation for its origin.

b) Decay of the Γ-point two $X_6^+ \times R_6^-$ excitons to the $X_6^+ \times X_6^-$ exciton.

In the insets of Figs. 1 and 2, the detailed line shapes of the emissions α, β and γ in TlCl and the emission α' in TlBr are shown. These emissions are observable with the excitation by the light whose energy is higher as well as lower than E $(X_6^+ \times X_6^-)$. The peak energies of the emissions α, β and γ are 3.0542 eV, 3.0525 eV and 3.0513 eV respectively. The peak energy of α' is 2.2812 eV. The three vertical lines given in each inset are the values of 2 E $(X_6^+ \times R_6^-)$ - E $(X_6^+ \times X_6^-)$. In the estimation of these three values, we use three energy levels of the $X_6^+ \times X_6^-$ exciton, where the lowest level is the spin triplet state and the other two higher levels are the singlet-triplet mixed states. Their energies at 4.2 K are 3.3984 eV,[5] 3.3997 eV[6] and 3.4040 eV[6] in TlCl, and 3.0079 eV,[7] 3.0084 eV[6] and 3.0100 eV[6] in TlBr. The values of E $(X_6^+ \times R_6^-)$ are 3.225 ± 0.001 eV in TlCl and 2.6444 ± 0.0005 eV in TlBr at 4.2 K.[1]

We observe that the energies of the three vertical lines fall almost within the line widths of the emissions α, β and γ in TlCl and of the emission α' in TlBr. This shows that the energies of these emissions are fairly close to 2 E $(X_6^+ \times R_6^-)$ - E $(X_6^+ \times X_6^-)$. Thus we conclude that the emissions α, β and γ in TlCl and α' in TlBr are due to the radiative decay of two $X_6^+ \times R_6^-$ excitons both at the same M-point with leaving one Γ-point $X_6^+ \times X_6^-$ exciton through a zero-phonon process. We are however unable to clarify whether or not two $X_6^+ \times R_6^-$ excitons form their excitonic molecule before their decay. This is because the accuracies of the values of E $(X_6^+ \times R_6^-)$ in TlCl and TlBr are the same order of the binding energies, E_m^b, of the molecule of $X_6^+ \times R_6^-$ excitons which are 1.5 ± 0.3 meV in TlCl and 0.7 ± 0.4 meV in TlBr. The above results show that the indirect $X_6^+ \times R_6^-$ exciton is converted to the direct $X_6^+ \times X_6^-$ exciton through the interaction of two indirect $X_6^+ \times R_6^-$ excitons in thallous halides.

REFERENCES

1) J. Nakahara, K. Kobayashi and A. Fujii: J. Phys. Soc. Japan 37 (1974) 1312; *ibid.* 37 (1974) 1319.

2) J. Nakahara and K. Kobayashi: J. Phys. Soc. Japan 40 (1976) 180.

3) K. Cho: Optics Commun. 8 (1973) 412.

4) O. Akimoto and E. Hanamura: J. Phys. Soc. Japan 33 (1972) 1537; *ibid.* 35 (1973) 973.

5) S. Kurita, K. Kobayashi and Y. Onodera: Progr. theor. Phys. Suppl. No.57 (1975) 10.

6) S. Kurita and K. Kobayashi: J. Phys. Soc. Japan 30 (1971) 1645.

7) The energy of the triplet state of the $X_6^+ \times X_6^-$ exciton in TlBr is not known. We assume that the energy separations between the three levels in TlBr are proportional to those in TlCl. We then

estimate the energy of the triplet state in TlBr from the observed
energy separations between three levels in TlCl and the observed
energy separation between two singlet-triplet states in TlBr.

BIEXCITON FLUORESCENCE LINE SHAPE IN CdS

R. Planel and C. Benoît à la Guillaume

Groupe de Physique des Solides[*]
de l'Ecole Normale Supérieure
Université de Paris VII.
2 place Jussieu, 75221
Paris Cedex 05

ABSTRACT

A careful experimental study of the M line in CdS is presented, with special attention to polarisation properties. The data can be fitted by a theoretical model of the line shape in both polarization ($E \perp C$ and $E // C$) taking into account the results of group theory, with the recombination of the molecule considered as giving two polaritons. A more precise value of the energy $E_B(o)$ of the ground state of the excitonic molecule is obtained $E_B(o)/2 = 2.5497$ eV ± 0.2 meV.

I. INTRODUCTION

Much work has been already done on the biexciton luminescence in wurtzite type semiconductors.[1,2] However, a complete agreement about the origin of the M line is far from being complete.[3,4] We present here additional experimental results on the polarization properties of the M line in CdS, which are compared to a theoretical model more elaborate than the model proposed earlier by E. Hanamura.[2] The good fit between the experiment and the theory provides a better proof of the existence of biexciton in CdS, and a more accurate value of its ground state energy $E_B(o)$.

II. EXPERIMENTAL SET UP

CdS platelets of good quality are immersed in pumped helium (T = 1.6 K). They are excited by the 4765 Å light of a cavity dumped A_r^+ laser, delivering 20 nsec pulses. At 15 W peak power focused in a spot of 60 μm in diameter, the excitation is about 0.5 MW/cm^2. The

[*] Laboratoire associé au C.N.R.S.

repetition rate was chosen in order to maintain the mean power below 10 mW. Emission spectra were taken in both polarization. A crucial point to select purely Γ_1 photon (E//C) is the use of a narrow vertical slit (the C axis of the crystal being horizontal) in order to select photons propagating in a narrow beam k⊥C. Another advantage of this geometry (k⊥C, E//C) is that one obtains a very good extinction of the I_2 line, which occurs in the same region as the M line. Since the M line is not so strongly forbidden in that geometry, one can observe it at a power density as low as 1 kW/cm^2 (see Fig. 3).

III. THEORETICAL MODEL[5]

We take here the point of view that a biexciton decays into two polaritons, connected with the lowest A, n=1 exciton state. A given event:

$$\text{Biexciton}(\vec{K}_B, \ E_B) \rightarrow P(\vec{K}_1, \ E_1, \ \vec{\pi}_1) + P(\vec{K}_2, \ E_2, \vec{\pi}_2),$$

where $\vec{K}$, E and $\vec{\pi}$ refers respectively to wave vector, energy and polarization of the particle, has to be characterized by a matrix element; in addition the usual rules of conservation of energy and momentum are taken into account. Fig. 1 shows that this point of view gives significant changes with respect to the simple theory[2] especially for biexciton of small momentum.

The matrix elements are determined in the approximation where one polariton is photon-like ($\vec{K}_1$ neglected) and the second is exciton-like ($\vec{K}_2 \sim \vec{K}_B$). We retain only the process allowed in the dipolar approximation.

Taking the z axis along C and the x axis along the momentum of the photon emitted we distinguish two cases:

1) E⊥C. The photon is of symmetry Γ_5. Since the biexciton is Γ_1, the emission of a Γ_5 exciton in the crystal is allowed (it can be purely transverse, or mixed). The emission of a Γ_6 exciton is forbidden. In

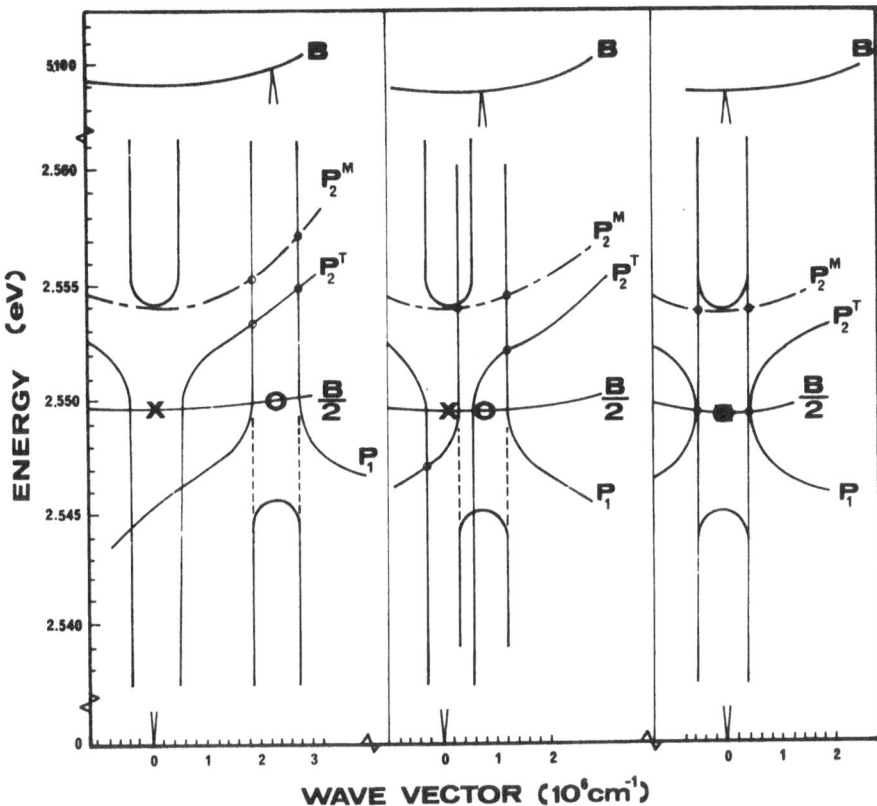

Fig.1. Schematic diagram showing the conservation of energy and momentum in the
dissociation of a biexciton of momentum $\vec{K}_B$ into two polaritons. Here,
the direction of momentum is restricted to a single direction perpendicular to the C axis.
In a) $\vec{K}_B$ is larger than the momentum of photons. One polariton is
photon-like, the other exciton-like (longitudinal or transverse).
In b) K_B is of the order of the momentum of a photon.
In c) $K_B = 0$.

fact, the emission of two polaritons with polarizations perpendicular

to each other is forbidden; this is why the matrix element in that case

is taken as proportional to $\cos \vec{\pi}_1 \cdot \vec{\pi}_2$.

2) E//C. The photon is of symmetry Γ_1. Then the process giving a Γ_5

(or a Γ_6) exciton is forbidden at the zone center. We have to take

into account the results of group theory away from $\vec{K} = o$. It can be

shown that the $\vec{K} \cdot \vec{P}$ theory can be applied to any eigenstate of a system

with translational invariance; so we can use it for exciton or biexciton states. The only mixing of interest occurring to first order in $\vec{K}$ is between Γ_1 and Γ_{5L} states and is proportional to $|\vec{K}_\perp|$ (Γ_{5L} is the state of polarization $\vec{\pi}$ parallel to $\vec{K}_\perp$). This means that for E//C, we can emit in the crystal only a mixed Γ_5 exciton (which becomes longitudinal if $\vec{K}_2 \perp C$), since in our approximation $\vec{K}_B = \vec{K}_2$. The matrix element is taken like $|\vec{K}_{B\perp}|$. The emission of Γ_5 transverse exciton is forbidden, as well as a Γ_6 exciton.

In addition, in the case E⊥C, there is a problem of probability P_e of escape of the polariton P_1 outside of the crystal. Since the total spectrum of the polariton created in the annihilation of a biexcit-on exhibits two peaks which are mirror like with respect to $E \sim E_B/2$, we assume a constant value of p_e for the lower energy peak and $p_e \sim 0$ for the upper peak.

Figure 2 gives the results of the line shape calculation, assuming Boltzmann distribution of the molecules with two different effective temperatures T_B. A few comments are of interest.

1) At low T_B, the line shape are much broader than expected when the simpler theory is used. This is due to the finite wave vector and the finite mass of both polaritons P_1 and P_2, and also the anisotropy of the exciton mass (related to the anisotropy of the valence band).

2) It is difficult to observe the longitudinal transverse splitting on the E⊥C spectrum, except at very low T_B. One reason is that in wurtzite crystal, the longitudinal mode is only a special case (k⊥C) of the mixed mode. The L. T. splitting is in fact more easily seen as a shift between the two curves for E⊥C and E//C.

3) Polaritons in the upper branch give negligible contribution to the spectrum.

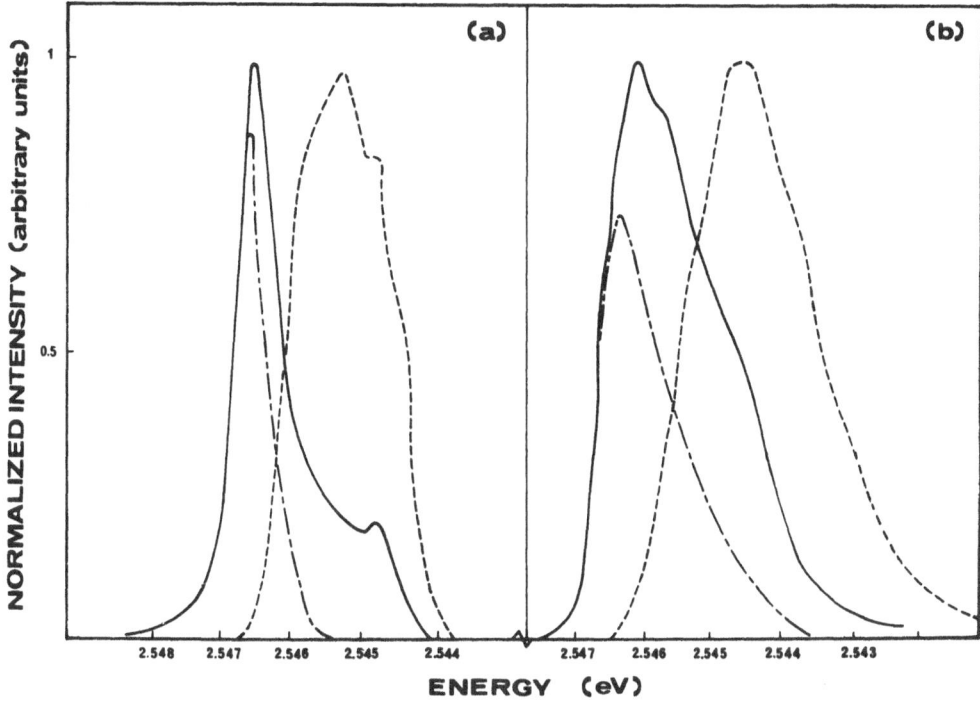

Fig.2. The theoretical M line shape for kT_b = 0.1 meV (a) and 0.5 meV (b) in
Γ_5 (——) and Γ_1 (----) observation.
The numerical values used for calculation are summarized in note (6) of
the text.
The two curves are arbitrarily normalized. In Γ_1 observation, only the
emission of a "mixed" polariton p_2 is allowed. In Γ_5 observation, there
are two contributions of equal integrated intensity. The transverse one
is shown by (— – —).

IV. DISCUSSION OF THE RESULTS

Figure 3 shows the results obtained at a rather low excitation level
for two different samples. A very good fit of the experimental line
shape for E//C is obtained without any additional broadening and for T_B
rather low. For E⊥C, it is just possible to verify that the theoretical
line shape is compatible with the experimental one, which is still
dominated by the I_2 line. A value of $E_B(o)/2$ = 2.5497 eV ±0.2 meV
gives the best fit to the data.[6]

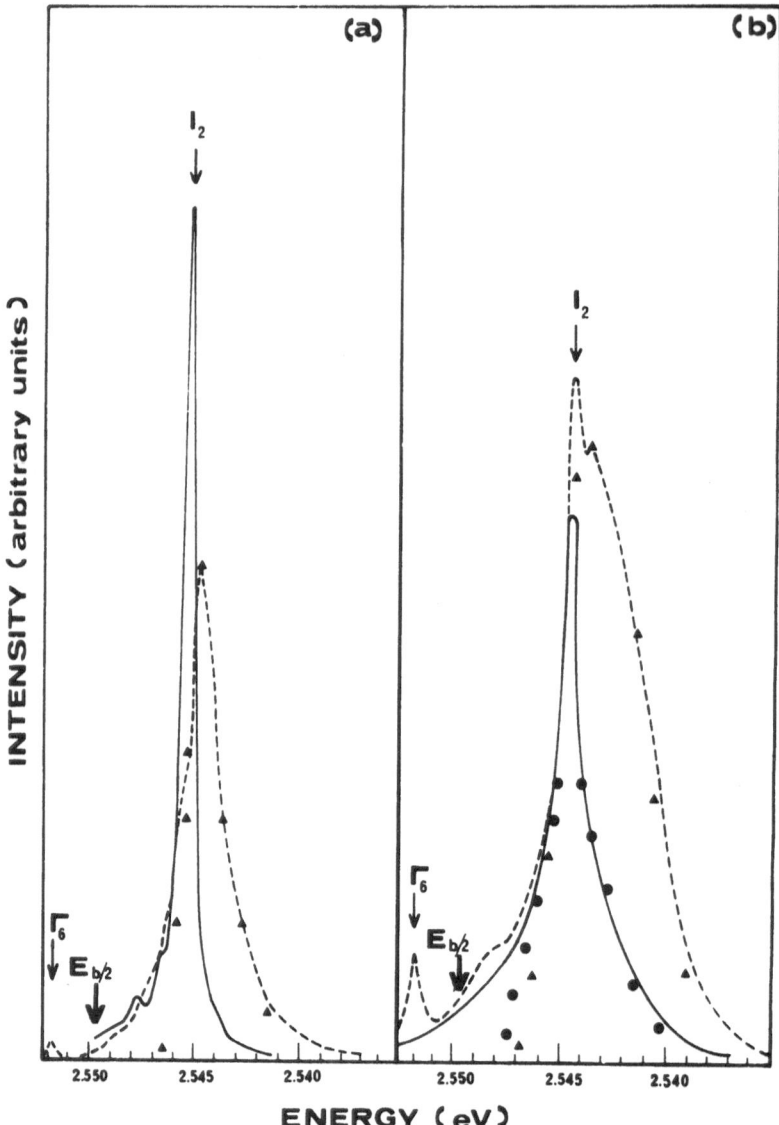

Fig.3. Experimental results in Γ_5 (——) and Γ_1 (----) observation of
the M line at rather low ((a): 1 kW/cm^2) and intermediate
((b): 6 kW/cm^2) excitation level. The spectra in a) are
taken from a purer sample than in b), so that the I_2 line is
more extinguished by the Γ_1 observation.
When the M line is clearly distinguished, the fit gives
satisfactory results ($\bullet$ in Γ_5 observation, $\blacktriangle$ in Γ_1 observation).
The obtained values for kT_b are 0.4 meV (a), and 1.2 meV (b).

Figure 4 gives data at higher excitation (150 kW/cm^2). In that

case, it is difficult to give a good fit for E//C, because in addition

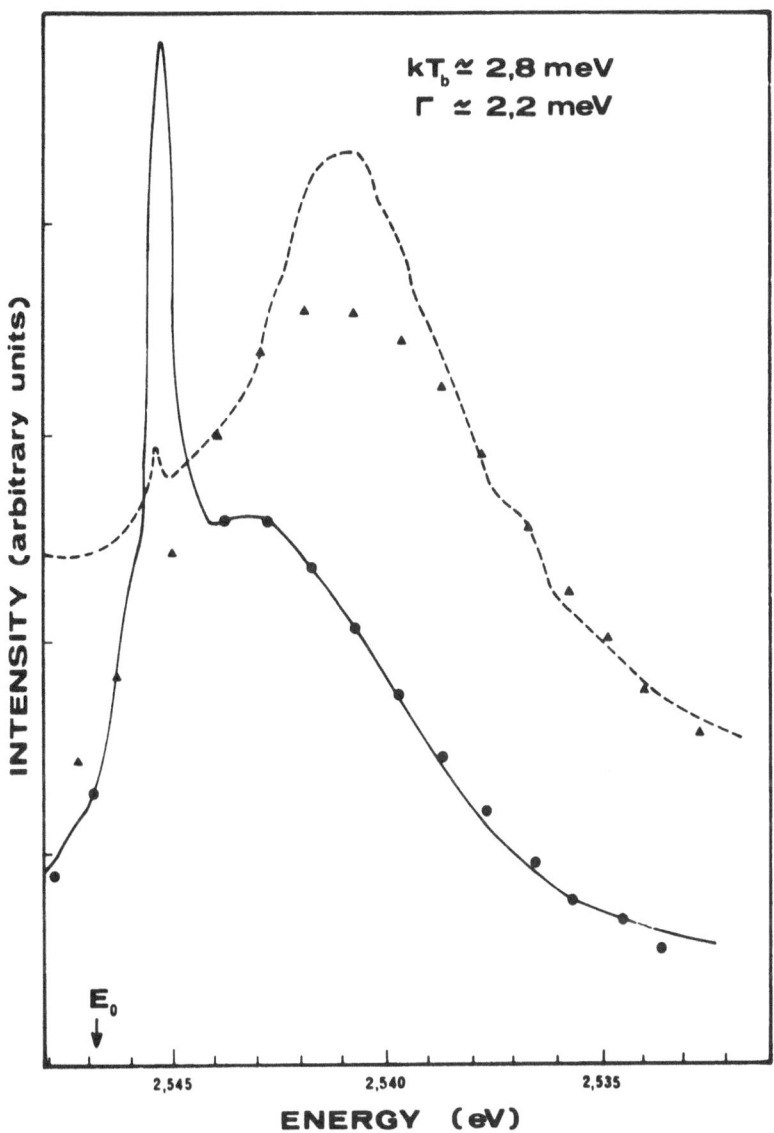

Fig.4. Experimental results in Γ_5 (——) and Γ_1 (----) observation of the M line, at rather strong excitation level (150 kW/cm^2). Here the theoretical fitting is possible using the eqs.(1)(●) and (3)(▲) derived from the model of Hanamura. Although the energy dependence obviously differs in Γ_5 and Γ_1 observation, the fitting of the "forbidden" line is made difficult by the irregular background growing at these excitation levels, and in this configuration of observation. A possible theoretical curve, coherent with the results of Γ_5 observation is however shown.

to the M line, a rather large background appears between 2.556 and

2.540 eV, which could correspond to interaction between excitons, involving at least one exciton of band B. The E⊥C spectrum can be fitted
by Hanamura's model, slighly modified to take into account the partici-
pation of both exciton modes (transverse and mixed) by an average exciton energy.

$$F(E) = (E_o-E)^{1/2} \exp \left[(E_o-E)/kT_B \right] \otimes \left[(E_o-E)^2 + \Gamma^2 \right]^{-1} \qquad (1)$$

where $\otimes$ means convolution product with

$$E_o = E_B(o) - E_{ex}^L(o) + 5/8 \, \Delta_{LT} \, . \qquad (2)$$

In that case, rather large biexcitonic temperature T_B and Lorentzian
broadening parameter Γ are required, as in ref. (1). It should be
noted that, by inspection of the 1 LO and 2 LO replica of the free
exciton line, the temperature of excitons is close to T_B. The line
E//C, after subtraction of the background emission is fitted by:

$$F(E) = (E_o^1-E)^{3/2} \exp \left[(E_o^1-E)/kT_B \right] \otimes \left[(E_o^1-E)^2 + \Gamma^2 \right]^{-1} \qquad (3)$$

with

$$E_o^1 = E_B(o) - E_{ex}^L(o) \, . \qquad (4)$$

The simultaneous fit of both line yields a value of $E_B(o)$ very close, to
the one obtained at low excitation level, but with a poorer precision.

V.   CONCLUSION

We think that these data give additional support to the existence
of the biexciton in CdS, and the same experimental technique and
theoretical analysis should be generalized to other wurtzite type
material. We can predict the spectrum emitted by biexcitons at k = 0
(Bose condensation). In that case, P_1 and P_2 have opposite momentum
and should have the same polarization, so they should have the same
energy. A single sharp line at $E_B(o)/2$ should be emitted (and not at

E_o given by eq.(2)).

REFERENCES

1) S. Shionoya, H. Saito, E. Hanamura and O. Akimoto: Solid State Comm. 12 (1973) 223.

2) E. Hanamura in *"Luminescence of Crystals, Molecules and Solutions"* ed. F. Williams (Plenum Press, N.Y., 1973) p.121.

3) A. F. Dite, V. I. Revenko, V. R. Timofeev and P. D. Altukhov: Zh ETF let. 18 (1973) 579 (JETP Lett. 18 (1974) 341).

4) J. Voigt and G. Mauersberger: Phys. Stat. sol. (b) 60 (1973) 679.

5) A more detailed paper on the model has been submitted for publication in Phys. Rev.

6) We have used the following parameter for CdS: $\varepsilon_\infty = 7.3$, exciton masses $m_{/\!/} = 5.2\ m_o$, $m_\perp = 0.9\ m_o$; $E_{ex}^L = 2.5540$ and $\Delta_{LT} = 1.9$ meV. Our value of photon energies take into account the refractive index of air, a point which has been usually omitted, as in the well known paper of J. J. Hopfield and D. G. Thomas, Phys. Rev. 122 (1961) 35.

STRESS-INDUCED SPLITTING OF EMISSION LINES FROM
EXCITONIC MOLECULES IN CdS AND ZnO

Yusaburo Segawa and Susumu Namba

The Institute of Physical and Chemical Research,
Wako-shi, Saitama, Japan

ABSTRACT

Splittings of spontaneous emission lines in laser excited wurtzite-type CdS and ZnO are observed under a uniaxial stress at 1.8 K. The relative intensity of the split components of the excitonic molecule emission line is accounted for by using Hanamura's wave function for the excitonic molecule.

In the past few years, there have been extensive studies on various aspects of high density excitation effects in solids. The nature of an excitonic molecule has been one of the main points of interest to be clarified. Recently, it has been proved that an experiment on a uniaxial stress effect is very suited for the study of the excitonic molecule.[1] Namely, we have observed the stress-induced splitting of the emission line which is ascribable only to the energy scheme of the excitonic molecule. This report deals with a detailed account of the experimental results and brief theoretical investigation of the uniaxial stress effect on the excitonic molecule in CdS and ZnO.

The experimental setup used has been already described.[2] Figure 1 shows the emission and the reflection spectra on the (0001) plane of CdS under the uniaxial compression σ perpendicular to the C axis of the crystal. Since the wavevector k of the emitted light is parallel to the C axis, the spectra are completely unpolarized in the strain free crystals. Under the uniaxial stress $\sigma \perp C$, they split into two components with $E \parallel \sigma$ and $E \perp \sigma$. The splitting patterns of the reflection spectrum and that of the M band are reversed as shown in Fig.1. In the reflection spectrum, the higher energy component is polarized with $E \parallel \sigma$. In the M band, on the contrary, the $E \perp \sigma$ component is on the higher

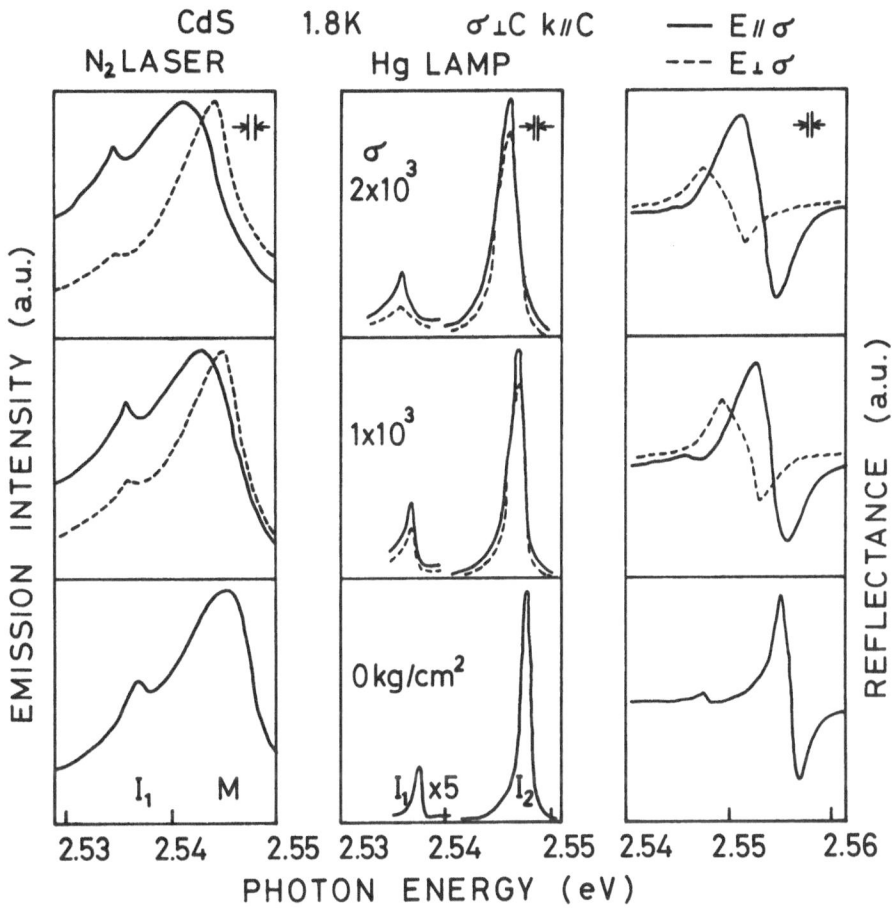

Fig.1. The change of the emission and the reflection spectra of CdS at 1.8 K. The intensity of the laser light on the surface of the sample is about 0.2 MW/cm^2. The intensities of the spectra are normalized with respect to the E//σ components.

energy side. This reversed splitting pattern of the M band is ascriba-ble only to the energy scheme of the excitonic molecule.[1] In this scheme, the initial state of the transition is the excitonic molecule ground state and the final state is the free exciton state. The split-ting of the M band results from the splitting of the free exciton state. The intensity change of the split components is also different. The relative intensities of the split components are plotted against the applied pressure for the reflection spectrum, the M band and the bound

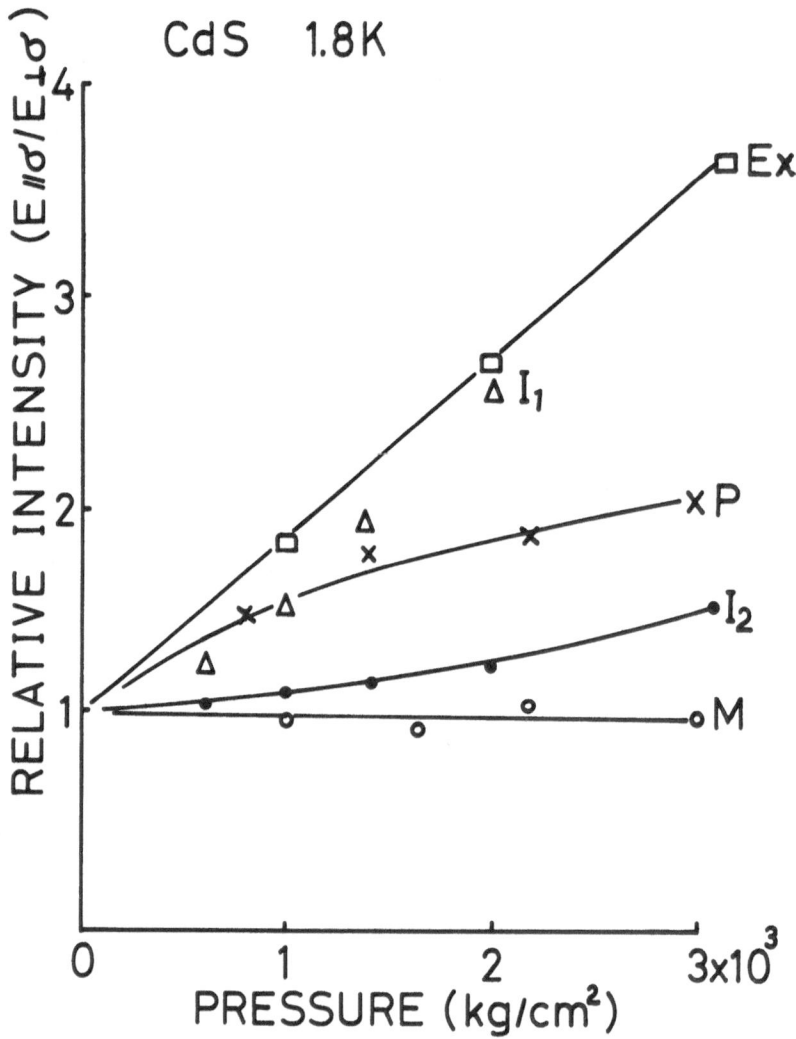

Fig.2. The relative intensity change of the E//σ and the E⊥σ components
of the free exciton (Ex), the M band (M) and the bound excitons
(I_1 and I_2). The stress effect of an emission from the free
exciton scattering process is also shown (P).

exciton I_2[2]) in Fig.2. As shown in the figure, all emission bands ex-

cept for the M band, as well as the reflection peak, show the decrease

in the E⊥σ components with increasing pressure. In the M band, however,

the relative intensity is nearly stress-independent. This is one of the characteristic features of the M band in CdS. From Fig.1, the energies

of the excitonic molecule ground state under the uniaxial stress can be

determined by adding the energies of the split components of the exciton

state and the M band. The resonant energies of the exciton state are

taken from the reflection spectra. The obtained energy shift of the

molecule ground state is shown in Fig.3. As is seen in the figure, the

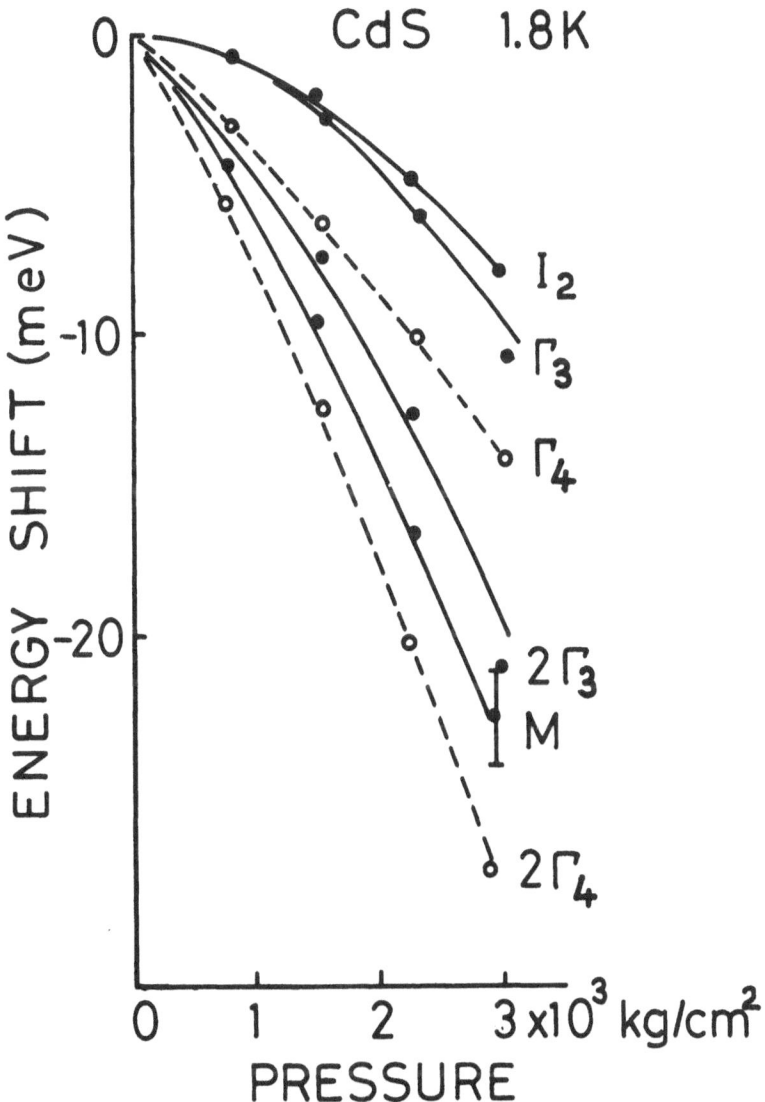

Fig.3. The energy shift of the Γ_3 exciton (Γ_{5x}, E//σ), Γ_4 exciton
(Γ_{5y}, E$\perp\sigma$) and the excitonic molecule ground state (M) against
the applied pressure.

molecule ground state does not show splitting within an experimental
error. As shown in Fig.4, splitting has been observed also in the M

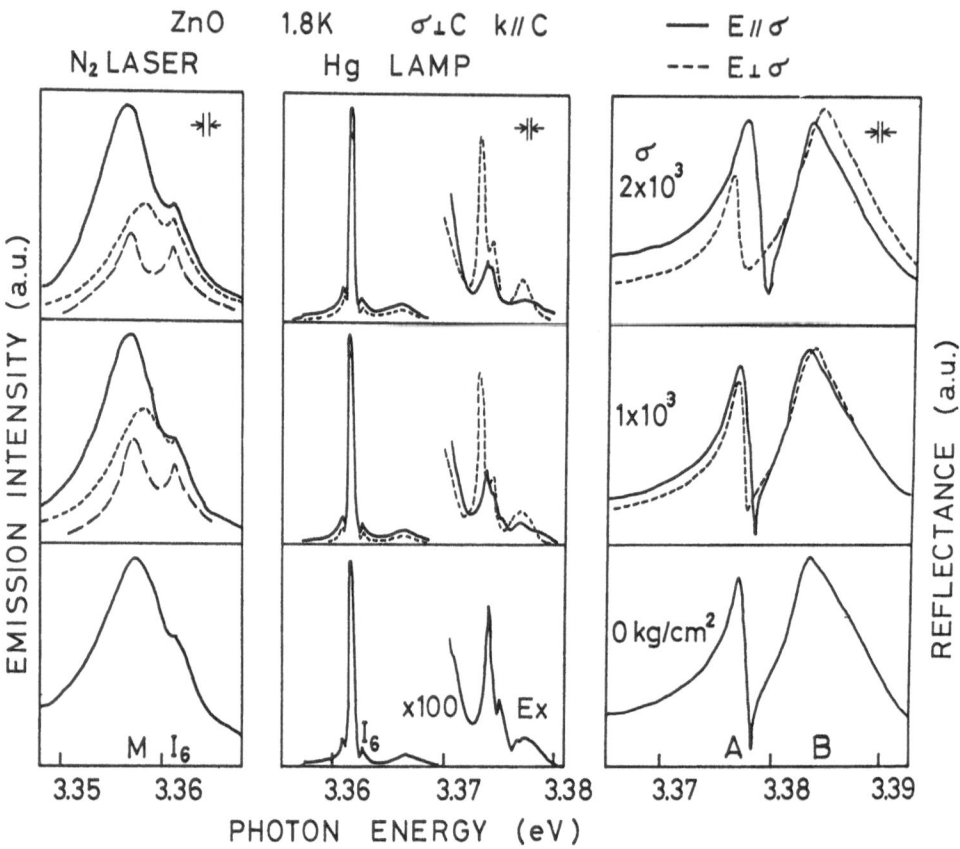

Fig.4. The changes of the emission and the reflection spectra of ZnO at 1.8 K.
The exciting power of the laser is about 1 MW/cm². The intensities of
the spectra are normalized to the E//σ components. In the spectra of
the M band, long dashed curves show the E//σ component under lower
excitation levels where the relative intensities of the M band and the
I_6 bound exciton are comparable with those of the E⊥σ component. In
these cases, the splitting of the M band is still observable.

band of ZnO in the same geometry of σ⊥C and k//C. The behaviour of this
band is qualitatively similar to that in CdS with regard to the polarization of the split components. However, the M band of ZnO show the
same intensity change as that of the reflection anormaly of the A exciton. In both the lower energy reflection peak of the A exciton and
the M band, the relative intensities of the E⊥σ components decrease with

increasing stress.

The splitting of the M band results from the splitting of the final state, and therefore, the relative intensity of the split components of the M band directly shows the transition probability between the excitonic molecule ground state and the free exciton state. In the stress-free crystals, CdS has C_{6v} symmetry at k=0 and irreducible representations of the A exciton are Γ_5 and Γ_6. The Γ_6 exciton state is purely triplet and dipole forbidden. The Γ_5 exciton is dipole allowed with E⊥C. In ZnO, the A exciton state is more complicated. According to the Hopfield's theory,[3] the A exciton consists of three states, Γ_1, Γ_2 and Γ_5. The Γ_2 exciton is dipole forbidden and the Γ_1 exciton has a very weak oscillator strength with E//C. Under the uniaxial stress perpendicular to the C axis, the Γ_{5x} exciton and the Γ_{5y} exciton split by the combined effect of the spin-exchange and the uniaxial stress.[4] In both CdS and ZnO, the Γ_{5x} state (E//σ component) has a higher energy and a stronger oscillator strength as observed in the reflection spectra of CdS and ZnO. However, on the Γ_6 state, the off-diagonal matrix elements of the spin-exchange Hamiltonian between the Γ_{6-} and Γ_{6+} states vanish. Therefore, we assume that the splitting of the Γ_6 states is negligible because it arises from higher order interactions. The diagonal matrix elements of the stress Hamiltonian have the same form for the Γ_6 and the Γ_5 states, respectively, and the center of the gravities of the split Γ_5 excitons and the Γ_6 excitons shift parallel with each other under the uniaxial stress.

The excitonic molecule wave function has a symmetry of

$$\Gamma_{5x} \cdot \Gamma_{5x} - \Gamma_{5y} \cdot \Gamma_{5y} + \Gamma_{6-} \cdot \Gamma_{6-} - \Gamma_{6+} \cdot \Gamma_{6+} \qquad (1)$$

as presented by Hanamura for CdS.[5] In the same manner, for ZnO, the excitonic molecule wave function has a form of

$$\Gamma_{5x} \cdot \Gamma_{5x} - \Gamma_{5y} \cdot \Gamma_{5y} + \Gamma_1 \cdot \Gamma_1 - \Gamma_2 \cdot \Gamma_2 \; . \qquad (2).$$

From these wave functions, we can obtain two results. First, the ground

state energy of the excitonic molecule should not split under the
uniaxial stress and stay between the double energies of the Γ_{5x} and the
Γ_{5y} states. This proposal is consistent with our experimental result
as shown in Fig.3. Second, the oscillator strength of the E//σ component (E⊥σ) of the M band shows the same intensity dependence as that of
the E//σ component (E⊥σ) of the free exciton. This proposal is consistent with the result of ZnO as shown in Fig.4. But on CdS, this is
not consistent. This fact seems to be attributable to the difference
between the binding energies of the molecules in CdS and ZnO. The
binding energy of the molecule in CdS is comparable with the splitting
of the Γ_5 exciton states at $\sigma = 3 \times 10^3$ kg/cm^2. Then the energy of the
Γ_{5y} state (low energy component) approaches to the $\frac{1}{2} \times$ (molecule ground
state energy) under the uniaxial stress. On the other hand, the Γ_{5x}
state shifts to the opposite energy side. Therefore, the weight of the
Γ_{5y} exciton state in molecule wave function may become larger than that
of the Γ_{5x} state and obscures the smaller oscillator strength of the Γ_{5y}
component. In ZnO, the binding energy of the molecule is large compared
with the splitting energy and this effect can be excluded.

The authores are very grateful to Professor E. Hanamura for valuable
discussions.

REFERENCES

1) Y. Segawa and S. Namba: Solid State Commun. 17 (1975) 489.

2) The excitons bound to neutral centers do not show the splitting
   under the uniaxial stress. J. P. Woerdman: Solid State Commun.
   13 (1973) 949; O. Goede, M. Blaschke and E. Hasse: Phys. Stat.
   sol. (b) 70 (1975) K41.

3) J. J. Hopfield: J. Phys. Chem. Solids 15 (1960) 97.

4) D. W. Langer, R. N. Euwema, K. Era and T. Koda: Phys. Rev. B. 2
   (1970) 4005.

5) E. Hanamura: J. Phys. Soc. Japan 39 (1975) 1506.

SPATIAL DIFFUSION OF HIGHLY CREATED EXCITONS IN CdS

Yasuo Oka and Takashi Kushida

The Institute for Solid State Physics
The University of Tokyo
Roppongi, Minato-ku
Tokyo 106, Japan

ABSTRACT

The exciton density is determined in CdS from the magnetic field dependence of the triplet-exciton emission intensity. The result shows that the exciton density is saturated markedly for very intense excitations by a nitrogen laser. This behavior is analyzed in terms of the diffusion process of optically created particles from the generated surface to the interior of the crystal. The diffusion length is found to be enhanced by two orders of magnitude for high excitations. This fast diffusion is ascribed to the high-density electron-hole plasma created at the crystal surface.

In the study of highly excited states of materials, one of the fundamental parameters is the density or the number of the optically created particles. This has been investigated extensively, for instance, in the case of the electron-hole drops in Ge. However, for the particles such as excitons and excitonic molecules in semiconductors, there exist only relatively poor estimations of their densities. In this paper we present a new method to determine the density and distribution of optically created excitons in CdS from the magnetic field dependence of the triplet-exciton emission intensity. The result of an analysis suggests the existence of electron-hole plasma state, which plays an important role in the spatial diffusion of excited particles.

High-purity platelets of CdS were excited by a pulsed nitrogen laser (λ = 337.1 nm, duration 10 nsec) and a cw argon laser (λ = 476.5 nm). Excitation power densities were 1 MW/cm^2 at the maximum in the case of the nitrogen laser and 10 W/cm^2 in the case of the argon laser. Fluorescence spectrum in the region of the free exciton line and its LO phonon sideband was measured at various excitation levels under the magnetic field H perpendicular to the crystal c-axis. Figure 1 shows

the emission spectrum under the excitation power of 10 kW/cm^2 obtained

in the configuration of the polarization vector of the emitted light

parallel to the c-axis. In this configuration, the longitudinal

EMISSION INTENSITY

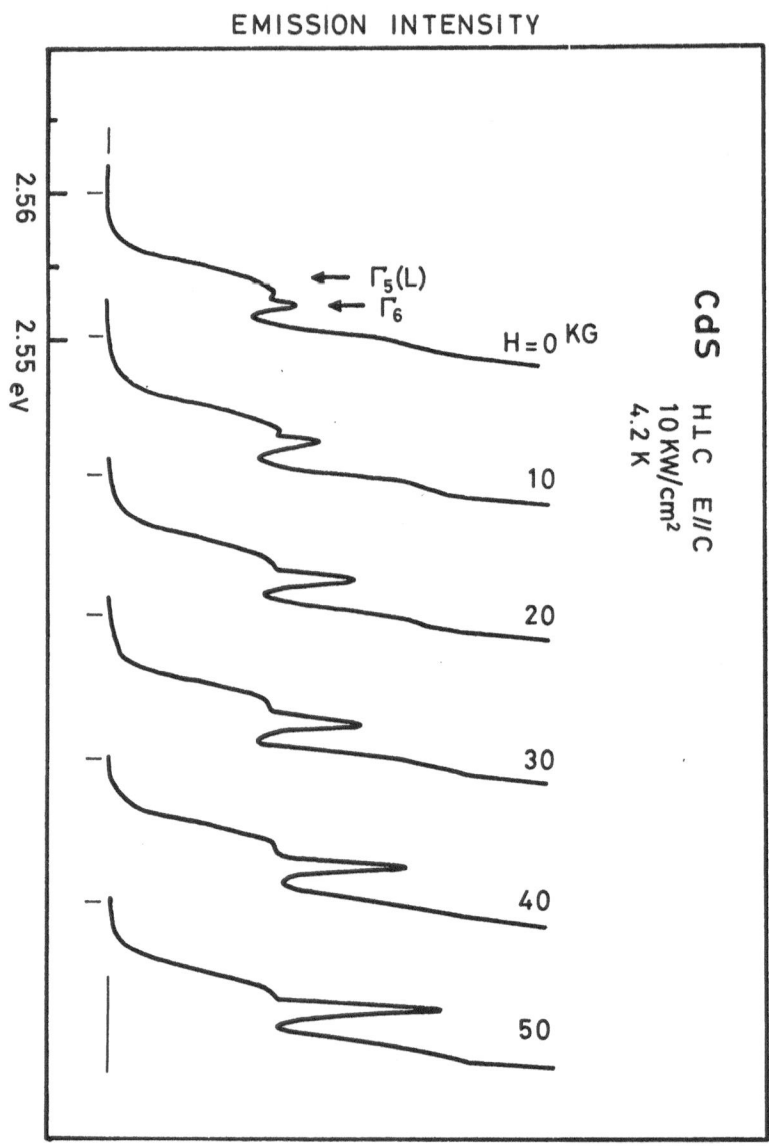

Fig.1. Emission spectra of CdS around the free exciton energy under
the excitation of 10 kW/cm^2 with various magnetic fields
perpendicular to the c-axis.

component of the singlet exciton emission (Γ_5) and the triplet line

(Γ_6) are observed. In Fig.1, we notice that the intensity of the triplet

exciton line is dependent on the magnetic field. This H-dependence
was found to vary with excitation power. In the case of weak excitation
by an argon laser, the triplet line varies as H^2. On the other hand,
the dependence is more gradual for intense excitations by a nitrogen
laser and the triplet emission intensity remains almost unchanged with
H at the maximum excitation intensity $(1 \ MW/cm^2)$.

The above result is explained well by the mixing of the triplet
state with the singlet due to the interaction between the excitons.[1]
The triplet state under the influence of the interaction among excitons
and the magnetic field (H⊥c) may be expressed as

$$| T > = | \Gamma_6 > + (\gamma + \beta H) | \Gamma_5 > \tag{1}$$

with $\qquad \beta = g_\perp \mu_B / \Delta E ,$

where γ is the mixing coefficient due to the high density effect, and
$g_\perp$, μ_B, and ΔE are g-value of the triplet state for H⊥c, the Bohr
magneton, and the exchange splitting energy between the Γ_5 and Γ_6 states,
respectively ($g_\perp = 1.72$ and $\Delta E = 1.3$ meV in CdS). The evaluation of the
mixing coefficient γ is made by using Hanamura's theory on high density
excitons.[2] For the interaction between the lowest state excitons, the
main contribution comes from the exclusion and exchange effects among
excitons. Then the mixing coefficient γ is expressed as

$$\gamma = \frac{\rho}{2} \frac{26}{3} \pi a_0^3 E_{ex}^b / \Delta E , \tag{2}$$

where ρ is the density of created excitons, a_0 and E_{ex}^b are the Bohr
radius and the binding energy of the exciton, respectively.

The strength of the electric dipole transition from the triplet
state f_T is expressed by renormalizing the wave function of eq.(1) as

$$f_T \propto \frac{(\gamma + \beta H)^2}{1 + (\gamma + \beta H)^2} . \tag{3}$$

In Fig.2, H-dependence of f_T is shown for several values of γ (or ρ).

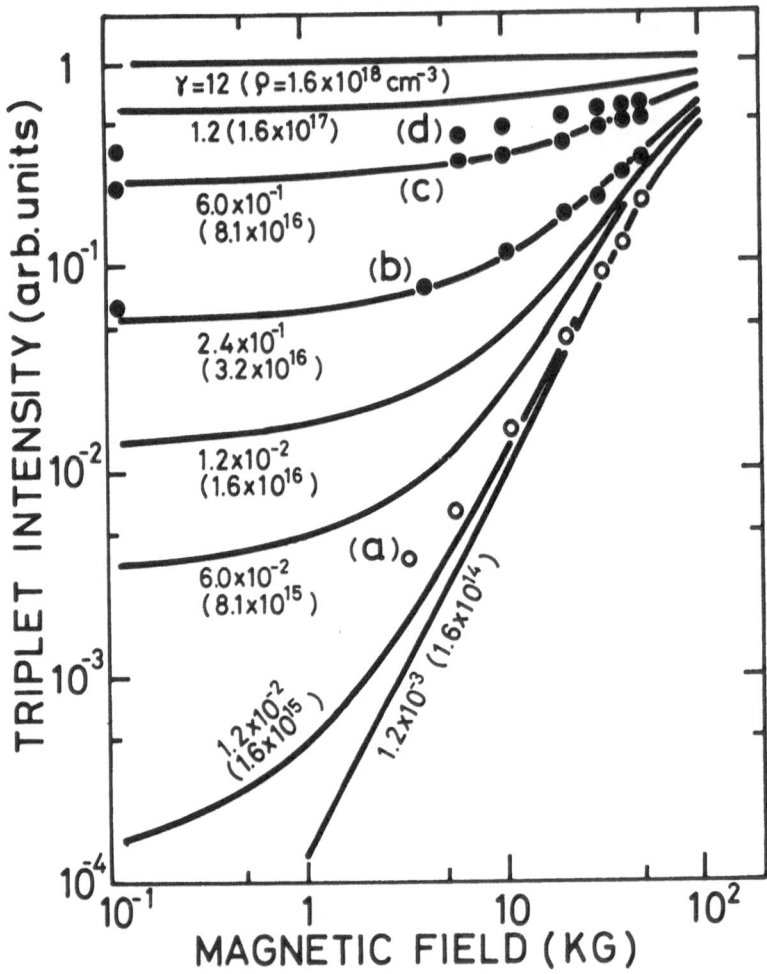

Fig.2. Theoretically derived magnetic field dependence of the electric dipole transition strength of the triplet exciton line (solid line) and the experimentally obtained dependence of the triplet emission intensity (circles). Here, (a), (b), (c) and (d) correspond to the excitation densities of 10 W/cm^2, 10 kW/cm^2, 100 kW/cm^2, and 1 MW/cm^2, respectively.

It is noted that f_T increases in proportion to H^2 for the small values of γ, namely for low density of excitons, while it is independent of magnetic field in large γ limit. Experimental points are also plotted in this figure. We see that the experimental result is explained well by the theoretical dependence. By fitting the experimental H-dependence of the triplet emission intensity with the theoretical curve, we can determine the exciton density ρ for each excitation level. The exciton density ρ thus obtained is shown in Fig.3 by a solid line. The exciton density is considered to increase linearly for the excitations below 10 kW/cm^2. In the region of higher excitations, however, the density ρ is saturated markedly and tends to a value of 10^{17} cm^{-3}. The critical density ρ_c of the transition from the insulating phase of the exciton to the metallic phase is about 6×10^{17} cm^{-3} in CdS. It is noted that the saturated value of the exciton density is of the same order as ρ_c.

Now let us study another aspect of the exciton emission. The intensity of the phonon-assisted emission line of exciton is considered to represent the total number of the excitons created in the crystal. Excitation power dependence of the intensity of the 2LO phonon sideband of the exciton line was measured simultaneously with the measurement of the zero phonon line of exciton described above. The intensity of the 2LO sideband was found to increase almost linearly with the excitation intensity. This result agrees with the observation reported by Leheny *et al.*[3)] The total number of created excitons can be estimated from the photon flux density of excitation under the assumption that the quantum efficiency for exciton generation is unity. An experiment under pico-second pulse excitation revealed that the intensity of the phonon-assisted exciton emission decays in a time of $1 \sim 5 \times 10^{-10}$ sec. We employ the value of 10^{-10} sec for the lifetime of the exciton. Then we obtain the total number of excitons per unit area of the crystal ρ^{tot} as shown in Fig.3 by a dashed curve.

In spite of the strong saturation of the exciton density, ρ^{tot} increases almost in proportion to the excitation power. This result clearly indicates that the diffusion of optically created particles is significant in the case of very intense excitations. Even in the weak excitation case, excitons are concluded to diffuse from the generated surface region into the interior because the diffusion length of exciton of 3.3×10^{-5}cm, which is determined from the experimental value of

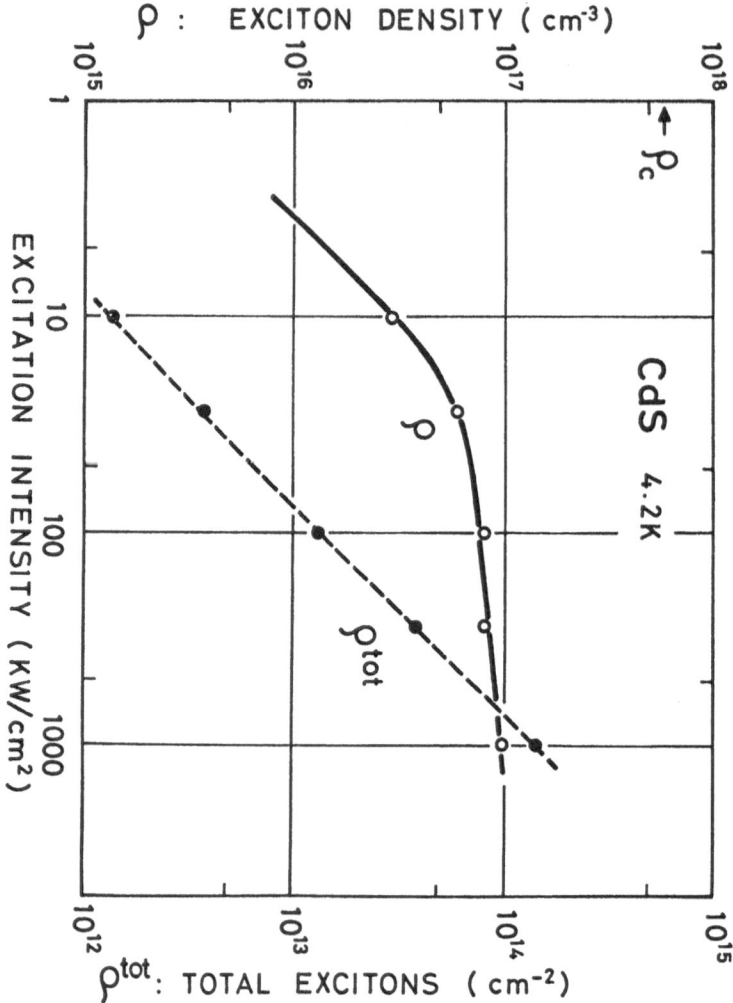

Fig.3. Excitation intensity dependence of the exciton density ρ and the total number of the created exciton per unit area ρ^{tot}.

$\rho\,tot/\rho$, is much larger than the reciprocal of the optical absorption coefficient. In the high excitation case, because the diffusion occurs much strongly, we conjecture that some new mechanism becomes dominant above the excitation density of 10 kW/cm^2. Since the saturated value of the exciton density is lower but close to the critical density ρ_c, the metallic electron-hole plasma state is considered to play an impor- tant role in this fast diffusion. In fact, if we use the absorption coefficient of incident photons of 5×10^5 cm^{-1} and the lifetime of the optically created particles of $10^{-10} - 10^{-11}$ sec, the density of created particles is estimated to be much higher than the critical density of exciton ρ_c under the intense optical excitations above 100 kW/cm^2 (10^{23} photons/cm^2·sec). Thus the state realized just after the excitation is probably the metallic phase of the exciton, i.e., the electron-hole plasma (EHP) state. However, we mainly observe emissions due to excitons such as phonon sidebands of exiton line and also the exciton-exciton collision emission. Thus we consider in the following way. The EHP diffuses very rapidly into the crystal and the plasma density is decreased. Then the excitons are formed from the EHP and luminescence related to the exciton is emitted. We shall calculate the distributions of EHP and excitons on the basis of this consideration. Since the duration of the excitation laser-pulse is 10^{-8} sec and the lifetimes of the excitons and EHP are of the order of 10^{-10} sec or less, the steady state analysis is considered to be a good approximation.

The density of the electron-hole pairs in the plasma in the steady state $N(X)$ is assumed to decrease exponentially with the depth from the surface X as

$$N(X) = Ae^{-X/L} + Be^{-\alpha X} , \qquad (4)$$

where L and α are the diffusion length of EHP and the absorption coefficient for the incident photons. This dependence is of the same form as that of the solution of the diffusion equation for the particles optically created within the penetration depth. The coefficients A and B determined from the boundary condition are given by

$$A = N(0) - B$$

$$B = \frac{1}{1 - \alpha^2 L^2} \eta (1 - R) F \alpha ,$$

where $N(0)$ is the density of EHP at the surface and η, R and F are the lifetime of EHP, the reflectivity and the flux density of incident photons, respectively.

Distribution of the exciton $\rho(X)$ in a steady state can be expressed by the following diffusion equation.

$$D \frac{d^2\rho}{dx^2} - \frac{\rho}{\tau} + \frac{N}{\eta} = 0 \quad , \tag{5}$$

where τ and D are the lifetime and the diffusion coefficient of the exciton. By substituting eq.(4) into eq.(5), $\rho(X)$ is solved as

$$\rho(X) = \left[\rho(0) - A' + B'\right] e^{-\frac{X}{\sqrt{D\tau}}}$$

$$+ A'e^{-X/L} - B'e^{-\alpha X} \quad , \tag{6}$$

in which $A'/A = A'/B = -\tau L^2 /\eta(D\tau - L^2)$. The integration of eq.(6) gives the total number of exciton per unit area:

$$\rho^{tot} = \int_0^\infty \rho(X)\, dX \quad . \tag{7}$$

If we use the values of ρ and ρ^{tot} in Fig.3 as $\rho(0)$ and the total number of exciton, the diffusion length L can be determined from the above relations for various excitation levels. We shall use the parameter values of $\tau = 10^{-10}$ sec and $\alpha = 5 \times 10^5$ cm^{-1}. Further, the diffusion coefficient D is calculated as 10 cm^2/sec from τ and the diffusion length of the exciton $(= \sqrt{D\tau})$ in the low excitation case. Under the excitation by intense pico-second pulses the formation time of exciton has been measured to be 10^{-11} sec in CdSe.[4] Since the lifetime of EHP is considered to be the same as the exciton formation time, we employ the value of $\eta = 10^{-11}$ sec.

Figure 4 shows the spatial distributions of the exciton and EHP determined from eqs.(6) and (4) by using the obtained value of L for each excitation level. The result may be summarized as follows. In the case of the excitation of 10 kW/cm^2, excitons remain within the range of about 3×10^{-5} cm, which is determined by the usual diffusion of

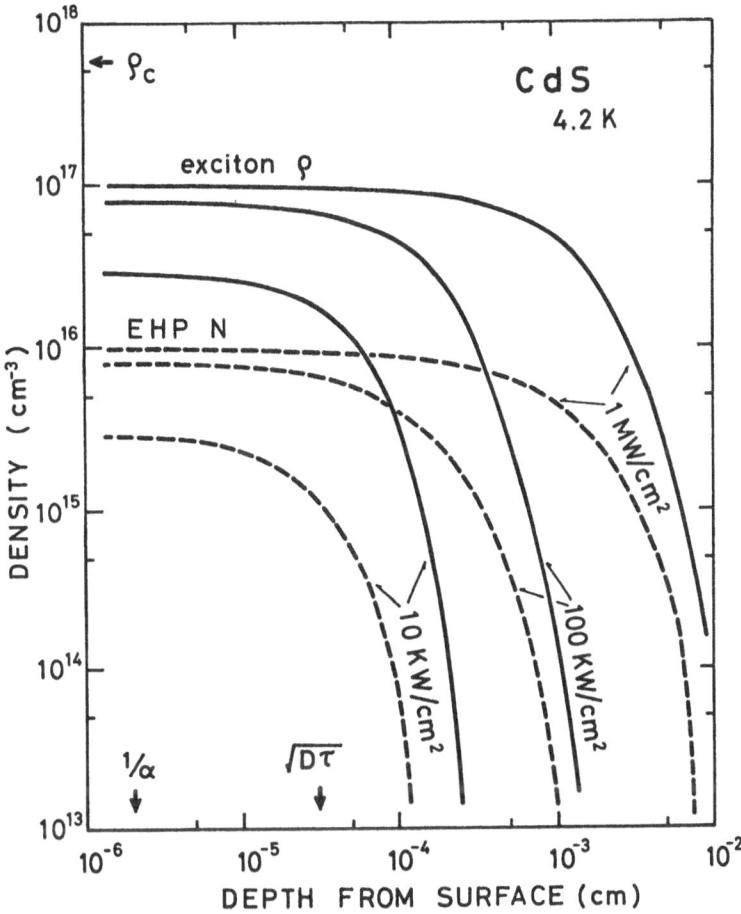

Fig.4. Spatial distribution of the density of the exciton and of
the electron-hole plasma N for three excitation intensities.

exciton in low density. When the excitation power is raised, the diffusion length of optically generated EHP increases considerably. The
distributed region of EHP reaches the depth of 10^{-3} cm from the surface
at the maximum excitation density of 1 MW/cm^2. Since the excitons are
formed from the EHP, the distribution of exciton follows that of EHP.
On account of the fast diffusion of EHP into the crystal, the density
of the exciton is suppressed within 10^{17} cm^{-3} even at the crystal surface. In the steady state, the density of EHP is always one order of
magnitude smaller than that of exciton. This is because the EHP has a
lifetime much shorter than the exciton.

In conclusion, the spatial distribution and diffusion of excited particles under the intense optical excitation were studied. An analysis suggests that the high density electron-hole plasma state is actually realized in highly excited crystals and plays an important role in the rapid diffusion. However, the emissions due to EHP has not been identified yet. The broad band nature of the EHP emission and also the existence of various intense emission lines near the band edge may cause the difficulty to observe this emission.

REFERENCES

1) Y. Oka and T. Kushida: Solid State Commun. 15 (1974) 1571.

2) E. Hanamura: J. Phys. Soc. Japan 29 (1970) 50, *ibid* 37 (1974) 1545.

3) R. F. Leheny, R. E. Nahory, and K. L. Shaklee: Phys. Rev. Letters 28 (1972) 437.

4) H. Kuroda and S. Shionoya: J. Phys. Soc. Japan 36 (1974) 467.

EFFECTS OF EXCITONIC MOLECULES ON EMISSION AND TRANSMISSION SPECTRA OF CdS SINGLE CRYSTALS

J. Voigt and F. Henneberger

Sektion Physik der Humboldt-Universität
zu Berlin, Bereich Halbleiteroptik

ABSTRACT

Based on new results of transmission and emission experiments on CdS single crystals in a very extended range of excitation intensities at 1.8 K, we discuss the annihilation and formation of excitonic molecules in a model involving A_{Γ_6}-excitons and A_{Γ_5}-polaritons. It is shown that the presented results as well as those by Saito and Shionoya can be consistently explained in this model. A theoretical foundation of the model and a way to calculate emission and absorption spectra due to excitonic molecules are presented.

I.   INTRODUCTION

In CdS single crystals, depending on the experimental conditions used, two excitonic molecule (EM) emission lines are reported in the literature differing in line shape and energetic position.[1,2] Saito and Shionoya[1] found, under 100 kW pulsed laser excitation, a new line in the emission spectrum of CdS at 2.546~2.544 eV whose position and half-width depend on the excitation intensity. We reported[2] a characteristic emission spectrum with a main peak at 2.5492 eV, appearing in very thin high-quality CdS single crystals at low temperature, under intense stationary excitation by a high pressure mercury lamp. To explain both lines on the basis of EM-annihilation, we extended the model[1] used up to that time for this process by including the polariton character of A_{Γ_5} ground state excitons. Additionally, A_{Γ_6} ground state excitons are taken into consideration. In this model, the line reported by Saito and Shionoya (M_s) is attributed to transitions at EM of wave vectors $k_M > k_{photon}$, whereas the line reported by us (M_1) corresponds to transitions at $k_M \approx 0$. In the following, this "polariton"-model of EM-annihilation was supported by results of transmission experiments.[3]

Besides an absorption peak (M_1) at E_1 = 2.5491 eV, which we attributed

to the direct formation of EM by optical absorption reverse to the EM-

annihilation involving A_{Γ_6}-excitons, we found a second (smaller) peak

(M_2) at E_2 = 2.5471 eV, which we ascribed to EM-formation from excitons

of the energy E_L = 2.5547 eV (*i.e.* exciton-like A_{Γ_5}-polaritons of the

upper branch as well as longitudinal excitons). Recently, Müller *et al.*[4]

observed a correlation between (M_S) and (M_1) EM-emission lines at dif-

ferent excitation intensities and concluded also that (M_S) and (M_1)

were due to EM having $k_M > k_{photon}$ and $k_M \approx 0$, respectively.

In this paper we report new experimental results of transmission

and emission experiments on CdS single crystals in a very extended

range of excitation intensities at 1.8 K, which confirm the proposed

"polariton"-model of EM-annihilation in detail, and, in particular,

explain the correlation between the different EM-lines (M_1), (M_2) and

(M_S). Furthermore, a theoretical foundation of the model is given and

formulas are derived for evaluating emission and transmission spectra.

II.   EXPERIMENTAL

The experimental arrangement used is described in detail in ref.

(5). The CdS single crystals investigated were grown from the vapour

phase with thickness ranging from 5×10^{-5} to 5×10^{-4} cm. The samples

were carefully prepared and mounted in such a way that no internal

strains occurred with decreasing crystal temperature. The measuring

temperature was 1.8 K (samples immersed in the liquid helium bath).

III.   RESULTS

At stationary excitation three characteristic emission lines lying

energetically below the free exciton emission are observed (Fig.1, curve

6): The EM-line (M_1) at E_1 = 2.549 eV (halfwidth H ≈ 1.5 meV) investi-

gated by us in detail in ref. (2), the line (M_2) at E_2 = 2.5469 eV (H ≈

0.5 meV), observed so far only in transmission,[3] and the line at E_{I_2} = 2.5460 eV ($H \approx 0.5$ meV), whose intensity strongly increases with increasing crystal thickness, whereas the ratio of intensities of the lines (M_1) and (M_2) is nearly constant in all crystals.

At lowest laser excitation*) only the lines (M_1) ($H \approx 1.4$ meV) and

Fig.1. Spectral dependence of the emission at different laser intensities ((1) 1.3% I_0, (2) 2.5% I_0, (3) 5.4% I_0, (4) 12% I_0, (5) 37% I_0) and at stationary excitation (6) $\vec{E} \perp \vec{C}$, T = 1.8 K.

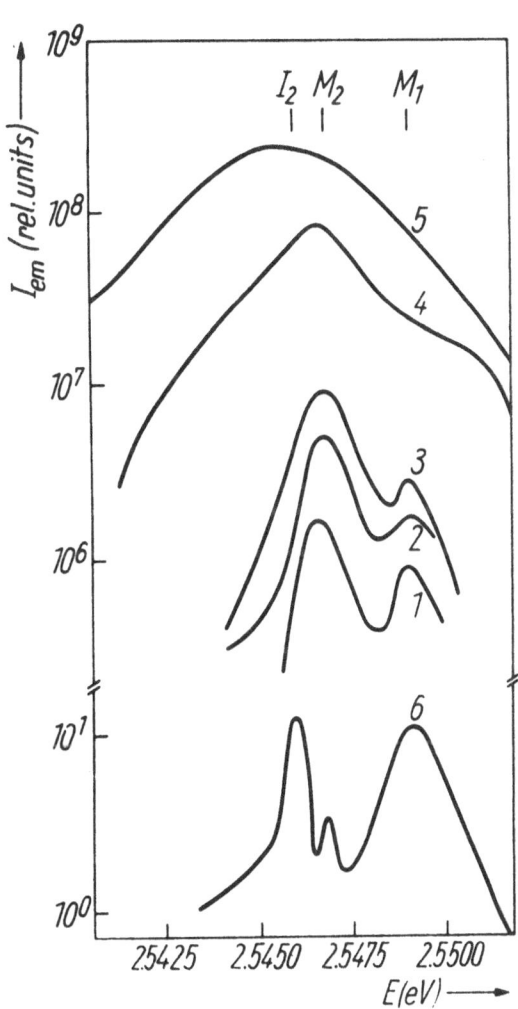

*) The lowest laser intensities used in our experiments to resolve (M_1) and (M_2) are more than two magnitudes smaller than the lowest one in the experiment of Saito and Shionoya.[1]

M_2 ($H \approx 1.4$ meV) appear (Fig. 1, curves 1 and 2). The essential difference in comparison to the spectra at stationary excitation consists
in the fact that (M_2) is the dominant line. With increasing laser excitation, (M_2) grows more strongly than (M_1) and this is accompanied by
some broadening (Fig. 1, curve 3). At still higher excitations, (M_2)
is transformed to the line (M_S) ($H \approx 5$ meV) whose energetic position
shifts to low energy side with increasing excitation intensity (Fig. 1,
curves 4 and 5). The line (M_1) contributes to a high energy tail of
(M_S), as seen from a shoulder (Fig. 1, curve 4) appearing at not too
high excitation intensities.

The transmission spectrum of the crystal investigated in emission
is given in Fig. 2. Besides the well pronounced spike at the longitu-

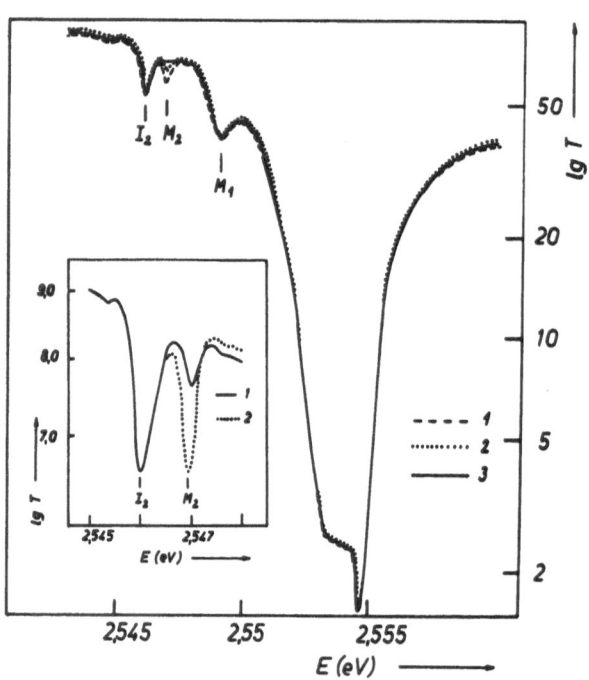

Fig.2. Spectral dependence
of the transmission
at different inci-
dent light inten-
sities ((1) I_0, (2)
22% I , (3) 5.6% I_0).
Insert: (1) without
additional light,
(2) HBO 500 $\lambda < 400$ nm.

dinal energy E_L = 2.5546 eV all structures found in the emission (M_1),
(M_2) and I_2 are reproduced in the transmission showing a very good
agreement with respect to the energetic positions as well as the half-
widths. With increasing intensity of the incident light a remarkable

change of the transmission spectrum is only found in the line (M_2)
(the spectra of Fig. 2 are normalized to constant incident intensity),
whose intensity increases. At low intensities (M_2) cannot longer be
detected in transmission. On the other hand, a strong additional excitation with $\lambda < \lambda_{ex}$ results in a strong enhancement of the line M_2
(see insert in Fig. 2).

IV. DISCUSSION

4.1 Qualitative Interpretation

The experimental results can be consistently explained in the
model for EM-annihilation proposed by us in ref. (2) which involves
both, A_{Γ_6}-excitons and A_{Γ_5}-polaritons (Fig. 3). The line at E_{I_2}
= 2.5460 eV is attributed to the known I_2 bound exciton emission[5)]
because of its dependence on sample thickness and its absence even
at the lowest laser excitation intensity. The lines (M_1) ,

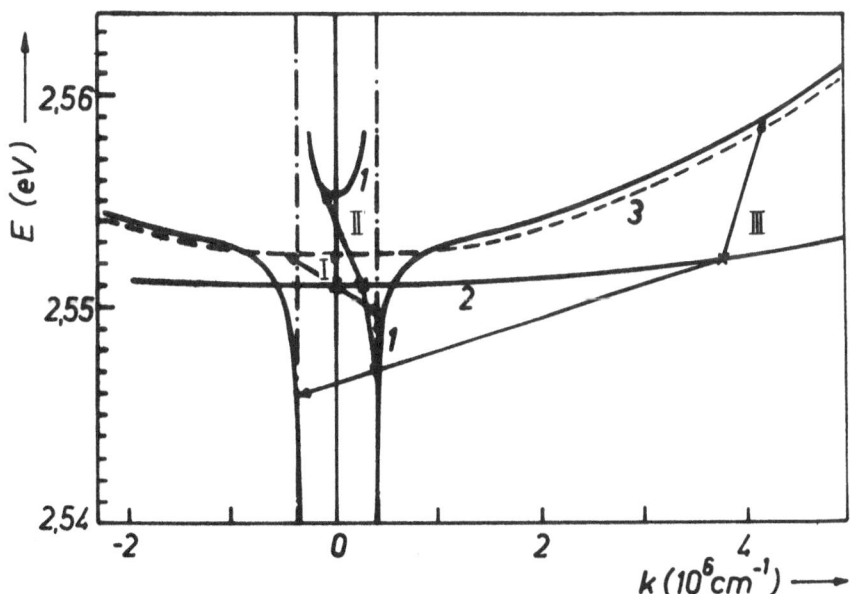

Fig.3. Dispersion curves of the A_{Γ_5}-polariton (1), EM (2), and
A_{Γ_6}-exciton (3).

(M_2) and (M_S) are interpreted as being due to EM-annihilation. Without
consideration of A_{Γ_6}-excitons, due to the splitting of the dispersion

curves of A_{Γ_5}-polaritons two EM-emission lines are expected in the model given in Fig. 3 (transitions II and III) leaving exciton-like polaritons in the upper and lower polariton branch, respectively. The transition II is only allowed (due to energy momentum conservation) for a very small range of k_M-values where the dispersion of EM can be neglected, resulting in a sharp emission line, whereas the main part of transition III is given by EM's with $k_M > k_{photon}$. The transition I designates the radiative recombination of EM's leaving a A_{Γ_6}-exciton. There are two reasons from which we conclude that (M_1) is correlated to the transition I rather than to the transition III. Firstly, (M_1) does not shift to low energy side with increasing laser intensity as one expects for transition III due to the heating of the EM gas. Secondly, the difference $E_1 - E_2 = 2.2$ meV is greater than the longitudinal-transverse splitting of the A_{Γ_5}-exciton, $E_{\Gamma_{5L}} - E_{\Gamma_{5T}} = 1.9$ meV, determined from the transmission measurement of high precision.[6] But excellent agreement is obtained by setting $E_1 - E_2 = E_{\Gamma_{5T}} - E_{\Gamma_6}$ and taking into account the electron-hole exchange splitting; $E_{\Gamma_{5T}} - E_{\Gamma_6} \lesssim 0.4$ meV.

The line (M_2) is attributed to the transition II, from its energetic position and halfwidth. The line (M_S) includes all transitions I, II and III, but the dominant transition is the transition III due to the distribution of EM in the band with a maximum at $k_M > k_{photon}$. On the other hand, at stationary excitation, where the temperature of the EM gas coincides with the lattice temperature, a narrow distribution of EM near $k_M = 0$ exists, resulting in dominant lines (M_1) and (M_2). Therefore, the redistribution of EM in the band with increasing laser intensity causes the observed changes in the emission spectra.

The results of transmission measurements, carried out at "normal" intensities, show that the processes I and II can take place also in reverse direction. But the intensity dependence one expects[7] is only verified for the line (M_2). The behaviour of the line (M_1) can be explained, taking into consideration the different relaxation times of the processes involved.

4.2 Theoretical Remarks

In earlier theoretical calculations[7,8] of the emission and absorption spectrum of EM, the polariton-effect of excitons as final and
initial states of the considered transitions has been neglected. Therefore, the theoretical results are only applicable to EM and excitons of
translational momentum large in comparison to that of photons. However,
our experimental results show that EM and excitons of small momentum
play a dominant role. Thus, the extension of the conventional models
of EM-annihilation and formation is necessary.

To include polariton-effects for these processes we start from the
model Hamiltonian H given by Hanamura.[8] (The interaction between EM's
is neglected because its smallness at the considered excitation densities). Following Hopfield[9] we transform H into the polariton representation, introducing annihilation and creation operators a_{ik}^+, a_{ik} of
polaritons at energy $E_i(k)$; i = 1,2, as linear combinations of the
corresponding exciton and photon operators. Then H can be written as

$$H = H_M + H_{POL} + H_{M-POL},$$ (1)

where H_M and H_{POL} describe non-interacting EM and exciton-polariton
systems, respectively. The interaction between both systems is given
by H_{M-POL}, its explicit form is written as

$$H_{M-POL} = \sum_{i,j=1}^{2} \sum_{\vec{k},\vec{q},\vec{q}'} C_{i1}(\vec{q})F(\vec{k})C_{i2}(\vec{q})\delta(\vec{k}-\vec{q}-\vec{q}')a_{i\vec{q}}^+ a_{j\vec{q}'}^+ C_{\vec{k}} + h.c.$$ (2)

($C_{\vec{k}}^+$ – creation operator of EM of momentum $\vec{k}$). The coupling constant
$F(\vec{k})$ is proportional to the oscillator strength calculated by Gogolin
and Rashba,[7] $C_{ij}(\vec{q})$ is the transformation matrix tabulated in ref.(9).
From eq.(1) the used transitions scheme (Fig. 3) can be justified immediately. In this picture the EM-annihilation (leaving a photon and an
exciton) and the reverse absorption process is replaced by the decay of

EM into two polaritons and the reverse fusion process due to the inter-action (2). Treating H_{M-POL} as perturbation, the optical spectra of EM can be calculated in the framework of first order perturbation theory. Basing on this concept we have derived formulas for emission and ab-sorption spectra of EM, which cannot be presented in this short paper (because their analytical form is very lengthy). Generally, they depend on

1. The ordinary oscillator strength of the optical transition[7] and, additionally, the polariton transformation coefficients[9] which charac-terize the exciton- and photon-like character of polaritons.

2. The density of states of the interband energy $E_M(k) - E_i(k)$ (instead of $E_M(k) - E_{ex}(k)$).[7,8]

3. The occupation numbers of EM (emission) and polariton states (ab-sorption).

To make a comparison of experimental and theoretical results it is necessary to carry out numerical computations using the occupation numbers of EM and polaritons as fitting parameters. This will be reported in a forthcoming paper.[10]

ACKNOWLEDGEMENTS

We are greatly indebted to Prof. E. Gutsche for critical remarks on the manuscript. We thank Mr. J. Puls for making the transmission experiments.

REFERENCES

1) H. Saito and S. Shionoya, *Proc. Internat. Conf. Luminescence, Leningrad* (1972) ed. F. Williams (Plenum, N. Y., 1973) p.104.

2) J. Voigt and G. Mauersberger, Phys. Stat. sol. (b) 60 (1973) 679.

3) J. Voigt and 1. Rückmann, Phys. Stat. sol. (b) 61 (1974) K85.

4) G. O. Müller *Proc. 12th Internat. Conf. Phys. Semicond., Stuttgart* (1974) ed. M. H. Pilkuhn (Teubner, Stuttgart, 1974) p.123.

5) J. Voigt, F. Mir, and G. Kehrberg, Phys. Stat. sol. (b) $\underline{70}$ (1975) 625.

6) J. Voigt, Phys. Stat. sol. (b) $\underline{64}$ (1974) 549.

7) A. A. Gogolin and E. I. Rashba, Zh, eksper. theor. Fiz. $\underline{17}$ (1973) 69.

8) E. Hanamura, *Proc. Internat. Conf. Luminescence, Leningrad* (1972) ed. by F. Williams (Plenum Press, N.Y., 1973) p.121.

9) J. J. Hopfield, Phys. Rev. $\underline{112}$ (1958) 1555.

10) F. Henneberger and J. Voigt, to be published.

LONG RANGE ORDER AND SUPERFLUIDITY
FOR BOSE CONDENSED EXCITONS

H. Haug

Institut für Theoretische Physik, Universität
6 Frankfurt/Main, Robert-Mayer-Str. 8,
Fed. Republic of Germany.

ABSTRACT

The problem whether a Bose-Einstein condensed system of Wannier excitons will exhibit superfluid properties is briefly reviewed. The long range order of such a system is shown to lead to a two-fluid model, which describes the superfluid **flow of excitation energy**. Qualitative remarks are given about the influence of perturbations, which will cause the exciton superfluidity to decay in time.

About 15 years ago the possibility of a Bose-Einstein condensation of excitons in highly excited semiconductors at low temperatures has been considered for the first time[1,2]. The analogy with He-II soon gave rise to speculations about the superfluid properties of the Bose condensed exciton gas. Moskalenko[2] argued on the basis of a two-fluid model that a crystal with a Bose condensed exciton gas would act as a thermal superconductor. However, the direct observations of these superfluid effects are rather difficult and it was suggested to search for evidence of a change in the excitation spectrum of the exciton system, which should alter at the inset of condensation from a quadratic to a linear spectrum for small momentum values. The consequences of superfluidity for the exciton transport has further been treated in terms of a Ginzburg-Landau equation.[3] The penetration depth of excitons is expected to increase considerably if the condensation takes place. Keldysh and Kozlov[4] as well as Hanamura[5] stressed the importance of the fact that excitons are not perfect Bosons. The deviations from the Bose nature leads to a repulsive interaction between two excitons if the spin is not taken into account. Both theories, the pairing theory of ref.4 and the Boson treatment of ref.5 confirmed the existence of a condensate and of a linear excitation spectrum, which have been assumed in the earlier treatments. All existing microsopic theories of the condensed exciton phase are limited to the low density limit $na_0^3 \ll 1$, where n is the exciton concentration and a_0 the exciton Bohr radius. The region in which the exciton density is close to the ionization limit ($na_0^3 \approx 1$) is up to now not accessible to theory. If the spins of the electrons and holes are taken into account, the exciton molecule can be shown to be stable. The interaction between excitation molecules is repulsive, so that a Bose-Einstein condensation of excitonic molecules rather than of excitons is to be expected.[6] In the following we disregard the formation of molecules (by omitting the spin variables), because the question about the possibility of superfluidity is fundamentally the same for a gas of excitons or excitonic molecules. The possibility of superfluidity has been questioned by Kohn and Sherrington[7] for all composite Bosons, which are built up from particles and holes (*e.g.* excitons or excitonic molecules). We reinvestigate this problem, following essentially the approach by Hanamura and the author[8]. We treat a system of Wannier excitons which is supposed to be in thermal quasi-equilibrium; *i.e.* we assume that the exciton relaxation time is much smaller than the exciton life time. Under these conditions one can describe a quasi-stationary state, in which the generation rate is balanced by the decay rate, by the introduction of a quasi-chemical potential. We do not treat the interactions of the electron-hole system with other perturbing fields such as photons, phonons and impurities, assuming that these interactions are weak as compared to the interactions within the system. Furthermore, a direct semiconductor is treated, but the generalization to an indirect one is straightforward.[8] The Hamiltonian for this system in the effective mass approximation is

$$H = \int d^3x_e \, \psi^+(\vec{x}_e) \, (-\frac{\hbar^2}{2m_{x_e}} \nabla^2_{x_e} - \mu_e + E_g) \, \psi(\vec{x}_e)$$

$$+ \int d^3x_h \, \phi^+(\vec{x}_h) \, (-\frac{\hbar^2}{2m_{x_h}} \nabla^2_{x_h} - \mu_h) \, \phi(\vec{x}_h) \qquad (1)$$

$$+ \frac{1}{2} \int d^3x_e \int d^3x'_e \, V(|\vec{x}_e - \vec{x}'_e|) \, \psi^+(\vec{x}_e) \, \psi^+(\vec{x}'_e) \, \psi(\vec{x}'_e) \, \psi(\vec{x}_e)$$

$$+ \frac{1}{2} \int d^3x_h \int d^3x'_h \, V(|\vec{x}_h - \vec{x}'_h|) \, \phi^+(\vec{x}_h) \, \phi^+(\vec{x}'_h) \, \phi(\vec{x}'_h) \, \phi(\vec{x}_h)$$

$$- \int d^3x_e \int d^3x_h \, V(|\vec{x}_e - \vec{x}_h|) \, \psi^+(\vec{x}_e) \, \phi^+(\vec{x}_h) \, \phi(\vec{x}_h) \, \psi(\vec{x}_e),$$

where ψ, ϕ and μ_e, μ_h are the field operators and the chemical potentials of the electrons and holes, respectively. $V(r)$ is the Coulomb
potential. The condensed state of the exciton system (eq.1) has been
analysed in ref.4 and ref.5. Using the results of these treatments,
one can show that the second reduced electron-hole density matrix is
of the form

$$< \psi^+(\vec{x}'_e) \, \phi^+(\vec{x}'_h) \, \phi(\vec{x}_h) \, \psi(\vec{x}_e) > = \, \varphi^*(r') \, \varphi(r) \, \rho_1(\vec{R};\vec{R}'),$$

$$\qquad (2)$$

$$\rho_1(\vec{R};\vec{R}') = \Psi^*(\vec{R}')\Psi(\vec{R}) + \tilde{\rho}_1(\vec{R};\vec{R}'),$$

where $\vec{r}$ and $\vec{R}$ are the relative coordinate and the center of mass coordinate of an exciton, respectively. φ is the wave function of the
internal exciton motion in the lowest state; we call $\rho_1(\vec{R};\vec{R}')$ the
first reduced exciton density matrix. ρ_1 shows the well-known off-
diagonal long range order (ODLRO) with the order parameter $\Psi(\vec{R})$ and the
non-condensate part $\tilde{\rho}_1$ which decays as $|\vec{R}-\vec{R}'| \to \infty$. The question now
is,[7] whether ODLRO of the electron-hole density matrix (which corres-
ponds to DLRO for a particle-particle density matrix) does lead to a
superfluid behaviour of the exciton system. In order to investigate
this question, we derive the equation of motion for the second electron-
hole density matrix by using the Heisenberg equations of the various

field operators. Because we confine ourselves to the region of the phase diagram in which only excitons exist, we pair each unpaired field operator which appears in the resulting equation. The product of two unpaired electron operators, *e.g.*, would be transformed into

$$< \psi^+(\vec{x}_9') \; \psi(\vec{x}_e) > \rightarrow \int d^3 r' \int d^3 r < \psi^+(\vec{x}_e') \; \phi^+(\vec{x}_h') \; \phi(\vec{x}_h) \; \psi(\vec{x}_e) > \varphi(r') \; \varphi^*(r) .$$

(3)

Finally, we eliminate the internal exciton motion and obtain an equation for the first reduced exciton density matrix

$$i\hbar \frac{\partial}{\partial t} \rho_1(\vec{R};\vec{R}') = - \frac{\hbar^2}{2M} (\nabla_R^2 - \nabla_{R'}^2) \rho_1(\vec{R};\vec{R}')$$

$$+ \int d^3 R'' \; \{W(|\vec{R}-\vec{R}''|) - W(|\vec{R}-\vec{R}'|)\} \; \rho_2(\vec{R},\vec{R}'';\vec{R}'',\vec{R}'),$$

(4)

where the exciton-exciton interaction is given by

$$W(|\vec{R}-\vec{R}'|) = \int d^3 r \int d^3 r' \; \{V(|\vec{R}-\vec{R}'|) + V(|\vec{R}-\vec{R}' + \vec{r}-\vec{r}'|)$$

$$- V(|\vec{R}-\vec{R}'-\vec{r}'|) - V(|\vec{R}-\vec{R}'+\vec{r}|)\} \; \{|\varphi(r)|^2 |\varphi(r')|^2$$

(5)

$$- \varphi(|\vec{r}+\vec{R}-\vec{R}'|) \; \varphi^+(r) \; \varphi(|\vec{r}'+\vec{R}'-\vec{R}|) \; \varphi^+(r')\}$$

$$\approx \delta^3(\vec{R}-\vec{R}') \; \frac{26}{3} \pi E_{ex}^b a_o^3 ,$$

where E_{ex}^b is the exciton binding energy and a_o the exciton Bohr radius. This result is valid in the low density limit $na_o^3 \ll 1$ and for $m_h \gg m_e$. The interaction potential is repulsive due to exchange effects. Using the same techniques as above, we further derive an equation of motion for the exciton order parameter

$$i\hbar \; \dot{\Psi}(\vec{R}) = [- \frac{\hbar^2}{2M} \nabla_R^2 + E_g - E_{ex}^b - (\mu_e + \mu_h)] \Psi(\vec{R})$$

(6)

$$+ \int d^3 R' \; W(|\vec{R}-\vec{R}'|) \; \rho_{3/2}(\vec{R},\vec{R}';\vec{R}') ;$$

where $\rho_{3/2}$ is the three-leg function.

Both equations (4) and (6) are of the same form as the corresponding ones for a system of interacting elementary Bosons.[9] As is known from the theory of He-II, eqs.(4) and (6) are sufficient to derive rigorously the following conservation laws

$$\frac{\partial}{\partial t} n + \vec{\nabla} \cdot \vec{j} = 0 \qquad \frac{\partial}{\partial t} E + \vec{\nabla} \cdot \vec{Q} = 0$$

$$\frac{\partial}{\partial t} \vec{j} + \vec{\nabla} \cdot \overleftrightarrow{\Pi} = 0 \qquad \frac{\partial}{\partial t} \vec{v}_s + \vec{\nabla}(\tfrac{1}{2} \vec{v}_s^2 + \frac{\mu}{M}) = 0,$$ (7)

where n, $\vec{j}$ and E are the densities of the particles, the particle current and the energy, respectively. $\overleftrightarrow{\Pi}$ is the stress tensor, $\vec{Q}$ the energy current. The superfluid velocity is defined as $\vec{v}_s = \vec{\nabla}\theta h/M$ where θ is the phase of the order parameter Ψ. In order to derive a hydro-dynamic two-fluid model from these equations, we have to assume that the system is in local equilibrium, *i.e.* we must be in a regime which is dominated by exciton-exciton collisions (the local equilibrium should not be established primarily by exciton-phonon collisions). If these conditions are fulfilled, we obtain a two-fluid model, which describes the superfluid flow of excitation energy (not of mass and charge). Because the interaction (5) is weak, one can use the Bogolubov theory to calculate all thermo-dynamic quantities explicitly.

The finite lifetime of excitons will make superfluidity a transient phenomena. Recently, Nagaoka[10] and Nakajima[11] stressed that in a real system the phase θ of the order parameter is most likely pinned by number-nonconserving processes. In order to see the essence of this mechanism, it is sufficient to add to a Hamiltonian of weakly interacting Bosons the following perturbation

$$\underset{k}{\Sigma} (g_k b_k^+ + g_k^+ b_k)$$ (8)

where b_k is a Boson operator, and g_k is assumed to be a classical amplitude. In the condensed state we get in addition to the Bogolubov Hamiltonian a term which depends explicitly on the phase. Writing $g_o = |g_o| \exp(i\alpha)$, we get $\Delta E = 2|g_o| n_o^{1/2} \cos(\alpha - \theta)$. The ground state is no longer continuously degenerate with respect to θ, thus $\vec{v}_s$ is no longer a hydrodynamic variable. Perturbations like eq.(8) could arise mainly from interactions of the excitons with the light field and impurities. Thus it remains an open question, whether real systems can be found and prepared in which number-nonconserving processes are so weak, that effects of a superfluid flow of excitation energy can be observed.

REFERENCES

1) I. M. Blatt, K. W. Boer and W. Brandt: Phys. Rev. 126 (1962) 1691.

2) S. A. Moskalenko: Sov. Phys.-Solid State 4 (1962) 199.

3) V. A. Gergel, R. F. Kazarinov and R. A. Suris: Sov. Phys. JETP 26 (1968) 354.

4) L. V. Keldysh and A. N. Kozlov: Sov. Phys. JETP 27 (1968) 521.

5) E. Hanamura: J. Phys. Soc. Japan 29 (1970) 50.

6) E. Hanamura: see the contribution in this volume.

7) W. Kohn and D. Sherrington: Rev. mod. Phys. 42 (1972) 1.

8) H. Haug and E. Hanamura: Phys. Rev. B11 (1975) 3317.

9) K. Weiss and H. Haug: *Cooperative Phenomena* ed. by H. Haken and M. Wagner (Springer, Berlin, 1973) p.219.

10) Y. Nagaoka: see the contribution in this volume.

11) S. Nakajima: see the contribution in this volume.

SUPERFLUID AND EXCITONIC STATES

Sadao Nakajima

Institute for Solid State Physics,
University of Tokyo,
Minato-ku, Tokyo 106 Japan

ABSTRACT

The fundamental difference between DLRO and ODLRO in relation to superfluidity is demonstrated by comparing the pseudospin model of Frenkel excitons with the quantum lattice model of ^{4}He and assuming Bose condensation in both models. The argument is extended also to Wannier excitons.

Since the classification of Yang,[1] we have been accustomed to associate superfluidity with the off-diagonal long range order (ODLRO). A formal proof was indeed given by Kohn and Sherrington[2] on the lack of superfluidity in the case of the diagonal long range order (DLRO). An explicit calculation of the energy transport in the excitonic phase of semimetals (or semiconductors) was done by Zittartz,[3] who concluded the lack of superheatconductivity.

However, Hanamura and Haug[4] have argued recently that the Bose condensation of Wannier excitons may result in the superflow of energy in contrast to the conclusion of Zittartz, though one has no super-fluidity of mass and charge in accordance with Kohn and Sherrington. Nagaoka[5] has pointed out, on the other hand, that in the case of DLRO characterized as a coherent state of a certain wave (charge, spin, strain *etc.*) it should always be possible to find some perturbation to destroy the "superflow" of the wave by clamping its phase. In the case of ODLRO, where the gauge symmetry of a basic matter field (*e.g.* the electron pair field in a superconductor) is broken, no other system than the superfluid itself can fix the phase[6] since any realistic perturbation should be gauge invariant.

The purpose of the present report is to point out that this fundamental difference between DLRO and ODLRO in relation to superfluidity may most clearly be seen by comparing the pseudospin model[7] of Frenkel excitons on the one hand with the quantum lattice model[8] of superfluid ^{4}He on the other. Thus, take a lattice of two-level atoms and let ground and excited levels of the j-th atom be represented by up and down states of the pseudospin σ_{jz}, respectively. We assume the Hamiltonian

$$H = - \sum_j E\sigma_{jz} - \frac{1}{2} \sum_{j \neq \ell} J_{j\ell} \{(1+\lambda)\sigma_{jx}\sigma_{\ell x} + (1-\lambda)\sigma_{jy}\sigma_{\ell y}\} , \qquad (1)$$

where E is the atomic level separation, $J_{j\ell}$ are transfer matrix elements, and $0 \leq \lambda \leq 1$ is the symmetry breaking parameter. Following Hanamura,[9] we may replace E by E-μ if we wish to deal with quasi-steady states of a highly excited insulator, where μ is the chemical potential of the exciton.

When $\lambda=0$, (1) is invariant under the rotation of pseudospins around the third axis (the XY isotropy) and has the same form as the quantum lattice model of ^{4}He. In this model, E is the chemical potential of the He atom and the XY isotropy means the number conservation of atoms, *i.e.*, the gauge invariance in the usual sense. The symmetry should therefore be preserved whatever perturbation (*e. g.* the interaction with walls) we may add. In the case of excitons, on the other hand, we always find some processes, in which the number of excitons is not con-served, so that $\lambda \neq 0$ in general. For instance $\lambda=1$ if J arises from the electric dipole interaction.[7]

We now suppose that the Fourier transform $J(k)=\Sigma J_{j\ell}\exp[ik \cdot R_{j\ell}]$ is a positive maximum at k=0, where $R_{j\ell}=R_j-R_\ell$ is the relative lattice vector. Thus pseudospins must be almost all aligned in a certain direction at low temperature. In the harmonic approximation, the direc-tion is determined by minimizing

$$\mathcal{Q} = - \frac{1}{2} \left[E\cos\theta + \frac{1}{4} J(0)\sin^2\theta \{1+\lambda\cos2\phi\} \right] . \qquad (2)$$

Here

$$\frac{1}{2} - <\sigma_{jz}> = \sin^2 \frac{\theta}{2}$$ (3)

is the number of excitons per site and ϕ is the phase common to all the atomic polarizations.

For $E > E_c$, where $2E_c = (1+\lambda)J(0)$, we have the normal ground state represented by pseudospins all parallel to the third axis ($\theta = 0$). The elementary excitation from this state, $i.\ e.$, the Frenkel exciton, is represented by the magnon of pseudospins. Strictly speaking, when $\lambda \neq 0$, the ground state contains some excitons[7] corresponding to the zero-point precession of pseudospins. We obtain $\phi = 0$ by minimizing this zero-point energy.

For $E < E_c$, we have the excitonic state, which is represented by pseudospins inclined with $\theta = \cos^{-1}(E/E_c)$. From (3), we see then that the number of zero-point excitons with k=0 is macroscopic (Bose condensation of Frenkel excitons). We also see from (2) that the phase ϕ is fixed to zero once $\lambda > 0$, however small it may be.

In the harmonic approximation, it is not difficult to obtain the elementary excitation energy

$$\varepsilon_k = [\ \frac{1}{2}\ (1+\lambda)\ \{\lambda J(0) + \frac{1}{2}\ (1-\lambda)(J(0) - J(k))\ \}$$

$$\times \{\sin^2\theta\ J(0) + \cos^2\theta\ (J(0) - J(k))\}\ ]^{\frac{1}{2}}\ .$$ (4)

It has the gap at k=0

$$\varepsilon_0 = [\ \frac{1}{2}\ \lambda(1+\lambda)\ \sin^2\theta\ ]^{\frac{1}{2}}\ J(0)$$ (5)

which vanishes for $\lambda = 0$. From (4) we then obtain the phonon-like excitation corresponding to the phonon in superfluid ^{4}He. It reflects the degeneracy of the ground state energy which is independent of the phase ϕ when $\lambda = 0$.

This degeneracy leads to the "Josephson effect". Thus the flow of excitons through a narrow boundary between two bulk subsystems is given by

$$C = \Sigma \ J_{\ell r} \langle \sigma_{\ell x} \ \sigma_{ry} - \sigma_{\ell y} \sigma_{rx} \rangle , \tag{6}$$

where sites ℓ and r are on the left and right of the boundary, respectively. Suppose that the phase ϕ has a jump at the boundary and is
uniform otherwise, so that $2 \langle \sigma_{\ell x} \rangle = \sin\theta\cos\phi_\ell$, $2\langle \sigma_{\ell y} \rangle = \sin\theta\sin\phi_\ell$, *etc.*
Ignoring the zero-point fluctuation, we obtain

$$C \simeq \sin^2 \frac{\theta}{2} \ \Sigma J_{\ell r} \ \sin(\phi_r - \phi_\ell) , \tag{7}$$

where the dependence on the sine of the phase difference is characteristic of the Josephson effect.[10] In order to satisfy the continuity
equation, the phase ϕ should not be quite uniform, but have a small
gradient in each subsystem. In fact, in the case of $\lambda=0$, the metastable
state corresponding to a relative minimum of Ω is possible with the non-
uniform phase $\phi_j = k \cdot R_j$ and $\theta = \cos^{-1}(J(k)/2E)$. Then (6) gives the uniform
superflow proportional to k. In the case of ^{4}He, this type of super-
flow is usually obtained by the use of the Galilean transformation.

These arguments can be extended to Wannier excitons. We assume
the simple model of semimetals

$$H = \Sigma_k \psi_k^+ \xi_k \tau_z \ \psi_k - \frac{1}{2} g \Sigma \psi_k^+ \vec{\tau} \psi_{k+q} \cdot \psi_\ell^+ \vec{\tau} \psi_{\ell -q} + \Sigma e^{i q \cdot X_i} u \psi_k^+ (1 - \lambda + \lambda \tau_x) \psi_{k+q} , \tag{9}$$

where ξ_k is the one-particle energy measured from the Fermi level, g
represents the short range interaction between electrons (up pseudospin)
and holes (down pseudospin), ψ_k is the two-component destruction opera-
tor, and $\vec{\tau} = (2\sigma_x, 2\sigma_y)$, $\tau_z = 2\sigma_z$. We have assumed the XY isotropy of the
interaction, through which the number of excitons is thus conserved.
We have destroyed this symmetry by including in (9) the interband
scattering due to impurities located at random sites X_i. Its importance
relative to the intraband scattering is measured by $0 \le \lambda \le 1$.

When $\lambda=0$, the order parameter

$$\vec{\Delta} = - g \ \Sigma_k \langle \psi_k^+ \vec{\tau} \ \psi_k \rangle \tag{10}$$

may be oriented in any direction on the XY plane of pseudospin space. We again have the flexibility of the order parameter as regards the phase ϕ and this may result in superfluidity, which is described by a GL type equation[11] near the transition point. Such a possibility is neglected in the argument of Zittartz.[3]

Once $\lambda \neq 0$, on the other hand, the direction of (10) is fixed along the Y-axis. Take the case $\lambda = 1$ for example. We apply the Hartree-Fock approximation to the interaction term in (9) and also the self-consistent Born approximation to the impurity scattering. The calculation is then the same as that of the superconductor with paramagnetic impurities,[12] whose spins are fixed all in one direction. Assuming constant density of states N_F near the Fermi level and small coupling constant gN_F, we obtain the self-consistency equation at temperature T

$$\vec{\Delta} = TgN_F \sum_{n=-\infty}^{+\infty} \left[\omega_n^2 + |\vec{\Sigma}_n|^2 \right]^{-\frac{1}{2}} \vec{\Sigma}_n \ . \tag{11}$$

Here

$$(1 + \Lambda_n) \Sigma_n^{(x)} = \Delta_x$$

$$(1 - \Lambda_n) \Sigma_n^{(y)} = \Delta_y \tag{12}$$

$$(1 - \Lambda_n) \omega_n = (2n+1)\pi T$$

and in terms of the life time τ due to the impurity scattering

$$\Lambda_n^{-1} = 2\tau \left[\omega_n^2 + |\vec{\Sigma}_n|^2 \right]^{\frac{1}{2}} \ . \tag{13}$$

For simplicity let us restrict ourselves to the transition temperature and linearize[11] as

$$\Delta_x \left[1 - TgN_F \sum_n \left[(2n+1)\pi T + \frac{1}{2\tau} \right]^{-1} \right] = 0 \ ,$$

$$\Delta_y \left[1 - TgN_F \sum_n \left[(2n+1)\pi T \right]^{-1} \right] = 0 \ . \tag{14}$$

Hence the excitonic phase with $\Delta_x = 0$, $\Delta_y \neq 0$ will appear at the same transition temperature as that of the pure system.

Finally we should mention that, in writing the interference factor as $\exp[iq \cdot X_i]$ in (9), we have tacitly assumed a direct gap. In general, we should write the scattering part as

$$H_{imp} = \Sigma \ u \ e^{iq \cdot X_i} \psi_k^+ \{(1-\lambda) + \lambda \vec{h}_i \cdot \vec{\tau}\} \psi_{k+q} \ , \tag{15}$$

$$\vec{h}_i = (\cos QX_i, \ \sin QX_i) \ . \tag{16}$$

Here Q is the wave vector pointing from the bottom of the electron band to the top of the hole band. The problem is thus similar to the super-conductor with randomly oriented impurity spins. After taking the average over X_i, no anisotropy in pseudospin space is left, so that the distinction between DLRO and ODLRO is by no means obvious. If we turn to the collective motion, instead of the individual excitations, however, we find that the phase of the excitonic condensate is again pinned down by impurity spins. Though we cannot go into the detail, the mechanism is similar to the one pointed out by Lee, Rice and Anderson[13] in the case of CDW produced by the Peierls instability. In the case of super-conductors with impurity spins, on the other hand, the phase of the Cooper pair remains free because of the gauge invariance.

ACKNOWLEDGMENT

The author wishes to thank Professor Y. Nagaoka for informing his idea prior to publication and also Professor H. Fukuyama and Dr. T. M. Rice for discussion on the last part of the present paper.

REFERENCES

1) C. N. Yang: Rev. mod. Phys. 34 (1962) 694.

2) W. Kohn and D. Sherrington: Rev. mod. Phys. 42 (1970) 1.

3) J. Zittartz: Phys. Rev. 165 (1968) 605 and 612.

4) E. Hanamura and H. Haug: Solid State Commun. <u>15</u> (1974) 1567.

5) Y. Nagaoka and J. Yamauchi: to be published in Solid State Commun.

6) P. W. Anderson: *Lectures on the Many-Body Problem* (Academic Press, New York, 1964) Vol.2, p.132.

7) P. W. Anderson: *Concepts in Solids* (W. A. Benjamin, New York, 1963) p.136.

8) T. Matsubara and H. Matsuda: Prog. theor. Phys. <u>16</u> (1956) 410.

9) E. Hanamura: J. Phys. Soc. Japan <u>37</u> (1974) 1545.

10) B. D. Josephson: Phys. Lett. <u>1</u> (1962) 251.

11) L. P. Gor'kov: Sov. Phys. JETP <u>37</u> (1960) 998.

12) A. A. Abrikosov and L. P. Gor'kov: Sov. Phys. JETP <u>12</u> (1961) 1243.

13) P. A. Lee, T. M. Rice, and P. W. Anderson: Solid State Commun. <u>14</u> (1974) 703.

DLRO, ODLRO AND SUPERFLUIDITY

Yosuke Nagaoka

Department of Physics,
Nagoya University
Nagoya, Japan

ABSTRACT

A necessary condition for superfluidity is discussed in connection with the classification of the long-range orders, DLRO and ODLRO, and it is concluded that it can take place only in systems with ODLRO. Based on this consideration, the possibility of superfluidity in a system of Frenkel excitons is examined. It is shown that it cannot occur in the excitonic phase, but that it can occur in the Bose-condensed phase of high-density excitons as a transient phenomenon.

According to Yang,[1] the long-range orders taking place in liquid helium and superconductors are sometimes called as the off-diagonal long-range order (ODLRO). Characteristic features of these orders are that the Gauge symmetry of the system is broken in the ordered state, and that the order parameter is off-diagonal with respect to the number of particles. We may call the other type of orders as the diagonal long-range order (DLRO), where the symmetry broken in the ordered state is the rotational or translational symmetry in the configurational or spin space and the order parameter is diagonal with respect to the number of particles. It is usually believed that superfluidity can occur only in systems with ODLRO in the above sense.[2]

From a mathematical point of view, however, it seems rather artifitial to discriminate the Gauge symmetry from the other type of symmetries. As is well known, there are various cases where ODLRO is mathematically equivalent to DLRO. As an example, let us consider the quantum lattice-gas model of bosons introduced by Matsubara and

Matsuda.[3)]   In this model, the Hamiltonian is given by

$$H = \frac{1}{2} \sum_{i>j} t_{ij}(a_i^+ a_j + a_j^+ a_i) + \sum_{i>j} U_{ij} n_i n_j - \mu \sum_j n_j , \tag{1}$$

$$n_j = a_j^+ a_j .$$

Here a_j^+ and a_j, the creation and annihilation operators of bosons at the lattice site j, obey the following commutation relations:

$$[a_i^+ , a_j^+]_- = [a_i , a_j]_- = [a_i , a_j^+]_- = 0 \qquad (i \neq j)$$

$$\tag{2}$$

$$[a_i , a_i]_+ = [a_i^+ , a_i^+]_+ = 0 , \qquad [a_i , a_i^+]_+ = 1 .$$

In eq.(1), the first term is the kinetic energy, the second term is the interaction between bosons, and μ denotes the chemical potential. The hard-core interaction between bosons is taken into account by the fermion-type commutation relation at the same lattice site.   If we introduce the pseudospin operators by

$$a_i \rightarrow S_i^x + iS_i^y , \qquad a_i^+ \rightarrow S_i^x - iS_i^y$$

$$\tag{3}$$

$$n_i \rightarrow \frac{1}{2} - S_i^z ,$$

they obey the usual commutation relations of spin operators

$$[S_i^x , S_j^y] = iS_i^z \delta_{ij} \qquad etc.$$

Using these operators, we can rewrite the Hamiltonian (1) as

$$H = \sum_{i>j} [J_{ij}^z S_i^z S_j^z + J_{ij}(S_i^x S_j^x + S_i^y S_j^y)] - h \sum_i S_i^z , \tag{4}$$

where

$$J_{ij}^z = U_{ij} , \qquad J_{ij}^\perp = t_{ij} , \qquad h = \mu + \sum_j U_{ij} .$$

In this example, we find an exact mathematical equivalence between a boson system and a spin system whose Hamiltonians are respectively given by eqs.(1) and (4). The mutual correspondence between
the two systems is given in Table 1. Then it is rather self-evident

Table 1. Correspondence between a boson system and a spin
system equivalent to it.

	boson system	spin system
symmetry	Gauge symmetry	Rotational symmetry around the z-axis
conserved quantity	The number of particles	The z-component of spins
order parameter	$\psi = \dfrac{1}{\sqrt{N}} \sum_j a_j$	$\vec{M}^{\perp} = \sum_j \vec{S}_j^{\perp}$
degeneracy of the ordered state	the phase of ψ	The direction of $\vec{M}^{\perp}$ in the x-y plane

that everything which occurs in the former system can occur in the
latter, too. Since the superflow of particles occurs in the boson
system, the superflow of the z-component of spins can occur in the
spin system. This implies that, if we generalize the concept of
superfluidity to general physical quantities, ODLRO is not a necessary
condition for the occurrence of superfluidity.[4] What is essential
for it is the continuous degeneracy of the ordered state with respect
to some degree of freedom as a result of the broken continuous
symmetry.

When we consider real systems, however, we find an essential
difference between a boson system and a spin system, or between the
Gauge symmetry and the rotational symmetry in the spin space. In
real spin systems, the rotational symmetry is only an approximate one

and we always find some sort of anisotropy energy. If it is sufficiently small, we may neglect it when we discuss thermodynamical
properties of the system. When we consider the direction of the
spontaneous magnetization, however, it plays an essential role, how
weak it may be. Due to the anisotropy energy, the symmetry of the
spin system reduces to a discrete one, and the direction of the
spontaneous magnetization is fixed to one of easy axes. Thus the
superflow of spin angular momentum cannot take place in real spin
systems. The Gauge symmetry is quite different from this. It is a
universal symmetry of physical systems, and is not destroyed by any
perturbation. This is the reason why we can find superfluidity only
in systems with ODLRO in the original sense of Yang.

Based on these general considerations, we shall next discuss the
possibility of superfluidity in an exciton system. For simplicity,
we consider Frenkel excitons in a lattice of two-level atoms.[5]
Assuming two atomic levels to be s- and p_z-levels, and taking only the
dipole interaction between atoms, we get the Hamiltonian as

$$H = \sum_j \frac{E}{2}(a^+_{jp}a_{jp} - a^+_{js}a_{js})$$

$$+ \sum_{i<j} \frac{p^2}{R^3_{ij}} \left(1 - \frac{3Z^2_{ij}}{R^2_{ij}}\right) (a^+_{ip}a_{is} + a^+_{is}a_{ip})(a^+_{jp}a_{js} + a^+_{js}a_{jp}) \quad ,$$

$$(5)$$

where $a^+_{j\alpha}$ and $a_{j\alpha}$ ($\alpha = s, p$) are respectively the creation and annihilation operators of electrons in the α-level of the atom j, E the atomic
energy splitting, p the matrix element of an electric dipole moment,
R_{ij} the distance between atoms i and j, and Z_{ij} its z-component. The
Hamiltonian can be rewritten again by using the pseudospin operators
defined by[6]

$$a^+_{jp}a_{js} \rightarrow S^x_j + iS^y_j \quad , \quad a^+_{js}a_{jp} \rightarrow S^x_j - iS^y_j \quad ,$$

$$\frac{1}{2}(a^{+}_{jp}a_{jp} - a^{+}_{js}a_{js}) \rightarrow S^{z}_{j} \quad ; \tag{6}$$

i.e. we have

$$H = -\sum_{i>j} J_{ij}S^{x}_{i}S^{x}_{j} - h\sum_{i} S^{z}_{j} \quad . \tag{7}$$

If we take into account more general interactions, then the Hamiltonian becomes

$$H = -\sum_{i>j} (J^{x}_{ij}S^{x}_{i}S^{x}_{j} + J^{y}_{ij}S^{y}_{i}S^{y}_{j}) - h\sum_{i} S^{z}_{j} \quad , \tag{8}$$

where $J^{x}_{ij} \neq J^{y}_{ij}$ in general. An essential difference of this system from the spin system described by the Hamiltonian (4) is the lack of the rotational symmetry around the z-axis.

If the exchange interaction is sufficiently strong compared with the magnetic field, spins are ordered in the x-y plane at low temperature. This is the so-called excitonic phase,[7] which corresponds to the order of electric dipoles in the real space. In contrast to the system with the Hamiltonian (4), in this case the direction of ordered spins is fixed depending on the relative magnitude of J^{x}_{ij} and J^{y}_{ij}. Therefore superfluidity cannot take place in the excitonic phase.

Next we consider the case where the magnetic field is strong and in the ground state all spins align in the z-direction. This corresponds to the usual insulating phase where all atoms are in the atomic s-level. Suppose the system is coherently excited to the state with $M_{x,y} \equiv \sum_{j} S^{x,y}_{j} \neq 0$. This is the Bose-condensed state of excitons discussed in detail by Hanamura and Haug.[8] Then in the molecular-field approximation, the magnetization obeys the equation of motion

$$\frac{dM_{x}}{dt} = (h - J_{y}M_{z})M_{y} ,$$

$$\frac{dM_{y}}{dt} = -(h - J_{x}M_{z})M_{x} , \tag{9}$$

$$\frac{dM_z}{dt} = (J_y - J_x)M_x M_y \quad ,$$

where $J_\lambda = N^{-1} \sum_j J_{ij}^\lambda$. These equations can be solved exactly by using elliptic functions. Though the anisotropy $(J_x \neq J_y)$ prevents a free precession of the magnetization and M_z is not a constant of motion, the motion is still a periodic one around the z-axis. The direction of the magnetization, or the phase of condensed excitons, is not fixed here. The situations are quite different between two cases, an excitonic phase and a Bose-condensed phase of excitons, though the Hamiltonian has the same symmetry. In the present case, we may expect the superflow of excitons, if the condensed excitons have a spatially inhomogeneous phase.

It should be emphasized here, however, that this is true only within the molecular-field approximation. In eq.(9) the magnitude of the magnetization $M^2 = \sum_\lambda M_\lambda^2$ is a constant of motion, but it is not in the original Hamiltonian. It implies that M^2 decays gradually in the course of time by the coupling with fluctuations.

The situation may be more explicit, if we rewrite the Hamiltonian again by using the boson operators B_j^+ and B_j defined by

$$S_j^x + iS_j^y = B_j^+(1 - n_j) \quad ,$$

$$S_j^x - iS_j^y = (1 - n_j)B_j \quad , \tag{10}$$

$$S_j^z = n_j - \frac{1}{2} \quad ,$$

where $n_j = B_j^+ B_j$. Then the interaction contains terms which do not conserve the number of bosons. Among them, quadratic terms, *i.e.* terms proportional to B^+B^+ and BB, can be eliminated by the Bogoliubov transformation, but higher order terms cannot. Even if the system is isolated from other systems, *e.g.* photons and phonons, and the total energy of the system is conserved, the number of excitons is not conserved.   Therefore in equilibrium excitons take the Planck

distribution, not the Bose-Einstein distribution.   It means that, if

the system initially has the macroscopic number of excitons in one

level, they decay gradually in the course of time, and that super-

fluidity can occur only as a transient phenomenon.   Whether you call

it as superfluidity or not depends on the time scale you are con-

sidering, and perhaps on your own taste.

REFERENCE

1)   C. N. Yang: Rev. mod. Phys. $\underline{34}$ (1962), 694.

2)   W. Kohn and D. Sherrington: Rev. mod. Phys. $\underline{42}$ (1970), 1.

3)   T. Matsubara and H. Matsuda: Progr. theor. Phys. $\underline{16}$ (1956), 569.

4)   We may generalize the definition of ODLRO to general conserved

quantities. For instance, the order of the spin system with the

Hamiltonian (4) is off-diagonal with respect to the z-component

of spins which is the conserved quantity in this system.   In

this sense ODLRO is a necessary condition for superfluidity.

5)   The possibility of superfluidity in this system was discussed by

S. Nakajima; preprint. See also Nakajima's paper in this

proceedings.

6)   P. W. Anderson: *Concepts in Solids* (Benjamin, N.Y., 1964), p.132.

7)   B. I. Halperin and T. M. Rice: Solid State Physics $\underline{21}$ (1968),

116.

8)   E. Hanamura: J. Phys. Soc. Japan $\underline{29}$ (1970), 50; *ibid.* $\underline{37}$ (1974),

1545.

E. Hanamura and H. Haug: Solid State Commun. $\underline{15}$ (1974), 1967.

See also Hanamura's and Haug's papers in this proceedings.

SEMICONDUCTOR-METAL TRANSITIONS

T. M. Rice

Bell Laboratories
Murray Hill, New Jersey 07974, U.S.A.

ABSTRACT

The theory of the semiconductor-metal transition is reviewed.
The study of electron-hole fluid has shed new light on the problem and
revived the old idea that the long-range Coulomb force makes the semiconductor-metal transition inrinsically first order. Recent theoretical work supports the idea of separate liquid-gas and metal-nonmetal
critical points in the electron-hole fluid in germanium.

I. INTRODUCTION

The theory of transitions between metallic and nonmetallic systems
has been the focus of increasing attention in recent years. It is now
some thirty years since the pioneering papers by Landau and Zeldovich,[1]
and by Mott[2] which posed the problem. Since that time a wide variety
of systems have been studied experimentally and theoretically. In this
brief review we shall not attempt to cover the diverse systems of interest and instead refer the interested reader to the excellent book on
the subject by Mott.[3] Rather, the focus is on the close relationship
between the study of a high density fluid of electrons and holes created by a nonequilibrium external source and the metal-semiconductor
transition as an indirect band gap passes through zero. The study of
the former system, which is the main topic of this conference, has shed
new light on some of the old questions on the latter topic. No attempt
will be made to review the electron-hole fluid in detail here. Such
reviews can be found elsewhere.[4-6]

In particular, in the original papers of Landau and Zeldovich[1] and
Mott[2] the role of the Coulomb force at the metal-semiconductor transi-

tion was stressed. Peierls (quoted in ref. 1) and Mott[2] argued that the long-range nature of the Coulomb force would cause a transition between the metallic and semiconducting phases to be intrinsically first order at zero temperature. The argument can be simply stated. In the semiconducting phase an electron and a hole attract each other with a long-range Coulomb potential which always has a bound state. On the other hand in a metallic phase with a finite density of free carriers, the long-range Coulomb force is screened and replaced by a short-range force which will not have a bound state for sufficiently strong screening. Therefore the metallic phase is stable only with a finite density of electrons and holes and it follows that the transition is first order.

In discussing the subsequent work it is helpful first to relate the two problems — the semiconductor-metal transition and the electron-hole fluid. This is the topic of the next section. In Section III the theory is presented for a low density fluid of electrons and holes and related to the earlier work on the excitonic insulator. In Section IV the theory of the metallic electron-hole liquid is reviewed and a general discussion of the high density regime is given. The transition between the two regions especially as a function of temperature is the subject of Section V. At present there are very interesting open questions on the transition from the low density of exciton gas which is weakly ionized to the high density plasma which is strongly ionized.

II.   THE SEMICONDUCTOR-METAL TRANSITION AND THE ELECTRON-HOLE LIQUID

We shall begin by discussing the relationship of the theory of the electron-hole liquid to the problem of the semiconductor-metal transition. In the first case one studies the energy of <u>fixed number</u> of electrons and holes interacting with Coulomb forces. The Hamiltonian of the system can be described using the effective mass approximation

$$H = - \sum_i \frac{\nabla_i^2}{2m_e} - \sum_j \frac{\nabla_j^2}{2m_h} + \frac{1}{2} \sum_{ij} \frac{e^2}{\kappa|\vec{r}_i^e - \vec{r}_j^e|} + \frac{1}{2} \sum_{ij} \frac{e^2}{\kappa|\vec{r}_i^h - \vec{r}_j^h|}$$

$$- \sum_{ij} \frac{e^2}{\kappa|\vec{r}_i^e - \vec{r}_j^h|} \qquad\qquad (2.1)$$

where m_e and m_h are the electron and hole effective masses and κ is the static dielectric constant. The assumption is made that in the steady state the rates at which electrons and holes are being generated and are recombining are sufficiently slow and may be neglected. Thus one assumes that there is true thermodynamic equilibrium of a model system described by eq. (2.1). The thermodynamic variables are the density and temperature.

In the semiconductor-metal transition the simplest case to study is that of an indirect gap semiconductor in which the energy gap is varied through zero. The background dielectric constant κ is due to vertical transitions and need not vary appreciably as the indirect band gap varies through zero. By varying the band structure the chemical potential, or the energy required to add an electron and hole to the system, is varied. In this system the number of electrons in the conduction band and the number of holes in the valence band are clearly not conserved. The relationship between the two systems was pointed out by Halperin and Rice,[7] namely that they are both described by the Hamiltonian (2.1) but with different thermodynamic variables.

There are, of course, other types of semiconductor-metal transitions involving direct gap materials or the overlap of Mott-Hubbard bands or disordered materials. The key assumption to which we will be restricted, is that the background dielectric constant κ and effective masses m_e and m_h remain finite at the transition. This can be true even in direct gap materials if the valence and conduction bands belong to different irreducible representations such that the interband matrix element of the momentum operator vanishes at the symmetry point. Disordered materials such as Si:P, where the P sites are frozen in fixed random positions in the lattice, are also excluded from this brief review.

III.   LOW DENSITY GAS OF EXCITONS: THE EXCITONIC INSULATOR

It was pointed out, some years ago, by Knox[8] that the binding energy of an exciton, E_x, could become greater than the band gap, E_{gap}, in an indirect gap semiconductor leading to an instability of the ordinary ground state of the crystal with respect to the spontaneous formation of excitons.   The value of E_x is given simply by a modified hydrogenic formula

$$E_x = \frac{\mu_o}{m\kappa^2} \text{ (rydbergs)} \tag{3.1}$$

where $\mu_o^{-1} = m_e^{-1} + m_h^{-1}$.   As discussed above, both μ_o and κ may depend sensitively on the direct energy gap, but not on an indirect energy gap.   The theoretical description of the new distorted phase, caused by the spontaneous formation of excitons was first developed by des Cloizeaux[9] and Keldysh and Kopaev[10] and subsequently discussed by many others.   It has been reviewed by Halperin and Rice[7,11] and by Kohn.[12] Consider the simplest case of a valence band with a single maximum at the zone center $(\vec{k}=0)$ and a conduction band with a minimum at the zone boundary $(\vec{k}=\vec{w})$.   If $E_{gap} < E_x$, excitons are present and the Hartree-Fock approximation was used to describe the distorted state and the expectation value

$$<b^+_{\vec{k}+\vec{w},\sigma} \, a_{\vec{k},\sigma'}> \neq 0 \tag{3.2}$$

for some σ and σ'.   The distorted state, generally known as the excitonic insulator, has a period in real space which is just double the period of the undistorted lattice.   In the Hartree-Fock picture, the one electron states of the distorted crystal are made up of linear combinations of wave functions of wave vectors $\vec{k}$ and $\vec{k}+\vec{w}$ from the

valence and conduction bands, respectively, of the nondistorted crystal.

If the expectation value (3.2) is real the distorted phase is a charge

density or spin density wave. If (3.2) is imaginary, orbital anti-

ferromagnetic or spin current states are possible.[7] The energies of

these four possibilities are split by interband scattering effects when

the electron and hole are on the same site. In the simplest model the

spin density wave state is lowest.

Let us examine these results from the point of view of a gas of

excitons. Let $A_{\vec{w}}^{+}$ be the creation operator for an exciton with wave

vector $\vec{w}$,

$$A_{\vec{w}}^{+} = \sum_{\vec{k},\sigma,\sigma'} f_{\sigma\sigma'}(\vec{k}) b_{\vec{k}+\vec{w},\sigma}^{+} a_{\vec{k},\sigma'} \qquad (3.3)$$

where $f_{\sigma\sigma'}(\vec{k})$ is an envelope wave function peaked around $\vec{k}=0$. A Bose

condensate of excitons means that $< A_{\vec{w}}^{+} > \neq 0$, which is identical to

(3.2). Treating the crystal in the Hartree-Fock approximation is equi-

valent to treating the gas of excitons as a repulsive Bose gas which

form a Bose condensate. The repulsion, in the Hartree-Fock approxima-

tion, arises from the Pauli exclusion principle on the electrons and

the holes.

The interaction between two excitons is not purely repulsive. At

large distances there will be a Van der Waals attraction between exci-

tons. The short-range repulsion due to the Pauli exclusion principle

will only apply to electrons or holes in a relative triplet state.

Thus there is a strong attraction between two excitons if both the

electrons and holes are in relative spin singlet states. Indeed it has

been shown that a bound state, an excitonic molecule, sometimes called

a biexciton can be formed in this case.[13-15] It is a very extended

state, however, and the binding energy of an excitonic molecule is only

a few per cent of E_x if the electron and hole masses do not differ by a large amount.   In the Hartree-Fock approximation the interaction is

averaged over space and spin configurations and the molecular correla-

tion ignored.   However, in the true dilute limit spatial molecular

correlations dominate.

For a nondegenerate band structure higher complexes are not bound

because of the large repulsive core in the exciton-exciton interaction

when either the holes or electrons are in a triplet state.   Brinkman

and Rice[16] have estimated the scattering length between two molecules by

treating each molecule simply as the superposition of two individual

excitons.   They obtained a value of $7a_x$ for the scattering length (a_s)

in the limit $m_e = m_h$ (a_x is the exciton Bohr radius).   One question of

interest is whether a molecular liquid or solid is stable.   This is

certainly true in the limit $m_e/m_h \rightarrow 0$, where the stable phase is a

molecular solid (*i.e.*, solid H_2).   In the opposite limit $(m_e = m_h)$

Brinkman and Rice[16] have argued, using the de Boer theory[17] of the quan-

tum corrections to equation of state, that no liquid-gas transition

will occur and the energy increases monotonically with the density, n.

They also propose that the molecular phase may be described as a weakly

interacting gas of molecules when the interparticle spacing $r_s a_x > a_s$

$[n = 3/4\pi r_s^3 a_x^3]$.   At higher density ($r_s \lesssim 7$ using the values of ref. 16)

the energy gained by the formation of molecules is overcome by the

repulsive interactions.   In this density regime it is possible that the

Hartree-Fock approximation, discussed above, could apply.   However, as

we shall see below, the energy of the metallic state will be very close

by and it is doubtful that there will be any region of validity of the

Hartree-Fock approximation, or equivalently of Bose condensed excitons.

It is interesting to study the limit of metastability of the low

density insulating phase.   [A first-order transition to a high density

metallic phase can occur at much lower density.]   Hanamura[18] considered

the criterion for the limit of metastability to be the density at which

the excitation energy (η) to create a spatially separate electron and hole goes to zero. If $\eta < 0$ there is no gap in the spectrum of current carrying excitations and one has a metallic state. Hanamura[18] has used the Hartree-Fock approximation to estimate $\eta(n)$. In fact, the calculation is similar to those made in the context of the theory of the excitonic insulator. He found $\eta(n)$ increases initially linearly with n. This repulsive interaction arises from the Pauli exchange repulsion between the electron and the exciton. The long-range interaction between an electron and an exciton is dominated by the attraction between the electron and the polarization induced on the exciton. Hanamura[18] included a polarization correction in lowest order and found it was sufficiently attractive to overcome the repulsion found in Hartree-Fock. He estimated that the critical value of metastability of the insulating phase occurs at a value of $r_s^{M-1} \simeq 3$. This calculation, however, does not adequately treat the spatial and spin-dependent correlations in the low density limit. Rice[19] has recently made a crude attempt to estimate r_s^{M-1} based on an effective electron-exciton potential and obtained a value $r_s^{M-1} \lesssim 2$.

So far, we have discussed the simplest case $m_e \approx m_h$ and nonorbitally degenerate electrons and holes. If $m_h \gg m_e$ (or *vice versa*) new possibilities arise. The stability of the molecular state is greatly enhanced and, as the experience of H_2 tells us, molecular solids and liquids can form. However, the zero point motion gives corrections which goes as $(m_e/m_h)^{\frac{1}{2}}$ and such states are confined to the limit $m_h \gg m_e$.

IV. THE METALLIC STATE: ELECTRON-HOLE LIQUID

At low densities excitons may be viewed as in the last section as well-defined entities with only weak interactions between them. Keldysh[20] first pointed out that a high density metallic state of electrons and holes could have a lower energy than the low density phase.

In such a metallic electron-hole liquid excitons cease to exist and any spatial correlation between a given electron and hole is extremely transitory. In this limit the exciton overlap is so strong that they lose their individuality and the e-h fluid must be viewed as two inter-penetrating Fermi fluids of electrons and holes. We will consider first the case of the idealized band structure with nondegenerate electron and hole bands and later in this section review the complica-tions introduced by the real band structure of Ge or Si.

In the high density limit the leading term in energy will be the kinetic energy,

$$E_K = \frac{3}{5} \left(\frac{k_F^2}{2m_e} + \frac{k_F^2}{2m_h} \right) = \frac{2.21}{r_s^2} \qquad (4.1)$$

expressed as the energy per pair in units of E_x. The kinetic energy gives a large positive (i.e., repulsive) contribution which increases with density as $n^{2/3}$.

The first correction is the exchange energy, E_{exch}, between the electrons and the holes separately. This is an attractive contribution arising from the reduction in Coulomb energy because of the spatial correlation imposed by the Pauli principle on electron (holes) in the same spin state.

$$E_{exch} = - \frac{3e^2 k_F}{2\pi\kappa} = - \frac{1.832}{r_s} \qquad (4.2)$$

The remaining term is known as the correlation energy, E_{corr}, which arises from the correlations primarily between particles in un-like quantum states. In the e-h liquid these are of two types. One is electron-hole correlations, i.e., between oppositely charged particles. The second is the electron-electron (or hole-hole) correlation between particles of like charge but opposite spin. These terms can be only approximately evaluated.

The first attack on the problem was by Hanamura[18] who used high density expansion to estimate this energy and obtained a local minimum in the total ground state energy-density curve $E_G(r_s)$ but with a minimum value at $r_s^o = 1.7$ of $E_G^o = -0.35E_x$. This value is substantially above the energy of separated excitons. It is known from experience with the single component electron gas that high density expansions in r_s are strictly limited to $r_s \leq 1$. In the single component electron gas a modified random phase approximation (RPA) is believed to give a good description of the energy even in the intermediate density regime ($r_s \sim$ 1-5). This approximation has been applied to the electron-hole liquid by Brinkman, Rice, Anderson and Chui.[21,16] These authors find a much lower minimum value ($-0.86\ E_x$) at a similar density.

The biggest question about the RPA is whether the correlation between electrons and holes are properly treated. Since it is just such correlations which give rise to the exciton bound state, they become increasingly important as the density is lowered. Within the RPA electron-hole correlations are included in lowest order and treated as opposite in sign to the electron-electron correlations. Recently Vashishta, Bhattacharyya and Singwi (VBS)[22] have applied the Singwi-Tosi-Land-Sjolander (STLS) method to the electron-hole liquid. In this method a set of self-consistent equations are obtained by approximating the equations of motion. The results of VBS for $m_e = m_h$ give a lower minimum value of $E_G^o = -0.99\ E_x$ but still $> -E_x$. Their approximation, however, builds in substantial more electron-hole correlation. A measure of such correlations is the enhancement factor $g_{eh}(o)$ which is the ratio of the probability of finding an electron at a hole to the mean density. Whereas, Brinkman and Rice[16] find a value of 2 for $g_{eh}(o)$ at $r_s = 1.7$, Vashishta $et\ al.$[22] report a value $g_{eh}(o) \approx 8$ at that density. A comparison of the two results for the ground state energy is shown in Fig. 1. Inoue and Hanamura[23] have attempted a variational calculation with a correlated wave function. They are forced to make approx-

imations in their evaluation so that
their answer is not truly variational.
Their final answer is very close to that
of VBS. It should be pointed out that
all of the above calculations involve
essentially uncontrolled approximations
and it is not possible to assess their
accuracy *a priori*. Nonetheless it appears that there is a metallic state
which is a local minimum of the energy
density curve but it is at an energy
slightly higher than that of separated

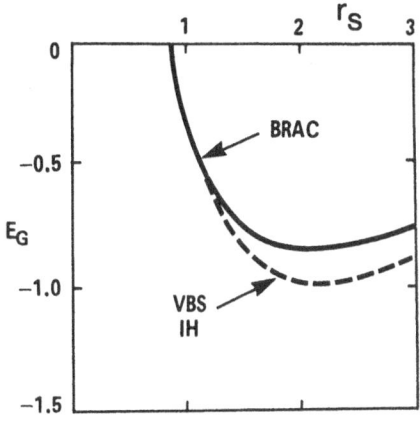

Fig.1 The ground state energy, E_G, of the isotropic nondegenerate e-h liquid in units of E_x as a function of the interparticle separation r_s in units of a_x.

excitons. This conclusion must be regarded as tentative, however,
since the energy differences are so small.

In the hydrogenic limit $(m_h \gg m_e)$ Wigner and Huntington[24] obtained
a ground state energy of $-1.05E_x$ for metallic hydrogen, slightly below
that of isolated hydrogen atoms but substantially above that of the
molecular solid (-1.17). Their calculation using the Wigner-Seitz
method included the electron-proton correlation to all orders. Several
calculations[16,25] have been performed at intermediate values of m_e/m_h and
it appears that for <u>isotropic nondegenerate bands</u> the low density molecular phase always has a lower energy than the high density state.

In general semiconductors have more complex band structures than
the ideal model discussed above. This can lead to big changes in the
energy. The cases of Ge and Si have been explored in detail by
several groups. The electron bands in both Ge and Si are highly ellipsoidal and have a high orbital degeneracy (ν_e) with $\nu_e = 4$ (Ge) and ν_e
$= 6$ (Si). In addition there are coupled hole bands which are four-fold
degenerate (including spin) at the zone center (Γ) and split away from
Γ. The effects greatly reduce the kinetic energy cost and to a lesser
extent the exchange energy. When the correlation energy is included a

greatly lowered ground state energy with a minimum at a higher density was found by several groups.[21,26,16,22,27]   These results shown in Fig. 2 are based on the RPA.   The results are fairly sensitive to the details of the band structure but the corrections to the RPA are proportionally much less important than in the ideal band structure.   The high degree of orbital degeneracy plays an important role both in stabilizing the metallic state and in minimizing the importance of the corrections to the RPA.   This can be seen by examining a diagrammatic expansion of the ground state energy.   In each order of an expansion in powers of e^2 the bubble diagrams, summed in the RPA (see Fig. 3), are the leading term and all corrections are down by a factor ν^{-1}, where ν is the degeneracy factor.   In Ge the degeneracy (including the holes and the spin) factor $\nu = 12$ and in Si, $\nu = 16$.   The first principles calculations are in very good agreement with experiments on both Ge and Si.[4,27,16]

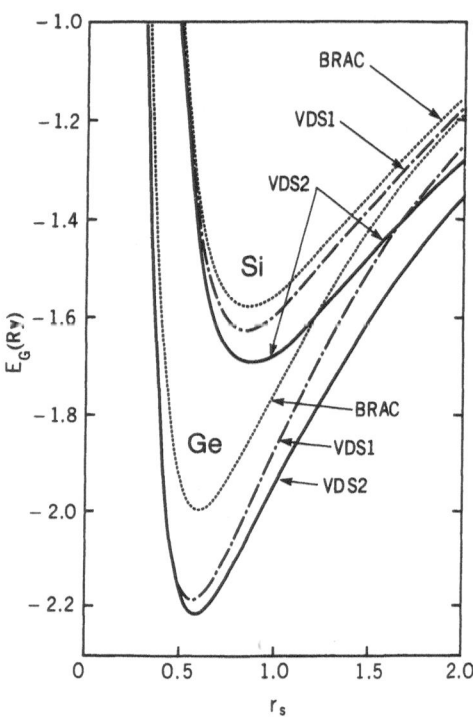

Fig.2   The ground state energy as function of r_s. The exciton rydberg and Bohr radius are defined with optical averaged effective masses [Ge; Ry = 2.65 meV; a_x = 177 Å; Si; Ry = 12.8 meV; a_x = 49 Å]. The curves BRAC[21] and VDS1[27] were both calculated within RPA and a more accurate band structure was used in VDS1. The curves VDS2[27] used the STLS approximation.

The implications of these results for the semiconductor-metal transition are clear.   Consider reducing E_{gap} from a large positive value.   Then before one reaches the condition $E_{gap} = E_x$ the condition $E_{gap} = -E_G^o$ will be satisfied and at that point a first-order transition

will take place directly from the
semiconducting phase to a metallic
phase with a finite density of
electrons and holes. This state-
ment is true only if $|E_G^o| > E_x$
but in general the energy bands
are anisotropic and have orbital
degeneracy both of which favor
the metallic phase.   Thus one
expects in a first-order transition which bypasses the excitonic
phases completely.

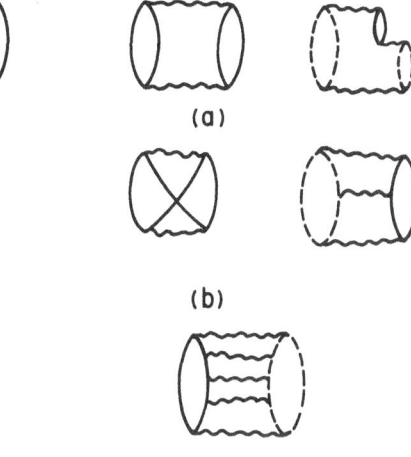

(a)

(b)

(c)

Fig.3 (a) The graphs included in
the RPA; (b) graphs omitted
in the RPA; (c) electron-hole
ladder graphs.   Solid and
dashed lines denote electron
and hole propagators.

What of the excitonic instability
in the metallic phase?   In the earlier
theory of the excitonic insulator, the excitonic instability arose as
an instability of the particle hole channel, *i.e.*, from repeated
electron-hole scattering [see Fig. 3(c)] .   The self-energy effects
on the electrons and holes were ignored and the chemical potential
taken as E_F, *i.e.*, as noninteracting electrons and holes.   The chemical
potential μ is given by

$$\mu = E_G(n) + n\partial E_G(n)/\partial n \tag{4.3}$$

so that at the minimum energy value, $\mu = E_G^o$.   Thus the chemical potential is much lower than E_F.   This lowering of the chemical potential
is due to the attractive potential seen by an electron or hole in the
metallic liquid.   The Pauli exclusion principle keeps apart electrons
of the same orbital and spin quantum numbers and thereby causes a net
positive charge in the vicinity of an electron.   Similarly correlations between electrons keep them apart and correlations between
electrons and holes enhance their overlap.   Explicit calculations of
the self-energy give a rigid shift of the band edges in the metallic

phase.[28]

If one reduces the density, n, below the equilibrium density, n_o, the electron-hole liquid is mechanically unstable towards a collapse. Nonetheless it is interesting to explore the question of the limit of metastability of the metallic phase ignoring the mechanical instability. This limit of metastability is set by the Mott criterion[2] which is the condition for the occurrence of an electron-hole or excitonic bound state. The metallic phase is stable if $q_{F.T.} > a_x^{-1}$ where $q_{F.T.}$ is the Fermi-Thomas screening wave vector. In a simple band structure this yields a condition $r_s < 10$,[16] which is four times larger than for a collection of hydrogen atoms. This increased metastability of the metallic phase is due to the increased screening and increased kinetic energy cost of localization when both positive and negative charges are light. The metastability is further enhanced in more complex band structures by orbital degeneracy, *etc.*

To summarize the phase diagram at zero temperature of the semi-conductor-metal transition is a first-order transition between the semiconducting phase for $E_{gap} > |E_G^o|$ to a metallic phase when $E_{gap} < |E_G^o|$ for a general phase diagram. For a simple band structure where the lowest electron-hole state is a molecular phase, this phase diagram may not change since the molecular phase does not require a change of crystal symmetry and will not be a distinguishable semiconducting phase.[11] If, however, the metallic phase has a band structure with 'nesting', *i.e.*, electron and hole Fermi surfaces which 'nest' when translated by some wave vector, then the distortions characteristic of the excitonic insulator, such as charge density waves or spin density waves can occur. An example of 'nesting' is the isotropic nondegener-ate band structure where the electron and hole Fermi surfaces are identical spheres. Under these circumstances the occurrence of charge or spin density wave states is more properly viewed as instability of a metal with a special band structure, as for example in Cr.

V.   THE SEMICONDUCTOR-METAL TRANSITION AT FINITE TEMPERATURE

In the previous sections we discussed the high density and low density limits and the transition between them at zero temperature. We turn now to finite temperatures and the interesting possibilities raised by the large region of metastability of the metallic and insulating phases.

At low temperatures, the calculation of the phase boundaries of the first-order transition discussed in the previous section is straightforward.   Taking temperature and density of electron-hole pairs as thermodynamic variables, at a density $n < n_o$ the system will break up into a high density liquid and a low density gas.   The density in the gas phase is very small at low temperatures and it will exert a negligible pressure on the liquid.   The Landau theory of Fermi liquids can be used to calculate the linear thermal expansion.

At low temperatures the gas outside will be composed of excitons, excitonic molecules and ionized electrons and holes.   The overall density will be very low and we can therefore ignore interaction effects and treat the equilibrium between the various species by classical statistical mechanics.   The chemical potential of each species is equal to that of the liquid.   The density of each species, i, is controlled by its work function, ζ_i.   This is the energy required to take each subspecies from the liquid to infinity at zero temperature.   Thus for i = exciton then $\zeta_{ex} = |E_G^o| - E_x$, the binding energy of the e-h liquid relative to the exciton.   For excitonic molecules

$$\zeta_{ex.mol} = 2\zeta_{ex} - E_{\beta mol} \tag{5.1}$$

where $E_{\beta mol}$ is the binding of the excitonic molecule with respect to dissociation into two excitons.   Since $E_{\beta mol}/\zeta_{ex}$ is estimated to be very small[13-15] ($\ll 1$) in Ge or Si it follows that there will be very few molecules in the gas outside.

The energies ζ for electrons and holes can be estimated as follows. First, the sum

$$\zeta_e + \zeta_h = \zeta_{ex} + E_x \ . \tag{5.2}$$

This follows at once from the fact that the energy to take out an electron and hole separately is equal to the energy cost of separating them from an exciton in the gas. Secondly, overall charge neutrality will set $\zeta_e = \zeta_h$ at finite temperature. Thus $\zeta_e = \zeta_h = \frac{1}{2} (\zeta_{ex}+E_x) > \zeta_{ex}$ in Ge or Si. The number of ionized e-h pairs will therefore be small relative to excitons at low temperatures. For simpler band structures, where $E_x \gg \zeta_{ex}$, this ionized fraction will be even smaller. The gas at low temperatures is predominantly composed of excitons.

As the temperature is raised the difference in densities between the liquid and the gas decreases. The simplest phase diagram possible would be to extrapolate the densities on both sides to form a single gas-liquid critical point (n_c, T_c) in the manner shown in Fig.4. This single critical point gives a single line for the phase diagram for the semiconductor-metal transition when we translate to the thermodynamic variables (μ, T).

In Ge and Si it has been found possible to make a reasonably accurate calculation of the gas-liquid critical point. This was first pointed out by Combescot[29] who used essentially the RPA for the free energy. As discussed in the previous section, the corrections to the RPA are greatly reduced by the high degree of orbital degeneracy in Ge and Si. The free energy per pair F can be written as

$$F(n,T) = F_o(n,T) + F_{xc}(n,T) \tag{5.3}$$

where F_o is the free energy of a noninteracting fluid of electrons and holes. The second term F_{xc} represents the exchange and correlation contributions

Within the RPA the effective masses are only slightly renormalized.[28] Thus there is very little change in the low lying excitations caused by the interactions. The principle effect of the interactions is to introduce the plasmon collective excitations at high energy. For densities $\sim 10^{17}$ cm^{-3} in Ge the plasmon energy $\hbar\omega_p$ ≈ 14 meV is very high and there will be no appreciable occupation of these modes until very high temperature. It is safe to assume that $\hbar\omega_p \gg k_B T$ near the critical point. This suggests replacing F_{xc} by $E_{exch}(n) + E_{corr}(n)$.

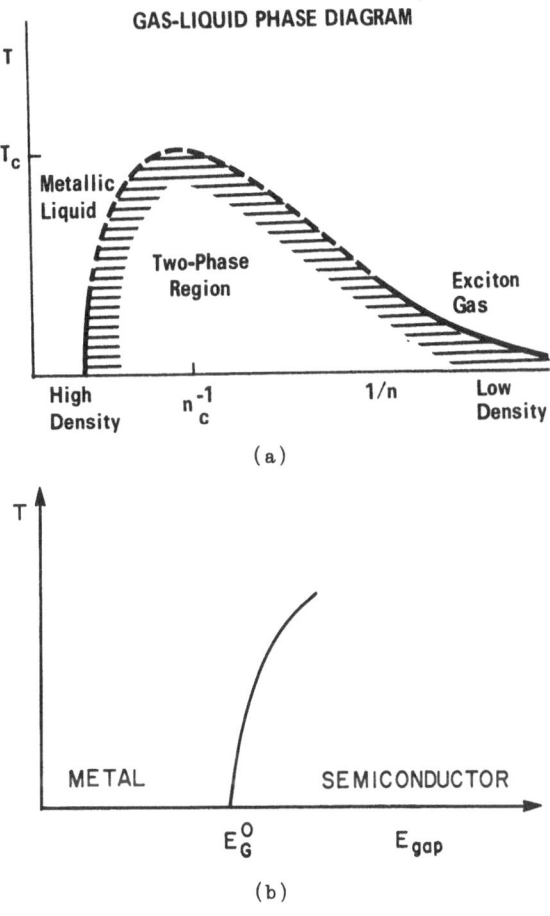

Fig.4 The simplest phase diagram of the e-h liquid obtained by extrapolating the low temperature behavior (a) temperature and density plane, (b) temperature and chemical potential plane.

The conditions for a critical point at temperature T_c and density n_c can be expressed simply in terms of the chemical potential μ as

$$\left.\frac{\partial\mu}{\partial n}\right|_{n_c,T_c} = \left.\frac{\partial^2\mu}{\partial n^2}\right|_{n_c,T_c} = 0 \tag{5.4}$$

where $\mu\,(=F+n\,\partial F/\partial n)$ is the chemical potential for an e-h pair. These conditions involve a knowledge of the third derivative of F with respect to n. This leads to some calculational difficulties. The

calculations have been performed by several groups now [29,19,27] and
the answers are in reasonable agreement. They predict a critical
point with $T_c \approx 7$ K and $n_c \approx 10^{17}$ cm^{-3} in Ge. Indeed a very different
approach by Reinecke and Ying[30] using the Fisher droplet model gave
fairly similar values. Further, in the experiments of Thomas *et al.*[31]
at critical point was found at $T_c = 6.5$ K and $n_c = 0.8 \times 10^{17}$ cm^{-3} in Ge.
These values are still well within the metallic regime by any criterion.
This assertion is also supported by the luminescence lineshape observed
near the critical point by Thomas *et al.* [31] We can conclude, with
some certainty, that in Ge (and presumably Si too) that there is a
liquid-gas critical point within the metallic regime.

As the temperature is lowered, there is a transition on the low
density side of liquid-gas curve between a metallic or strongly ionized
plasma in the vicinity of the liquid-gas critical point to a low
density weakly ionized exciton gas. This transition is a form of
metal-to-nonmetal transition. In previous sections we discussed
briefly some of the criteria that can be used at zero temperature to
determine the boundary between metallic and insulating behavior. At
zero temperature the limits of metastability of the metallic and insulating phases were very different implying a large first-order
transition if one could envisage a uniform expansion of the e-h liquid.
Such a first-order transition could persist to a high enough temperature to be visible along the coexistence curve. This possibility was
suggested by the present author[19] who revived the early suggestion by
Landau and Zeldovich[1] for fluid Hg. This would lead to a phase diagram as shown in Fig. 5.

The phase transition between weakly ionized and strongly ionized
plasma has been studied by several authors. Reviews have been written
recently by Norman and Starostin[32] and by Ebeling, Kraeft and Kremp.[33]
In their approach the ionization equilibrium is determined by an
equation of the form

$$n_{ex} = n_e^2 \left(\frac{2\pi\hbar^2}{kT} \frac{M_{ex}}{m_e m_h} \right)^{3/2} \frac{\nu_{ex} e^{\beta E_x - \beta \delta E_x}}{\nu_e \nu_h} \qquad (5.5)$$

where n_e and n_{ex} are the density of electrons (or holes) and excitons, respectively, M_{ex} is the translational mass of an exciton and ν_{ex}, ν_e and ν_h are the degeneracy factors of the excitons, electrons and holes including spin and orbital degrees of freedom. (In principle excited states of the exciton should also be included but in practice they are ignored.) The correction δE_x to the exciton binding energy includes the Debye-Huckel screening by the electrons and holes. So far the expression is purely classical but then the lowest order quantum correction to δE_x is added in, giving

$$\delta E_x = e^2/\kappa r_D \left[1 - C\Lambda/r_D \right]$$

$$(5.6)$$

where Λ is the thermal de Broglie wavelength for electrons (Λ

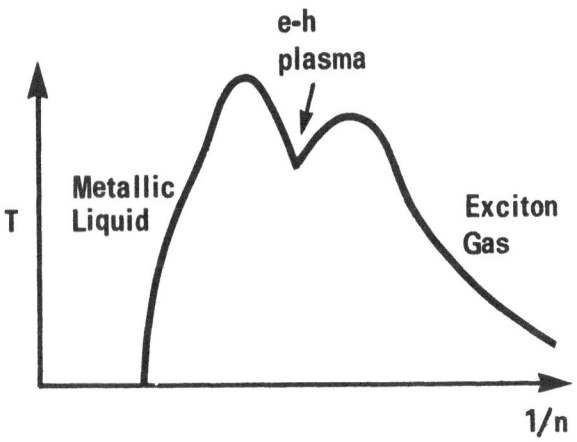

(a)

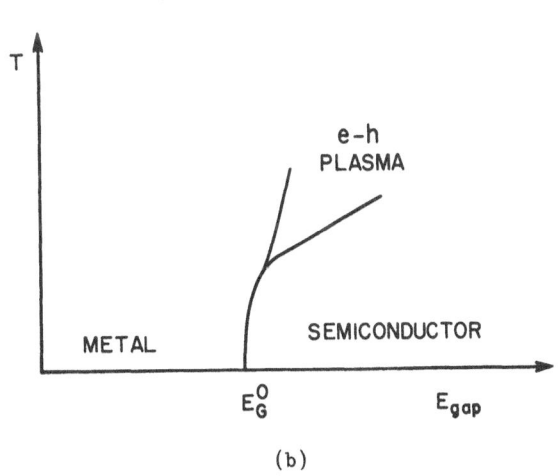

(b)

Fig.5 Phase diagram expected if a second critical point associated with the metal-nonmetal transition occurs (a) temperature and density plane and (b) temperature and chemical potential plane.

$= h(2\pi m_e kT)^{-\frac{1}{2}})$ and r_D is the Debye-Huckel screening radius $(r_D$

$= (8\pi\beta n_e e^2/\kappa)^{-\frac{1}{2}}$ and $\beta^{-1} = k_B T)$. The constant C is a numerical constant

≈ 0.1 A further approximation can be made which has similar accuracy

to namely

$$\delta E_x = \frac{e^2}{\kappa r_D} \frac{1}{1+C\Lambda/r_D} \qquad (5.7)$$

Norman and Starostin point out that the structure of (5.7) is typical

for the theory of strongly nonideal electrolytes.   Thus it will hope-

fully be more valid than (5.6) at higher densities.

The condition for mechanical stability can be expressed as

$(\partial n_{ex}/\partial n_e)_T > 0$.   If this condition is violated, then as one adds more

electrons and holes to the system the number of excitons would decrease,

clearly leading to an ionization catastrophe.   The critical point is

determined by the conditions

$$\left.\frac{\partial n_{ex}}{\partial n_e}\right|_{T_c} = \left.\frac{\partial^2 n_{ex}}{\partial n_e^2}\right|_{T_c} = 0 \qquad (5.8)$$

Kremp, Ebeling and Kraeft[34] have applied this theory (using 5-7 with

C=1/8) to the electron-hole fluid in Ge and estimate a critical point

temperature T = 9 K and density n $\simeq 10^{15}$ cm^{-3}.   This temperature and

density are still within the classical regime.   These authors point

out that their values are very different to that obtained above for the

liquid-gas critical point and support the possibility of two separate

critical points.

At present, the question of two critical points in Ge is an open

one and this represents one of the outstanding experimental questions

which are unresolved at the present time.   In the few metals that

have been studied near their critical point, Hg and Cs, no evidence of

two critical points has been found.   However, in Ge and Si the orbital

degeneracies and anisotropy in the band structure greatly stabilize metal and separate the liquid-gas critical point from the region of the metal-to-nonmetal transition, thus affording an opportunity to see two critical points. The observation of the second critical point would be a striking confirmation of the early ideas discussed in the introduction on the metal-insulator transition.

REFERENCES

1) L. D. Landau and G. Zeldovich: Acta Phys.-Chim. USSR 18, (1943) 194 [Eng. Trans. in Collected Works of L. D. Landau, ed. by D. TerHaar (Pergamon Press) 1965].

2) N. F. Mott: Proc. Phys. Soc. (London) A62, (1949) 416.

3) N. F. Mott: *Metal-Insulator Transitions* (Taylor and Francis, Ltd., 1974).

4) Ya. E. Pokrovskii: Phys. Stat. sol. (a) 11, (1972) 385.

5) L. V. Keldysh: in *"Eksitony v Polupnovodnikah"* (Excitons in Semiconductors) (Nauka, Moscow) 1971, p. 5.

6) T. M. Rice, J. C. Hensel, G. A. Thomas and T. G. Phillips (to be published).

7) B. I. Halperin and T. M. Rice: in *Solid State Physics*, ed. by F. Seitz, D. Turnbull and H. Ehrenreich (Acad. Press, NY, 1968) 21, 115.

8) R. S. Knox: Solid State Phys. Suppl. 5, (1963) 100.

9) J. des Cloizeaux: J. Phys. Chem. Solids 26, (1965) 259.

10) L. V. Keldysh and Yu. V. Kopaev: Fiz. Tverd. Tela 6, (1964) 2791. [Eng. Trans., Sov. Phys.-Solid State 6, (1965) 2219].

11) B. I. Halperin and T. M. Rice: Rev. Mod. Phys. 40, (1968) 755.

12) W. Kohn: in *Many-Body Physics*, ed. by C. deWitt and R. Balian (Gordon and Breach, NY, 1968), p. 351.

13) E. A. Hylleraas and A. Ore: Phys. Rev. 71, (1947) 493.

14) A. Akimoto and E. Hanamura: Solid State Comm. 10, (1972) 253.

15) W. F. Brinkman, T. M. Rice and B. J. Bell: Phys. Rev. B8, (1973) 1590.

16) W. F. Brinkman and T. M. Rice: Phys. Rev. B7, (1973) 1508.

17) J. deBoer, in *Prog. in Low Temp. Phys.*, ed. by C. J. Gorter (North-Holland, Amsterdam 1957), II.1.

18) E. Hanamura: in *Proc. of the Tenth Int. Conf. on the Physics of Semiconductors*, Cambridge, Mass., 1970, ed. by S. P. Keller, J. C. Hensel and F. Stern (USAEC, Div. of Tech. Info., 1970), p. 504.

19) T. M. Rice: in *Proc. of the Twelfth Int. Conf. on the Physics of Semiconductors*, Stuttgart, 1974, ed. by M. H. Pilkuhn, (Teubner, Stuttgart), p. 23.

20) L. V. Keldysh: in *Proc. of the Ninth Int. Conf. on the Physics of Semiconductors*, Moscow, 1968, ed. S.M. Ryvkin (Nauka, Leningrad, 1968) p.1303.

21) W. F. Brinkman, T. M. Rice, P. W. Anderson and S.-T. Chui: Phys. Rev. Lett. 28, (1972) 961.

22) P. Vashishta, P. Bhattacharyya and K. S. Singwi: Phys. Rev. Lett. 30, (1973) 1248.

23) M. Inoue and E. Hanamura: J. Phys. Soc. Japan 35, (1973) 643.

24) E. Wigner and H. B. Huntington: J. Chem. Phys. 3, (1935) 764.

25) P. Vashishta, P. Bhattacharyya and K. S. Singwi: Phys. Rev. B10, (1974) 5108.

26) M. Combescot and P. Noziéres: J. Phys. C 5, (1972) 2369.

27) P. Vashishta, S. G. Das and K. S. Singwi: Phys. Rev. Lett. 31, (1974) 911.

28) T. M. Rice: Il. Nuo. Cim. B23, (1974) 226.

29) M. Combescot: Phys. Rev. Lett. 32, (1974) 15.

30) T. L. Reinecke and S. C. Ying: Phys. Rev. Lett. 35, (1975) 311.

31) G. A. Thomas, T. M. Rice and J. C. Hensel: Phys. Rev. Lett. 33, (1974) 219.

32) G. E. Norman and A. N. Starostin: Tep. Vys. Temp. $\underline{8}$, (1970) 413. [Eng. Trans. High Temperature $\underline{8}$, (1970) 381].

33) W. Ebeling, W. D. Kraeft and D. Kremp: *Theory of Bound States and Ionization Equilibrium in Plasmas and Solids* (Akademie-Verlag, Berlin, 1975).

34) D. Kremp, W. Ebeling and W. D. Kraeft: Phys. Stat. sol. (b) $\underline{69}$, (1975) K59.

THE KINETICS OF DECAY OF ELECTRON-HOLE
DROPLETS IN GERMANIUM

J. C. Hensel

Bell Laboratories
Murray Hill, New Jersey 07974, U.S.A.

ABSTRACT

We briefly review several time-resolved microwave experiments
which have been used to study the kinetics of electron-hole droplet
decay.

I.   INTRODUCTION

It is now well-known that excitons in Ge can condense at low temperatures into a metallic, liquid state having the form of electron-
hole droplets (e-h droplets). The study of time-dependent phenomena
has played an important role in the achievement of our present stage of
understanding of this interesting system. In this paper we shall review microwave experiments,[2-4] which have elucidated the kinetics
of the decay of e-h droplets in Ge.

II.   THE KINETIC RATE EQUATIONS

In studying the decay kinetics we address ourselves to the following features of the e-h drop system: the binding energy of the e-h
condensate φ (relative to the energy of the free exciton), the phase
diagram of the two-phase system of an exciton gas coexisting with the
e-h liquid condensate, and the decay modes of e-h droplets. Interest-
ingly enough these topics have a great deal in common, and all are
amenable to investigation via decay kinetics. An important fact to re-
cognize is that the rates of drop decay are intimately related to the
thermodynamics of the system. This is explicitly seen from the rate
equation for an e-h droplet containing ν e-h pairs

$$\frac{d\nu}{dt} = - \frac{\nu}{\tau_O} - a\nu^{2/3}T^2 e^{-\varphi/k_B T} + bn_{ex}\nu^{2/3} \; . \tag{1}$$

On the r.h.s. the first term is the bulk decay rate in the drop given by a total lifetime τ_O; the second is a surface decay akin to thermionic emission with the work function φ; and the third term represents the "backflow", the capture of free excitons (whose density is n_{ex}) by droplets.

At very low temperatures the thermionic emission rate becomes vanishingly small; however, there is evidence[4] for a residual temperature independent surface evaporation which can be described analytically by including the term $-c\nu^{2/3}$ on the r.h.s. of eq. (1).

If a quantitative description of the exciton density n_{ex} and free-carrier density $n_{e,h}$ is required, then two additional rate equations must be written for these variables and integrated simultaneously with eq. (1).

An important class of experiments use pulse excitation and monitor the time dependence of ν after the pumping source has been turned off, whereupon the uncondensed carriers and excitons die rapidly by fast recombination in $\gtrsim 1\,\mu$sec. However, the conventional decay mechanisms are bottlenecked inside the droplets by the comparatively high densities of e-h pairs which saturate the traps. The enhanced recombination time is a sign of the existence of drops and makes a time analysis experiment possible. In the experiments described in the following sections it is assumed that the decay takes place into a "vacuum" ($n_{ex} \approx 0$) which permits one to neglect the backflow term. In this case the rate equation (1) for ν can be integrated at once to give

$$\nu(t) = \alpha^3 \tau_O^3 \left[e^{(t_c - t)/3\tau_O} - 1 \right]^3 \tag{2}$$

with $\alpha = aT^2 \exp(-\varphi/k_B T)$. (The inclusion of the extra surface term $c\nu^{2/3}$ in (1) results simply in the constant c being added to the r.h.s. of this equation defining α). This solution is highly non-exponential at long times and has a finite cut-off time t_c where $v(t_c) = 0$,

$$t_c = 3\tau_0 \ln \left[1 + \frac{v^{1/3}(0)}{\alpha\tau_0} \right] . \tag{3}$$

As regards to the behavior of n_{ex} or $n_{e,h}$, in this regime they follow $v(t)$, simply because the time scale for changes in $v(t)$ is much longer than the recombination times of excitons and e-h pairs.

III. d.c. CONDUCTIVITY

We turn now to several kinds of experiments that serve to monitor the time dependence of the system. By way of introduction we consider a measurement of the pulsed d.c. photoconductivity of a Ge sample. These experiments are most simply interpreted if one operates at low excitation where the photoconductivity is unquestionably dominated by free carriers from the decay of drops. Here the conductivity is governed by the rate equations and could be used as a tool to study the droplet decay kinetics.

An example of low pumping level d.c. conductivity data is shown in Fig. 1. The initial, fast pulse in both traces is the photoconductivity due to the laser pulse. The subsequent tails with much longer lifetimes are a sign of droplets; this conductivity results from the decay carriers. Since the main features in this

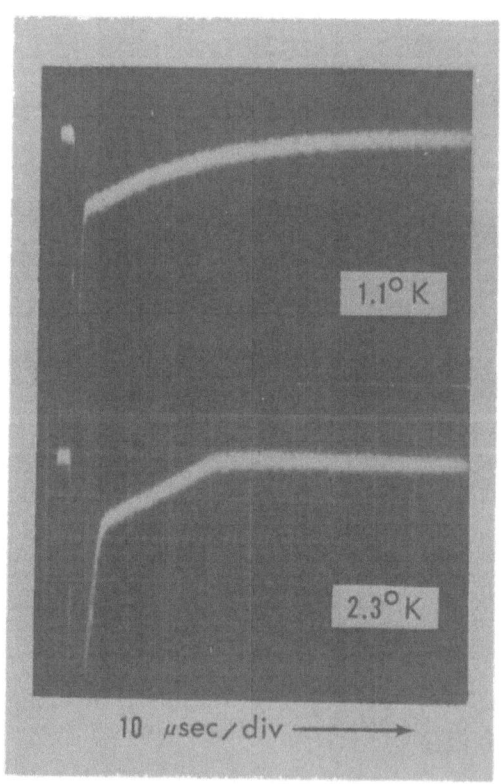

Fig.1. Oscilloscope traces of the d.c. conductivity signal following excitation by 1.2 nJ laser (1.064 μm) pulses of 0.3 μsec width.

data are identical in form to those seen by pulsed cyclotron resonance, we shall pass on forthwith to that subject.

IV. MICROWAVE CYCLOTRON RESONANCE

From cyclotron resonance experiments[3] done at very low pumping levels the main decay mechanisms for e-h drops have been established. The cyclotron resonance detects free carriers outside drops, and free carriers arise from two sources. The excitons, which evaporate from the surface, decay principally by an Auger process liberating free carriers. Likewise intrinsic Auger recombination takes place within the drop creating free carriers with sufficient kinetic energy to escape. The free carrier signal, a measure of $n_{e,h}(t)$, can be interpreted in terms of the rate equation for $\nu(t)$ as pointed out earlier.

Cyclotron resonance measurements of time decays were obtained at 53 GHz with the magnetic field fixed on resonance and the free carrier population $n_{e,h}(t)$ monitored from the change of the absorption signal amplitude as a function of time. In the inset in Fig. 2 is a typical recorder tracing taken at 2.5 K showing a fast signal coincident with the laser

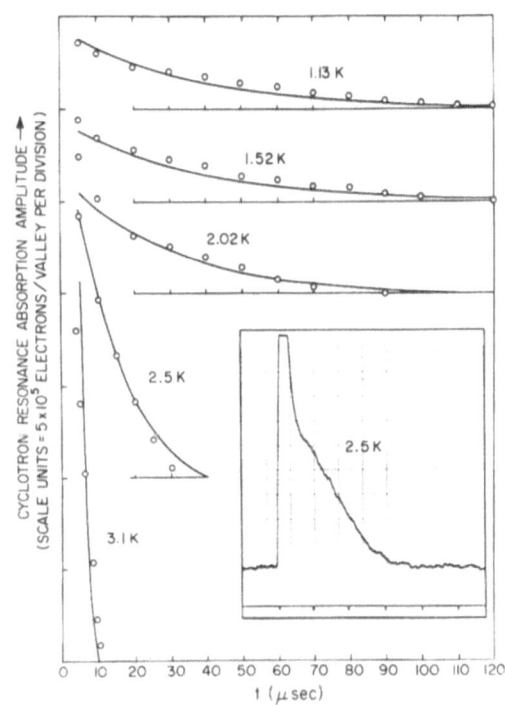

Fig.2. Decay profiles of the cyclotron-resonance absorption signal subsequent to excitation by 1.2 nJ laser (1.064 μm) pulses of 0.3 μsec width.

pulse followed by a much slower, nonexponential decay related to the e-h drops. From the data (circles) in the main figure two regimes can be recognized. Below 2 K where thermionic emission is negligible the decays are quasi-exponential in agreement with the simple solution e^{-t/τ_0} of eq. (1) with the lifetime $\tau_0 = 40 \pm 5$ μsec. Above 2 K a qualitative change in the decays is apparent in the profiles. The decays become distinctly nonexponential and terminate at a well defined cutoff time t_c marking the extinction of the drops. It is evident that t_c, a measure of the surface decays, depends very strongly on temperature. This effect is predictable from eq. (3) which gives, when $t_c \ll 3\tau_0$, the approximate result $t_c \approx 3\nu^{1/3}(0)\ a^{-1}T^{-2}\exp(\varphi/k_B T)$ where $\nu(0)$ is the number of e-h pairs per drop at t=0. Thus, $\ln t_c$ depends linearly on T^{-1}, and φ acts as an activation energy. The data are plotted in the form $\ln t_c$ vs T^{-1} as shown in Fig. 3 and fitted, dashed curve, by eq. (3). For the purposes of curve fitting the "cutoff" in the low tempera- ture regime 2 K is defined to be "signal-to-noise equal to one". From the fit in the linear region one gets $\varphi = (16^{+3}_{-2})K$. Similar results for φ have been ob-

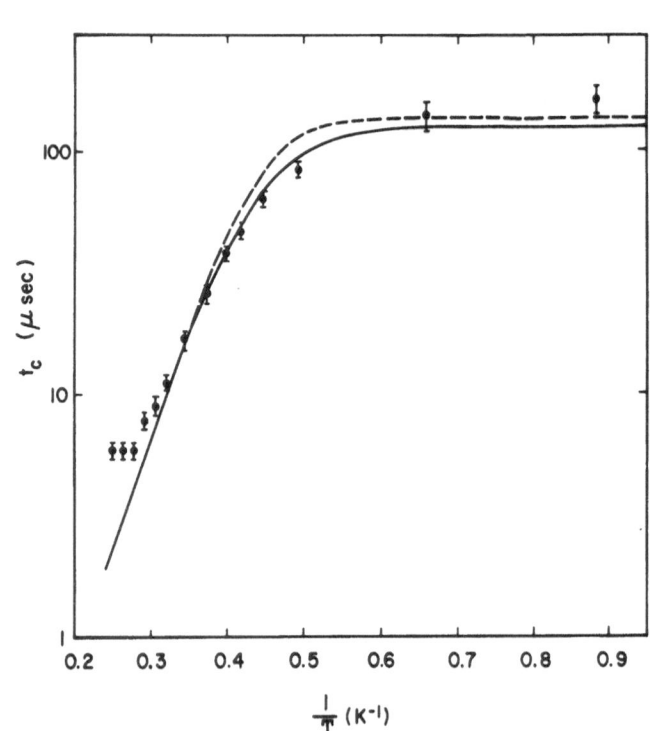

Fig.3. The cutoff times t_c of the decay profiles of Fig. 2 plotted vs T^{-1}.

tained from other kinetic experiments by Benoit à la Guillaume, *et al.*[5)]

and by Westervelt, *et al.*[6)]

In Fig. 2 there is good agreement between the simple theory just

outlined (solid curves) and experiment above 2 K, but below 2 K the

deviation from exponential decays is greater than predicted by the

rate equations. While such a discrepancy could arise from self-

heating of the drop relative to the lattice temperature, there is

reason to believe that it arises from direct emission of excitons

(expressed by the rate $cv^{2/3}$) due to excitations created in the e-h

liquid near the surface by radiative recombination of e-h pairs. The

extra surface contribution corresponding to the direct emission of one

exciton for each volume recombination within a surface layer of 0.3 μm

(this corresponds to a value of $c = 1.1 \times 10^{6} sec^{-1}$) is included in the

curves in Fig. 2 and the solid curve in Fig.3. More direct experi-

mental evidence for this process is discussed in Sec. V.

It is tacitly assumed that the number of drops does not change

during decay which permits an analysis to be made based on the rate

equation for a single drop. While the successful fit of the data is

in itself some justification, there is one independent piece of evi-

dence from the experiments. For temperatures when $t_c \ll 3\tau_0$, the ex-

panded form of eq. (3) shows that t_c depends upon $v^{1/3}(0)$, or equi-

valently for fixed number of drops, on the cube root of the pump power.

Data in Fig. 4 taken at 3.0 K obey this power law for $t_c > 15$ μsec. One

discomforting note, however, is the recent light scattering results[7)]

which show that the droplet cloud expands with power and not drop

size. This would imply that the lengthening of t_c with pumping is

associated with <u>cloud</u> size instead of *bona fide* growth of drops.

However, the light scattering results are representative of the strong

pumping regime and, hence, not in direct conflict with the cyclotron

resonance results in the low pumping regime.

We go on now to another kind of time-resolved cyclotron resonance

experiment –
the detection
of onsets. If
the excitation
intensity is
steadily in-
creased, start-
ing from a
very low level,
then at some
point a signal
appears, often
quite abruptly,[2]
signifying that
the phase bound-
ary has been
crossed and
nucleation of
drops has taken
place.   A plot[3]
of these "onsets"

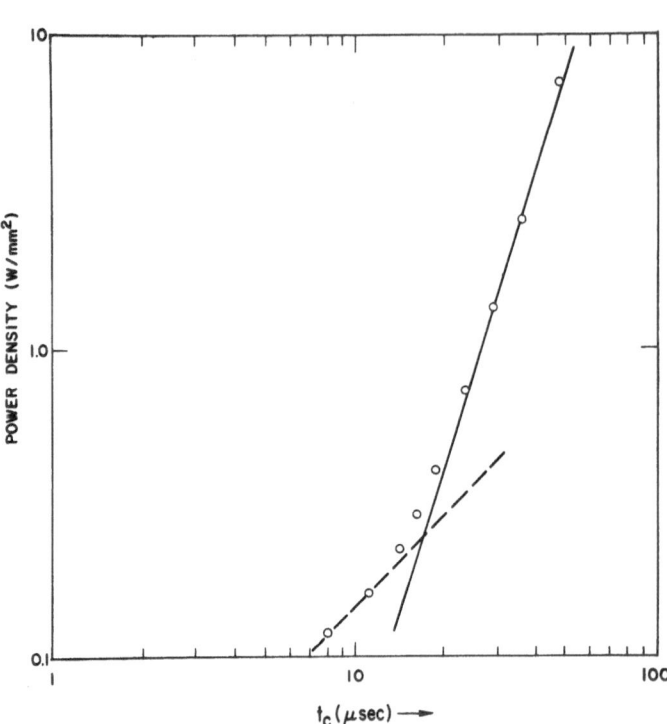

Fig.4. The cutoff times t_c of the decay profiles at 3.0K
plotted vs excitation power. (A power density of
10^{-1} W/mm^2 corresponds to 0.8 nJ pulses).   The
solid line has a slope of 3; the dashed line has
unity slope.

as a function of temperature maps out the edge of the phase boundary
as shown in Fig. 5.   The fit of the data above 2 K in Fig. 5 by the
expression

$$n_{ex} = g \left(\frac{M_o k_B T}{2 \pi \hbar^2} \right)^{3/2} e^{-\varphi/k_B T} \tag{4}$$

for the density of an ideal exciton gas in coexistence with e-h liquid
gives $\varphi = 16$ K in good agreement with the value obtained from the t_c's.
(In eq. (4) g is the degeneracy of the ground state of the exciton,
and M_o is its translational mass).   Below 2 K the data deviates

seriously from the form (4) for reasons that are not altogether clear. One strong possibility is the departure from ideal thermodynamic behavior below 2K due to the finite lifetime of the drops. Analogous results have been obtained by Lo, *et al.*[8] and McGroddy, *et al.*[9] by other onset methods.

Finally we note a persistent and as yet unexplained discrepancy between spectroscopic and thermodynamic values of φ as seen, for example, by comparing the cyclotron resonance value $\varphi = (16^{+3}_{-2})K$ (not unrepresentative of all thermodynamic measurements), and the most refined spectroscopic value $\varphi = 21 \pm 2K$ recently obtained.[10]

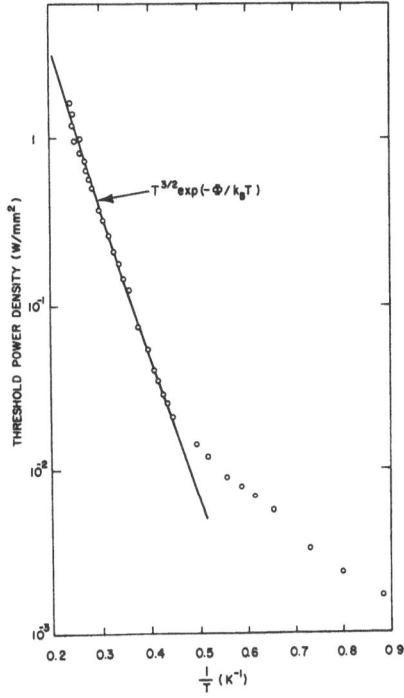

Fig.5. The optical power threshold for onset of the long-lived cyclotron resonance signal, detected at a delay time of 10 μsec.

V. MICROWAVE ELECTRIC SUSCEPTIBILITY

The detection of the microwave electric susceptibility of e-h drops affords a direct way to monitor their decay kinetics. For a dielectric medium, *i.e.* Ge, containing metallic inclusions, *i.e.* drops, there will be an enhancement of the **effective** dielectric constant at microwave frequencies, the measurement[2,4] of which can provide additional information about the decay processes of drops.

At low drop concentrations we can regard the drops as noninteracting spheres of radius R and dielectric constant ε_1. In a uniform electric field $\vec{\mathcal{E}}$ each sphere contributes a polarization

$$\vec{p} = \varepsilon_2 \frac{\varepsilon_1 - \varepsilon_2}{\varepsilon_1 + 2\varepsilon_2} R^3 \vec{\mathcal{E}} \tag{5}$$

to the surrounding medium (Ge) whose dielectric constant is $\varepsilon_2 (=16)$. We next assume that the dielectric properties of the drops are described by the Drude expression,

$$\varepsilon_1 = \varepsilon_2 \left[1 + \frac{i\omega_p^2 \tau^2}{\omega\tau(1 - i\omega\tau)} \right], \tag{6}$$

where ω $(= 3 \times 10^{11} \mathrm{sec}^{-1})$ is the microwave frequency, ω_p $(\approx 2 \times 10^{13} \mathrm{sec}^{-1})$ is the plasma frequency of the e-h liquid, and τ $(\sim 10^{-9} \mathrm{sec})$ is the carrier scattering time. The response of a dielectric to a microwave electric field is conventionally described by the dielectric suscepti-bility $\chi = \chi' + i\chi''$. Under the above described conditions, $\omega\tau \sim 10^3 \gg 1$ and $\omega/\omega_p \sim 10^{-2} \ll 1$, the components of χ derived from eqs. (5) and (6) simplify to

$$\chi' = \frac{3}{4\pi} \varepsilon_2 v, \qquad \chi'' = \frac{9}{4\pi} \left(\frac{\omega}{\omega_p}\right)^2 \frac{1}{\omega\tau} \varepsilon_2 v, \tag{7}$$

where v is the fractional volume occupied by the drops. Thus, the droplets give rise to a positive, real component χ' to the suscepti-bility (a dispersion signal, in microwave parlance) and to a negligibly small imaginary component χ'' (absorption signal).

Measurements at 1.1 K and 53 GHz show χ' to be a <u>positive</u> signal, as predicted, characteristic of e-h drops; on the other hand the χ'' signal is found to be essentially zero. The decay of the χ' signal after pulsed excitation shown in Fig. 6 exhibits several peculiar features. The decay is faster than 40 μsec, is markedly nonexponential and, in fact, reverses sign and remains negative for $t > 110$ μsec. The

dashed line serves to emphasize the extent to which the data depart
from the usual 40 μsec lifetime for volume decay.

The anomalous time dependence of χ' can be satisfactorily accounted for by the surface term $c\,v^{2/3}$. Note that in this experiment, at
1.1 K with pulsed pumping, the thermionic emission and backflow in eq.
(1) are negligible. The free carriers which arise from the surface
decay, as well as from bulk Auger decay, contribute a negative susceptibility. As the droplet decays this negative χ' which stems
partly from the surface will compete with and eventually exceed the
positive χ' which is proportional to the drop volume. The (+) and
(-) contributions to χ', obtained by solving the rate equations are
added in an arbitrary
ratio. The ratio is
fixed from the "com-
pensation time" as
seen in Fig. 6 where
the (+) and (-) con-
tributions are equal
and χ' vanishes. The
resultant fit shown by
the solid curve in
Fig. 6 is quite con-
vincing evidence of
the low temperature
surface emission con-
firming the implica-
tion from cyclotron
resonance.

In closing we
remark that various
kinetic experiments

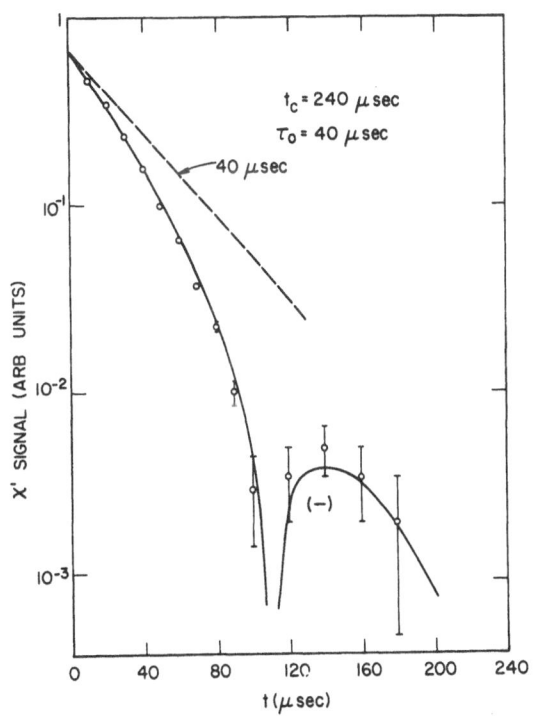

Fig.6. Time dependence of the χ' signal for 9 nJ
laser (1.064 μm) pulse excitation and 1.14K.

expose different aspects of the decay problem, as illustrated in the above work, and shed more light on the e-h system when considered in concert rather than individually.

It is a pleasure to acknowledge the collaboration of T. G. Phillips and T. M. Rice in the work summarized here.

REFERENCES

1) For a review see Ya. Pokrovskii: Phys. Stat. solidi (a) 11 (1972) 385.

2) J. C. Hensel, and T. G. Phillips: *Proceedings of the Eleventh International Conference on the Physics of Semiconductors, Warsawa,* 1972 (PWN, Warsawa, 1972), p.671.

3) J. C. Hensel, T. G. Phillips and T. M. Rice: Phys. Rev. Letters 30 (1973) 227.

4) J. C. Hensel, and T. G. Phillips: *Proceedings of the Twelfth International Conference on the Physics of Semiconductors, Stuttgart,* 1974, ed. M. H. Pilkuhn (Teubner, Stuttgart, 1974), p.51.

5) C. Benoit à la Guillaume, M. Capizzi, B. Etienne, and M. Voos: Solid State Comm. 15 (1974) 1031.

6) R. M. Westervelt, T. K. Lo, J. L. Staehli, and C. D. Jeffries: Phys. Rev. Letters 32 (1974) 1051, 32 (1974) 1331 (E).

7) J. M. Worlock, T. C. Damen, K. L. Shaklee and J. P. Gordon: Phys. Rev. Letters 33 (1974) 771.

8) T. K. Lo, B. J. Feldman and C. D. Jeffries: Phys. Rev. Letters 31 (1974) 224.

9) J. C. McGroddy, M. Voos and O. Christensen: Solid State Comm. 13 (1973) 1801.

10) G. A. Thomas, A. Frova, J. C. Hensel, R. E. Miller and P. A. Lee: Phys. Rev. B13 (1976) 1692.

EFFECT OF THE SURFACE ENERGY ON THE ELECTRON-HOLE DROP LUMINESCENCE IN Ge

C. Benoît à la Guillaume, B. Etienne and M. Voos

Groupe de Physique des Solides*
de l'Ecole Normale Supérieure,
Université de Paris VII - 75005
Paris, France.

ABSTRACT

The effect of surface energy on the luminescence line shape of electron hole droplets is considered theoretically. A shift $2\sigma/rn_o$ of the high energy cut off of the line is predicted, and two contributions to the variation of the line width are discussed. A differential method is described which allows the measurements of these effects. The radius of EHD near threshold can be fitted by a simple model which takes into account the surface energy and the internal EHD recombination.

I. INTRODUCTION

Bulk properties of electron hole drops (EHD) in germanium are now well understood.[1] Several authors have published recently theoretical calculation of the surface energy of the electron hole liquid.[2] In this paper we investigate the dependence of the EHD luminescence on their radius r, through the effect of surface energy. We show how small changes of luminescence line shape can be detected and used to measure EHD radius near the threshold of EHD condensation[3] and thus obtain a deeper insight into the nucleation processes.

II. THEORETICAL MODEL

We consider a droplet of radius r containing N electron hole pairs. We restrict our consideration to T = 0 for simplicity. The total energy of the drop is the sum of the volume and surface energy:

$$\xi = NE(n) + 4\pi r^2 \sigma(n) \tag{1}$$

* Laboratoire associé au Centre National de la Recherche Scientifique

.where $\sigma(n)$ is the EHD surface energy and $E(n)$ is the mean energy per pair as a function of the density n, which has a minimum value μ_∞ at $n = n_o$. Near n_o we use the development $E(n) = \mu_\infty + a (n-n_o)^2$. We can minimize ξ by a small change of n (and r). The condition $d\xi/dn = 0$ gives:[4)

$$n - n_o = \frac{1}{r} \frac{\sigma(n_o) - 3/2\, n_o (d\sigma/dn)_{n\,=\,n_o}}{a\, n_o^2} . \tag{2}$$

which is the change of the density caused by the pressure due to surface tension.

The corrected value of ξ is now

$$\xi = N\mu_\infty + 4\,\pi\,r^2\,\sigma(n_o) - \frac{N}{r^2} \frac{\left[\sigma(n_o) - 3/2\, n(d\sigma/dn)_{n\,=\,n_o}\right]^2}{a\, n_o^4} . \tag{3}$$

Then, one can obtain the chemical potential

$$\mu = \frac{d\xi}{dN} = \frac{1}{4\,\pi\,r^2\,n} \frac{d\xi}{dr}$$

$$\mu(r) = \mu_\infty + \frac{2\sigma}{rn_o} + \frac{1}{3r^2} \frac{\left[\sigma - 3/2\, n(d\sigma/dn)_{n\,=\,n_o}\right]^2}{a\, n_o^4} . \tag{4}$$

In fact, we will use only the first order contribution to $\mu(r)$, i.e. $\frac{2\sigma}{rn_o}$.

We consider now the effect on the luminescence line of the droplet of radius r. The high energy cut-off of the line corresponds to a transition of a droplet with N pairs to a droplet with (N-1) pairs, both in their ground state: This is just the definition of $\mu(r)$. As shown in figure 1, the high energy cut-off of the luminescence line depends now on r through the high energy shift $2\sigma/r\, n_o$. This shift is small, since its value is 6×10^{-3} meV at $r = 1$ μm if one takes $\sigma = 10^{-4}$ erg/cm^2.[2) The next step is to see if there is a change of the width $\Delta(r)$ of the luminescence line, which for $r \rightarrow \infty$ is equal, in the crudest

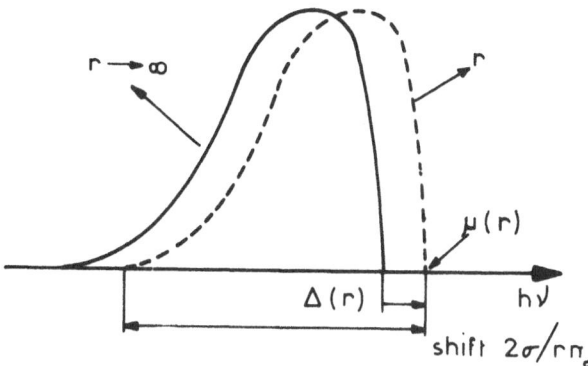

Fig.1. Recombination radiation line shape for
a droplet of radius r. The solid line
corresponds to r → ∞ and the dashed
line to r finite.

approximation, to $E_F(n_o)$ the sum of the Fermi energy of the electrons
and the holes. Since there is no possibility to push the development
$\xi(r)$ in eq.(3) to the term of order r^{-1}, we try to evaluate the difference $\Delta(r) = \xi^*(n^*) - \xi(n_r)$ between the energy of a droplet containing N pairs in its ground state and in an excited state with one pair
excited from the bottom of the bands to the Fermi levels. On can write:

$$\xi^*(n) = \xi(n) + E_F(n) + 4 \pi r^2 \Delta \sigma(n)$$

where $\Delta \sigma(n) = \sigma^*(n) - \sigma(n)$.

Here $\sigma^*(n)$ is the surface energy in the excited state. $\xi^*(n)$ should be
in principle minimized by allowing a small change of n to n^*, but
(n^*-n) turns out to be like r^{-3}, and such a change can be neglected.
Using the value of n in eq.(2), one gets:

$$\Delta(r) = E_F(n_o) \left[1 + \frac{2}{3r} \frac{\sigma(n_o)-3/2 \, n_o (d\sigma/dn)_{n \, = \, n_o}}{a \, n_o^3} \right] + 4 \pi r^2 \Delta\sigma. (5)$$

There are two causes of change to $\Delta(r)$; the second term in eq.(5)
corresponds to the compression (eq.2) and the third one gives also a r^{-1}

contribution if $\Delta\sigma$ is like r^{-3}.

We need now some evaluation of $d\sigma/dn$ and $\Delta\sigma$. In the most usual way of computing σ,[2)] one minimizes the quantity:

$$S = \int_{-\infty}^{+\infty} \left[m\ E(m) + \frac{\alpha}{m} \left(\frac{dm}{dx}\right)^2 \right] dx \qquad (6)$$

where $m(x)$ is the density profile in a direction x perpendicular to the surface, which is supposed to be the same for electrons and holes (no dipole layer). A further simplification is to assume that $m(x)$ varies according to a given function with just one parameter d, the thickness of the surface layer. With scaling arguments, S can be transformed to:

$$S = A(n)\ \frac{d}{n} + B(n)\ \frac{n}{d}\ .$$

Then, σ is obtained by minimizing S with respect to d:

$$\sigma(n) = 2\ \sqrt{A(n)B(n)}$$

$d\sigma/dn$ is probably small, because if one takes for $m(x)$ a linear function of x, then $(dA/dn)_{n_0} = 0$ and if one takes $m(x) = n\ (1+\exp(-x/d)^{-1}$ then $dB/dn = 0$. So we assume that $d\sigma/dn$ can be neglected in eq.(5).

In an attempt to obtain $\sigma^*(n)$ in eq.(6) we can modify the first term in the integrand

$$m\ E(m) \rightarrow m\ E(m) + m/nV\ [E_F(m) - E_F(n)]$$

but, we do not know how to modify the second one. Then, taking $m(x)$ as a linear function of x in the interval $(0,d)$ one can compute $A^*(n)$

$$A^*(n) = \int_0^n m\ E(m)\ dm + \frac{1}{nV} \int_0^n m\ [E_F(m) - E_F(n)]\ dm$$

$$A^*(n) = A - \frac{1}{8} \frac{n}{V}\ E_F(n).$$

In this approximation, B is unchanged. Thus

$$\Delta \sigma = \sigma \frac{\Delta A}{2A} = - \frac{\sigma n\ E_F}{16\ AV}\ .$$

The contribution to the width $\Delta(r)$ (last term of eq.(5)) is:

$$4\ \pi\ r^2\ \Delta\sigma = -\frac{3}{2}\ \frac{\sigma n\ E_F}{16A} = -\frac{3}{8}\ \frac{d}{r}\ E_F \qquad (7)$$

since $A = \frac{\sigma}{2}\ \frac{n}{d}\ .$

Taking $d = 60$ Å, this contribution (7) is $- 23\ E_F/r(r$ in Å) which has

to be compared to $E_F(n) - E_F(n_o) = \frac{40}{r}\ E_F$, taking $\sigma = 10^{-4}\ erg/cm^2$ and

$a = 1.2 \times 10^{-35}$ meV cm^{-6}. With the same units

$$\mu_r - \mu_\infty = + \frac{10}{r}\ E_F\ .$$

We must emphasize that this evaluation of $\Delta\sigma$ is a rough approximation. It proves that $\Delta\sigma$ is like r^{-3}. We hope that the negative sign
in (7) is correct, but we cannot have any confidence in the order of
magnitude. However, one can take it as an indication that a rather
strong compensation occurs between the two r^{-1} contribution to $\Delta(r)$ in
eq.(5).

III. EXPERIMENTAL SET UP

The idea is to measure, near the threshold of EHD condensation,
the shift of the luminescence line of EHD when their size changes.[3]
As shown below, EHD size depends on the pump level and also because of
hysteresis,[5] on the pump level history. A pure Ge sample $(N_A - N_D \sim
2 \times 10^{10}$ cm^{-3}) is immersed in liquid He and excited by a very stable
tungsten halogen lamp monitored by a special chopper, as shown in Fig.
2, so that both levels of pump J_1 and J_2 $(J_1 \sim 10\ J_2)$ have opposite
phases. The corresponding luminescence signals, analyzed through a
grating spectrometer followed by a cooled PbS cell is sent to a lock in
amplifier. The duration of J_1 is adjusted in order to get zero at the
maximum of the EHD line. The resulting signal is proportional to the

Fig.2. (a)(b) – Time and phase
dependence of the excitation
in the differential experiment.
The chopping frequency is 75
Hz.
(c) – Luminescence signal of
the 709 meV EHD line versus
excitation J_1 when J is continuously increased from zero
to 1000 (off scale) and back
to zero ($J_{th\ B} \sim 2\ mW/cm^2$).

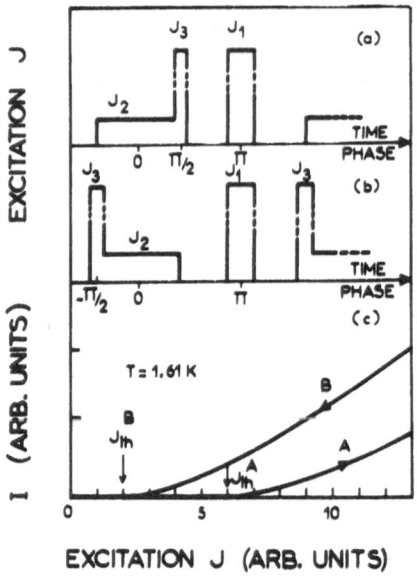

EXCITATION J (ARB. UNITS)

difference between the line shapes for excitation levels J_1 and J_2.
This differential method allows a great amplification of small changes
in line shape. Besides, adding just before J_2 and excitation J_3 equal
to J_1, and in phase quadrature with J_1 and J_2 (see Fig. 2(b)), we can
make measurements on the descending branch (curve B in Fig. 2(c)) of
the optical hysteresis. The excitation scheme of Fig. 2(a) gives data
on the rising branch (curve A); here J_3 is not functional, but is left
for convenience. One can switch from A to B just by changing the
rotation of the chopper from clockwise to counter clockwise. Fig. 3
gives typical data obtained on branch B at $J_2 \sim 10\ J_{th\ B}$. The peaks at
713,6 meV and 705.2 meV are due to free excitons.

In the region of the LA phonon assisted emission of EHD (~ 709 meV)
one can see a signal resembling the derivative of the EHD line. The
differential signal can be analyzed as follows. If f (hν) is the line
shape at J_1, we suppose that the change at J_2 can involve a shift δ and
a dilatation of the line ε giving f $[(1 + \varepsilon)h\nu + \delta\ ]$. Then the differential signal is:

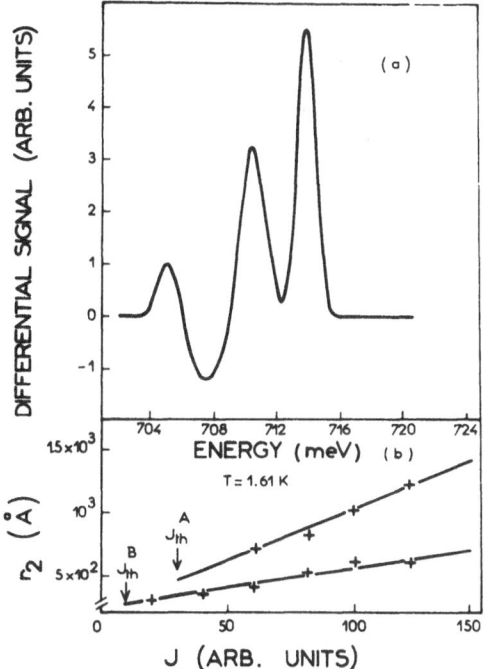

Fig.3. (a) – Typical differential
luminescence signal at 1.61K
for $J_2 \sim 10\ J_{th\ B}$.
(b) – Variation with the excitation of the radius of droplets
on branch A and B of Fig. 2(c).

$$\Delta f = (\delta + \varepsilon h \nu)\ df/dh\nu \qquad (8)$$

Here, if $h\nu = 0$ at the high energy cut-off of the line, then the shift
of the high and low energy cut-off are respectively δ and $(\delta - \varepsilon E_F)$.

IV. DISCUSSION OF THE EXPERIMENTAL RESULTS

a) Line Shape

Within the experimental uncertainties, the shape of the differential signal (Fig. 3(a)) is quite insensitive to the experimental conditions. This is consistent with the assumption that both δ and ε are
like r^{-1}. Using the analysis sketched at the end of Chapter III, one
obtains $\varepsilon E_F = 0.3\ \delta$. In fact, if r_1 and r_2 are the radii of EHD

corresponding to J_1 and J_2, all variations are indeed like $(r_2^{-1} - r_1^{-1})$.

$$\delta = 2\sigma/n_o \ (r_2^{-1} - r_1^{-1}) \qquad \text{and} \qquad \varepsilon E_F = \Delta(r_2) - \Delta(r_1).$$

As shown at the end of chapter II, if the only contribution to ε was the change of Fermi energy caused by the drop contraction, then εE_F would be equal to about 4 δ. This experimental result, $\varepsilon E_F = 0.3$ δ, shows that indeed a strong compensation occurs between the two contributions considered in eq.(5).

b) Measurement of Droplet Radius Near Threshold

When $J_1 = 10 \ J_2$, the r_1^{-1} term can be neglected because r is a rather rapid function of J as shown in figure 3(b), which gives the value of r as a function of excitation along the two branches A and B of Fig. 2(c). These data are of great importance for the study of nucleation since from the data of Figs. 2(c) and 3(b) one can deduce the variation of the density of droplets. In the remaining of this paper, we want just to compare the value of r obtained on branch B at the threshold with a simple theory.[3]

Taking into account the surface energy, the EHD work function is now[6] $\phi(r) = \phi_\infty - 2\sigma/rn_o$. Therefore, the well known Pokrovskii eq.(1) relating the density in the exciton gas n_{ex} to the drop radius r becomes:

$$n_{ex} = \alpha r + n_{ex,o} \exp(2\sigma/rn_o kT) \qquad (9)$$

where $\alpha = 4 \ n_o/3 \ v_{ex}\tau_o$ (v_{ex} thermal velocity of excitons, τ_o EHD lifetime) and

$$n_{ex,o} = g(2\pi \ m^* \ kT/h^2)^{3/2} \exp(-\phi_\infty/kT)$$

(g and m^*: degeneracy and mass of exciton).

Relation (9) is shown in Fig. 4(b). There is now a value r^* which is the radius of the smallest stable droplet.

Fig.4. (a) – Experimental temperature
dependence of ΔE near threshold
$J_{th\ B}$. On the right scale, the
corresponding values of r^* have
been plotted taking $\sigma = 10^{-4}$
erg/cm^{-2}. The solid line is a
fit using eq.(10) as explained in
in the text.
(b) – The free exciton density
n_{ex} as a function of droplets
radius r. The dashed line
corresponds to Pokrovskii's
model, and the solid line to the
results obtained when the effect
of surface energy is added.

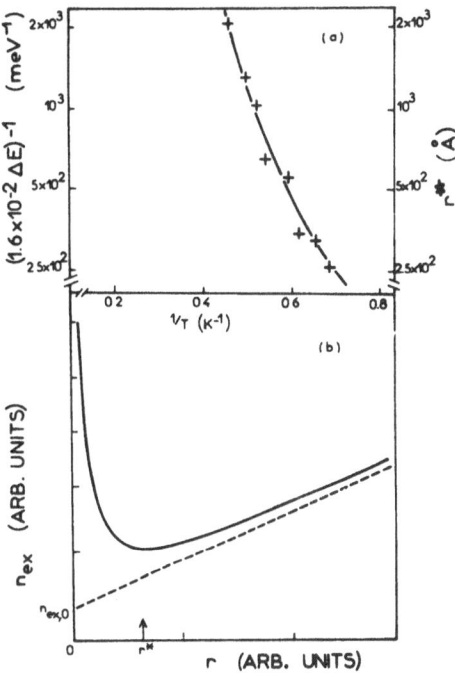

$$r^* = \beta(A\sigma T)^{1/2} \exp(-\phi_\infty/2kT) \exp(\sigma/r^* n_o kT) \qquad (10)$$

with $\beta^2 = 2\tau_o(3/4\ \pi)^{1/3}/k\ n_o^{4/3}$ and A is the coefficient of the

Richardson law for thermal emission; r^* is identified with the value of

r obtained at the threshold of branch B. The fit of experimental data

with eq.(10) is shown in Fig. 4(a). This fit depends in fact on two

parameters, ϕ_∞ and A/σ since what is measured primarily is the quantity

$\Delta E^* = 2\sigma/r^* n_o$; This is clear if eq.(10) is written:

$$\frac{1}{\Delta E} \exp(-\ \Delta E/2kT) = \frac{\beta n_o}{2}\ (\frac{A}{\sigma})^{1/2}\ T^{1/2}\ \exp(-\phi_\infty/2kT)\ .$$

Fig. 4(a) shows that a good fit is obtained taking $\phi_\infty/k = 23°K$ and A/σ

$= 1.6\times10^{14}\ s^{-1}K^{-2}erg^{-1}cm^2$. This value of A/σ is close to the calculated

value, taking $A = 3.2\times10^{10}sec^{-1}K^{-2}$ (with $g = 16$ and $m^* = 0.33\ m_o$) and

$\sigma = 10^{-4}erg/cm^2$. If $\phi_\infty/k = 16°K$ is used, A/σ is reduced by a factor of

120, a result which seems unreasonable. One can remark that the result

$\phi_{\infty}/k = 16^{\circ}K$ has always been obtained through the analysis of threshold data related to branch A[1,7] and that these data should be reanalyzed taking into account supersaturation effects.[8]

To conclude, let us recall the importance of these measurement of droplets radius and density near the threshold of condensation for a detailed study of the nucleation processes.

REFERENCES

1) See for example: Y. A. Pokrovskii: Phys. Status solidi(a) 11 (1972) 385; T. M. Rice: in *Proceedings of the XIIth International Conference on the Physics of Semiconductors, Stuttgart, Germany* (1974) edited by M. H. Pilkuhn (Teubner, Stuttgart, Germany, 1974) p.23; M. Voos, *ibid.* p.33; and references therein.

2) L. M. Sander, H. B. Shore and L. J. Sham: Phys. Rev. Lett. 31 (1973) 533; H. Buttner and E. Gerlach: J. Phys. C 6, L (1973) 433; T. M. Rice: Phys. Rev. B 9, (1974) 1540; T. L. Reinecke and S. C. Ying: Solid St. Comm. 14, (1974) 381; P. Vashishta, R. Kalia and K. S. Singwi: *Oji Seminar "Physics of Highly Excited States in Solids", Tomakomai, Japan,* Sept. 1975; B. Etienne, C. Benoît à la Guillaume and M. Voos: Phys. Rev. Lett. 35 (1975) 536.

4) An explicit dependence of σ upon r could be added. Then the term $\frac{1}{n}\frac{d\sigma}{dn}$ in eq.(2) would be changed into $(\frac{1}{n}\frac{\partial\sigma}{\partial n} + \frac{1}{r}\frac{\partial\sigma}{\partial r})$.

5) T. K. Lo, B. J. Feldman and C. D. Jeffries: Phys. Rev. Lett. 31 (1973) 224.

6) R. N. Silver: Phys. Rev. B 11 (1975) 1569.

7) J. C. Hensel, T. C. Phillips and T. M. Rice: Phys. Rev. Lett. 30 (1973) 227; J. C. McGroddy, M. Voos and O. Christensen: Solid State Comm. 13 (1973) 1801.

8) B. Etienne, C. Benoît à la Guillaume and M. Voos: to be published.

SURFACE PROPERTIES OF ELECTRON-HOLE LIQUID IN GERMANIUM*

P. Vashishta

Argonne National Laboratory, Argonne, Illinois 60439 U.S.A.

R. K. Kalia and K. S. Singwi

Physics Department, Northwestern University,
Evanston, Illinois 60201 and Argonne National Laboratory,
Argonne, Illinois 60439 U.S.A.

ABSTRACT

Using density-functional formalism and the exponential density
profiles with variational parameters, we show that inclusion of first
gradient correction to the exchange-correlation energy has a signifi
cant quantitative effect on the surface energy of the electron-hole
liquid (EHL) in germanium. The surface energy of EHL in Ge is also
calculated by self-consistently solving the Kohn-Sham equations in a
local density approximation. The effect of the exchange-correlation
gradient correction is also included. The surface energy of EHL in Ge
under very large <111> uniaxial stress is found to be a factor of six-
teen smaller than the value in Ge at zero stress. The question of
charge on an electron hole drop is also discussed.

The subject of the electron-hole liquid (EHL) in germanium and
silicon has become one of considerable experimental and theoretical
interest in the last several years.[1-5] Much effort has been devoted
to an understanding of the ground state energy and thermodynamics of
the bulk EHL.[4,5] To date there are only a few experimental and
theoretical investigations of surface properties, such as surface
energy, σ, and charge on an electron-hole drop (EHD) in Ge.[6-9]
The problem of the EHD surface is of special interest from a
theoretical standpoint. Until now, most of the theoretical investigations of surface properties of EHL have been based upon the Hohenberg-
Kohn-Sham (HKS) formalism for an inhomogeneous electron gas.[10] The

---

* Based on work performed under the auspices of the U. S. Energy Research and
Development Administration, the Materials Research Center of Northwestern
University, and the NSF under Contract No. GH-39127.

reason why a study of the surface properties of EHL is so interesting

is because of the fact that the EHL in semiconductors can be regarded

as a uniform multicomponent plasma, free of ionic effects which are

always present even in simple metals. EHL thus provides an ideal test-

ing ground for HKS density functional formalism.

In this paper we shall present calculations of the surface energy

of EHL in three systems—Ge under zero stress Ge(4:2); Ge under inter-

mediate stress (≈ 3 kg/mm^2) along <111> direction, Ge(1:2); and Ge

subject to a large uniaxial stress along < 111> direction Ge(1:1).[11]

For Ge(4:2), Ge(1:2) and Ge(1:1) our calculated equilibrium densities,

respectively, are 2.2×10^{17} cm^{-3} ($r_s = 0.58$), 6.9×10^{16} cm^{-3} ($r_s =$

0.86), 1.1×10^{16} cm^{-3} ($r_s = 1.57$). Our calculated values of the equi-

librium density, the binding energy, the derivatives of the ground state

energy with respect to the density (related to the isothermal compress-

ibility), Fermi energy, the temperature dependence of the equilibrium

density, Fermi energy and chemical potential, and the gas liquid tran-

sition temperature and critical density at the transition temperature

are in excellent agreement with experiments on EHL in Ge.[5]

Using the multicomponent generalization of HKS density-functional

formalism for an electron gas, the ground state energy E [{n}] which is

a unique functional of densities {n} can be written as

$$E[\{n\}] = \frac{1}{2} \sum_{ij} \int d\vec{r}' d\vec{r}'' \ v_{ij}(\vec{r}' - \vec{r}'')n_i(\vec{r}')n_j(\vec{r}'')$$

$$+ \sum_i \int d\vec{r}' \ v_i^{ext}(\vec{r}')n_i(\vec{r}') + T^0[\{n\}] + E^{xc}[\{n\}] \tag{1}$$

where $n_i(\vec{r})$ is the density of the ith component, v_{ij} is the Coulomb

interaction potential, v_i^{ext} is the external potential acting on the ith

component, $T^0[\{n\}]$ is the noninteracting kinetic energy and $E^{xc}[\{n\}]$ is

the exchange-correlation energy of the system. For slowly varying

density $n_i(\vec{r})$ retaining only the lowest order gradient correction to the

noninteracting kinetic energy and exchange-correlation energy, one can write

$$T^{o}[\{n\}] = \sum_i \int d\vec{r} \; t_i(n_i(\vec{r})) + \sum_i C_i \int d\vec{r} \; \frac{|\nabla n_i(\vec{r})|^2}{n_i(\vec{r})} \tag{2}$$

$$E^{xc}[\{n\}] = \sum_{ij} \int d\vec{r} \; \varepsilon_{ij}^{xc}(\{n_i(\vec{r})\}) - \frac{1}{2} \sum_{ij} \int d\vec{r} \; d_{ij}(\{n_i(\vec{r})\})$$

$$\times \; \nabla n_i(\vec{r}) \cdot \nabla n_j(\vec{r}) \tag{3}$$

where t_i is the kinetic energy of the ith component and ε_{ij}^{xc} is the ex-change-correlation energy contribution per unit volume. Substituting (2) and (3) in (1), the ground state energy (including the lowest order gradient correction in the absence of an external potential) can be written as

$$E = \int d\vec{r} \; \varepsilon(n_e, n_h) + \int d\vec{r} \; c(n_e)|\nabla n_e|^2 + \int d\vec{r} \; c(n_h)|\nabla n_h|^2$$

$$+ \sum_{ij} \int d\vec{r} \; d_{ij}(\{n_i\}) \; \nabla n_i \cdot \nabla n_j$$

$$+ \frac{e^2}{2\varkappa} \iint d\vec{r} d\vec{r}' \; \frac{(n_e(\vec{r}) - n_h(\vec{r}))(n_e(\vec{r}') - n_h(\vec{r}'))}{|\vec{r} - \vec{r}'|} \tag{4}$$

where $\varepsilon(n_e, n_h)$ is the energy per unit volume in the local density ap-proximation, and is a function of electron and hole densities, $n_e(\vec{r})$ and $n_h(\vec{r})$. $\varepsilon(n_e, n_h)$ consists of kinetic, exchange and correlation contributions. The kinetic and exchange energies in the local density approximation contain the anisotropy of conduction bands as well as valence band coupling (VBC) between light and heavy hole bands. The correlation energy that we use has been calculated in a fully self-consistent (FSC) approximation, and it includes the effect of anisotropy of the conduction bands. The last term in (4) constitutes the electro-static contribution. The expression for $\varepsilon(n_e, n_h)$ reads,

$$\varepsilon(n_e, n_h) = \frac{3}{10} \frac{\hbar^2}{m_{de}} \left(\frac{3\pi^2 n_e(\vec{r})}{v_e}\right)^{2/3} n_e(\vec{r}) + \frac{3}{10} \frac{\hbar^2}{m_H} \left(\frac{3\pi^2 n_h(\vec{r})}{1 + \gamma^{3/2}}\right)^{2/3} n_h(\vec{r})$$

$$+ \alpha(n_e^{4/3}(\vec{r}) + n_h^{4/3}(\vec{r})) + n_e(\vec{r}) \varepsilon_c(n_e(\vec{r})) + n_h(\vec{r}) \varepsilon_h(n_h(\vec{r})). \quad (5)$$

In (5), m_{de} is the density of states mass for the electron, ν_e is the number of conduction valleys, m_L and m_H are, respectively, the light and heavy holes masses, and $\gamma = \dfrac{m_L}{m_H}$. Following Rice, we have divided the bulk part of the exchange and correlation energy per particle equally between electrons and holes. The correlation energy per pair is fitted to a polynomial in inverse powers of r_s.

The expression for the coefficient of the first gradient correction to the kinetic energy of electrons and holes, respectively, are given as,

$$c(n_e) = \frac{\hbar^2}{72 \, m_{oe}} \times \frac{1}{n_e(\vec{r})} \quad , \qquad (6)$$

$$c(n_h) = \frac{\hbar^2}{72 \, m_H} \times \frac{F(\gamma)}{n_h(\vec{r})} \quad , \qquad (7)$$

where m_{oe} is the optically averaged mass of an electron and the expression for $F(\gamma)$, as derived by Reinecke and Ying,[6] is

$$F(\gamma) = \frac{(1 + \gamma^{1/2})}{(1 + \gamma^{3/2})} \times \left[7 - \frac{24 \, \gamma^{1/2}}{(1 + \gamma^{1/2})^2} \right] . \qquad (8)$$

If one ignores VBC then $F(\gamma) = (1 + \gamma^{1/2})/(1 + \gamma^{3/2})$.

The coefficients of the exchange-correlation gradient correction, $d_{ij}(\{n_i\})$, are obtained from the local field factors, $G_{ij}(Q)$, which include effects of multiple scattering of (e-e), (e-h), and (h-h).[4,5] d_{ij} are related to the Q^4 coefficients of $G_{ij}(Q)$. In accordance with the fact that the exchange-correlation energy in the local density approximation is divided equally between electrons and holes, the contribution arising from the first gradient correction to the exchange-correlation energy is also equally distributed between electrons and holes.

I. VARIATIONAL CALCULATION OF SURFACE ENERGY

In our model for the EHD with a surface we assume that the drop fills half the space, $z < 0$, has a boundary at $z = 0$, and the space b beyond $z = 0$ is assumed to be vacuum. Parameterizing the densities of electrons and holes by β_e and β_h, we have

$$
n_{e,h}(z) = \begin{cases} n_o[1 - \frac{1}{2} \exp(\beta_{e,h}z)] & z < 0 \\ \\ \frac{n_o}{2} \exp(-\beta_{e,h}z) & z > 0 \end{cases} \tag{9}
$$

where n_o is the density of e-h pairs deep inside the liquid. Surface energy σ is defined by

$$
\sigma = \frac{(E - \tilde{\mu}N)}{S} . \tag{10}
$$

In (10), $\tilde{\mu}$ is total bulk energy per pair, N is the number of e-h pairs, and S is the surface area perpendicular to z. We substitute (9) in (10) and minimize σ numerically with respect to β_e and β_h. Having obtained $n_e(z)$ and $n_h(z)$ which minimize σ, we substitute them back in (10) to evaluate the surface energy. The results are given in Table I. The coefficients of the exchange-correlation gradient correction have an uncertainty of about twenty per cent which will introduce an uncertainty of no more than fifteen per cent in our values of the surface energy, calculated using a variational principle, for EHD in the three systems.

Our values of β_e and β_h for Ge (4:2) are 2-3 times smaller than those of Rice, Reinecke and Ying.[6] Also, our value for the surface tension is about four times larger than those calculated by Sander et al. and Rice, and three times the value of Reinecke and Ying.[6] Silver calculates the surface tension of EHD in normal Ge to be 10^{-3} erg/ cm^2.[7] It is about six times larger than the experimental value of Alekseev et al.[8] In a recent calculation, Reinecke and Ying[7]

Approximation	System	σ erg/cm^2
Variational calculation	Ge (4:2)	3.5×10^{-4}
with exponential density	Ge (1:2)	1.1×10^{-4}
profiles: Section I	Ge (1:1)	0.2×10^{-4}
Solution of Kohn-Sham	Ge (4:2)	4×10^{-4}
equation: Section II		

Table I. Surface energy, σ, of EHL in germanium at zero stress and < 111 > uniaxial stress. The band structure of Ge (4:2) consists of four conduction valleys and a heavy and a light hole band which are degenerate at Γ point. Ge (1:2) has the band configuration (for < 111 > uniaxial stress $\approx$ 3 kg/mm^2) where the splitting between the bottom of the lowest conduction valley along the direction of stress and the three other conduction valleys is equal to the Fermi energy of electrons. Band configuration of Ge (1:1), consisting of a single conduction minimum and a valence minimum, is achieved under very large < 111 > uniaxial stress.

theoretically fit the experimental measurements of the phase diagram for Ge (4:2) and estimate the droplet surface tension to be ~10^{-4} erg/cm^2. Etienne *et al.'s* experimental estimate of σ for EHD in Ge is 2×10^{-4} erg/cm^2.[9] The experimental value of σ, as found by the Berkeley group, is 2.9×10^{-4} erg/cm^2.[12]

We have thus shown that in a calculation of surface energy of EHD which is based on the density functional formalism for an inhomogeneous electron gas, it is important to include the first gradient correction to the exchange and correlation energy. It is interesting to note that the value of σ for EHD in Ge (4:2) is an order of magnitude larger than in Ge (1:1).[13]

II. SURFACE ENERGY FROM THE SOLUTION OF KOHN-SHAM EQUATION

The main shortcoming of the variational calculation is that the treatment of the noninteracting kinetic energy is inexact; it is approximated by a local density term and the first gradient correction to the kinetic energy of electrons and holes. Since the density profiles used in the variational calculation are of exponential form with variational parameters in the exponents, the effect of quantum density oscillations (Friedel oscillations) is lost.

The variational calculation presented above can be greatly improved at the expense of much more work by treating the noninteracting kinetic energy exactly.[10,14] This implies solving a set of coupled Kohn-Sham equations, which will then yield self-consistent density profiles for electrons and holes. In this section we shall present such a calculation for EHL in Ge (4:2) under the assumption of local charge neutrality. The results of surface energy of EHL in Ge and Si under uniaxial stress will be published elsewhere.[15]

Using the fact that the energy $E[\{n\}]$ in (1) is a unique functional of densities, n_i's, one can derive a set of coupled K-S equations which for the ith component, when expressed in reduced units, reads

$$(-\frac{m_r}{m_i}\frac{d^2}{dz^2} + v_i^{eff}[n_e,n_h;z]) \psi_i(\varrho,z) = \frac{m_r}{m_i} (\varrho^2 - \varrho_{Fi}^2) \psi_i(\varrho,z) . \qquad (11)$$

When i stands for electrons, m_i is taken to be the density-of-states mass, m_{de}, thereby bringing in the effect of anisotropy of the conduction bands; ϱ_{Fi} is the Fermi wave vector of the ith component and m_r is the reduced mass used in defining the excitonic Bohr radius and excitonic Rydberg. The expression for the effective potential of the ith component is:

$$v_i^{eff}[n_e,n_h;z] = 8\pi\xi_i\int_{-\infty}^{z}dz'(z - z')[n_e(z') - n_h(z')]$$

$$- [\bar{\mu}_i^T(n_{oi}) + \bar{\mu}_i^{xc}(n_{oi})] + \mu_i^{xc}(n_i) , \qquad (12)$$

where $\xi = +1, -1$ for holes and electrons, respectively. In addition to containing the Hartree term, v_i^{eff} contains a potential, $\mu_i^{xc}(n_i)$ which arises from the exchange-correlation energy in the local density approximation, as well as the value of $\mu_i^{xc}(n_i)$ deep inside the liquid, which is $\bar{\mu}_i^{xc}(n_{oi})$, and $\bar{\mu}_i^T(n_{oi})$, the contribution of the noninteracting kinetic energy to the chemical potential.

Expressing the density of the ith component in terms of the solution $\psi_i(Q,z)$ of (11) we have

$$n_i(z) = \frac{\nu_i}{\pi^2} \int_0^{Q_{Fi}} dQ(Q_{Fi}^2 - Q^2) | \psi_i(Q,z) |^2 , \tag{13}$$

where ν_i is the number of equivalent bands.

Equations (11) - (13) constitute a set of equations to be solved self-consistently. Starting with a trial density profile for electrons and holes one constructs an effective potential from (12), substitutes it in (11) and solves the latter for the $\psi_i(Q,z)$'s. With these $\psi_i(Q,z)$ one constructs $n_i(z)$ from (13) and the above procedure is continued until the density of every component converges to a satisfactory degree. From our variational calculation as well as those of others,[6] it is well established that the value of β_e and β_h are very close and the electrostatic contribution to the surface energy is $\leq 1\%$. Thus in a self-consistent calculation for the surface energy of the kind mentioned above, it is a very good approximation to take $n_e(z) = n_h(z)$. With this approximation the first term in V_i^{eff} vanishes identically and the self-consistent equations for each component are uncoupled. Taking $n_e(z) = n_h(z)$ and examining the K-S equations, it is easy to infer that (11), when solved for electrons should involve a mass $\bar{m}_{de}$, where

$$\bar{m}_{de} = \frac{1}{\nu_e^{2/3}} \left[\frac{1}{2} \left\{ \frac{1}{\nu_e^{2/3} m_{de}} + \frac{1}{m_H(1 + (m_L/m_H)^{3/2})^{2/3}} \right\} \right]^{-1} . \tag{14}$$

Using $\bar{m}_{de}$ in (11), (12), and (13) we solve them self-consistently for $n_e(z)$. The degree of convergence achieved for the density, in our calculation, is about 1% of the bulk density. A plot of $n_e(z) = n_h(z)$ for Ge (4:2) is shown in Fig. 1. For the sake of comparison we have

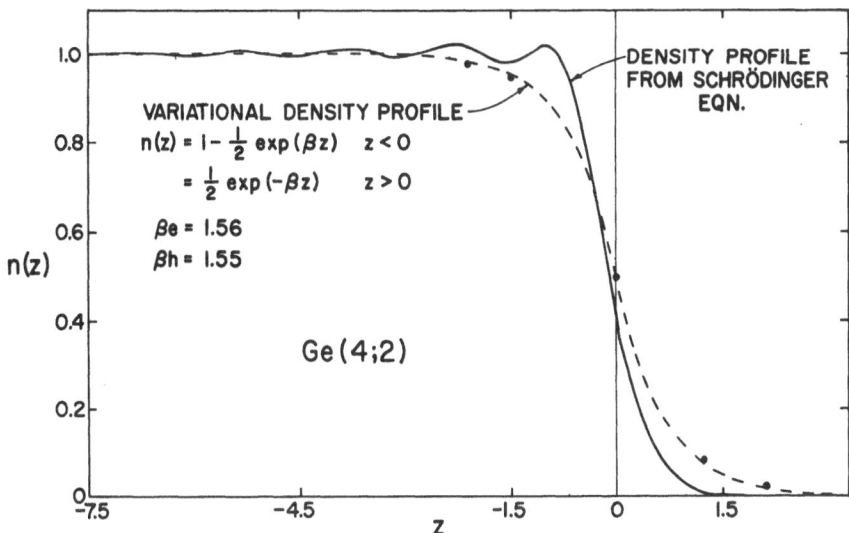

Fig.1. Normalized density profile $n(z) = \dfrac{n_e(z)}{n_0} = \dfrac{n_h(z)}{n_0}$ from a self-consistent solution of equations (11), (12), and (13) with local charge neutrality. The dashed curve is the normalized exponential density profile for electrons and solid points are for the holes.

also plotted in Fig. 1 the exponential profiles used in Section I which minimize the energy. It is obvious from Fig. 1 that the self-consistent density profile contains Friedel oscillations. The greater the value of r_s the larger the amplitude of these oscillations.

Since, as mentioned in Section I, in a variational calculation the first gradient correction to the exchange-correlation energy, $\Delta E_{xc}^{(2)}$, has a significant effect on the value of the surface energy, we have included its contribution to σ. The expression for $\Delta E_{xc}^{(2)}$ is,

$$\Delta E_{xc}^{(2)} = \sum_{ij} \int d\vec{r}\, d_{ij}(\{n_i\})\, \nabla n_i \cdot \nabla n_j \quad . \tag{15}$$

On evaluating (15) with the self-consistent density profile we find that $\Delta E_{xc}^{(2)}$ constitutes a small correction to the exchange-correlation energy in the local density approximation.

The value of $\sigma = 4 \times 10^{-4}$ erg/cm^2 for EHL in Ge (4:2) based on a self-consistent solution of the K–S equation is quite close to $\sigma = 3.5$

$\times\,10^{-4}$ erg/cm^2 calculated using the variational principle with exponential density profiles. It leads us to believe that as far as the calculation of surface energy is concerned a non-self-consistent calculation, such as presented in Section I, is reasonable provided the first gradient correction to the exchange-correlation energy is included.

In conclusion, our value of the surface energy, $\sigma = 4 \times 10^{-4}$ erg/cm^2 for EHL in Ge (4:2), based on a self-consistent solution, is in good agreement with the experimental value of 2.9×10^{-4} erg/cm^2 obtained by the Berkeley group from a systematic analysis of the hysteresis data on EHL in ultra-high purity Ge.[12]

III. CHARGE ON THE EHD

Besides the surface energy, another interesting surface property is the charge on the EHD. Pokrovsky and Svistunova[8] have shown experimentally that the EHD is "negatively charged in normal Ge" and that it is "positively charged in Ge subject to $<111>$ uniaxial stress (≈ 3 kg/mm^2)". It is easy to infer from a simple argument of Rice that the sign of charge on EHD will be in agreement with the findings of Pokrovsky and Svistunova, provided the dipole layer $\Delta\phi = 8\pi \int_{-\infty}^{+\infty} dz\, z \times (n_e(z) - n_h(z))$ is less than half the difference between the chemical potentials of electrons and holes. Thus the sign of the charge depends sensitively on the density profiles $n_e(z)$ and $n_h(z)$.

We shall now briefly discuss the sign of the charge on EHD in Ge under $<111>$ uniaxial stress. The kinetic energy per particle of an electron, T_e, and a hole, T_h, for Ge (4:2) ($\nu_e = 4$, $\nu_h = 2$) and Ge (1:2) ($\nu_e = 1$, $\nu_h = 2$) can be written as,[4]

$$T_e = \frac{\hbar^2\, q_F^2}{2\, m_{de}\, \nu_e^{2/3}}\,, \tag{16}$$

$$T_h = \frac{\hbar^2\, q_F^2}{2\, m_H}\, \frac{1}{(1 + (m_L/m_H)^{3/2})^{2/3}}\,. \tag{17}$$

Substituting m_{de} = 0.22, m_L = 0.042 and m_H = 0.347 in (16) and (17), we get

$$\frac{T_e}{T_h} = \frac{1.6}{v_e^{2/3}} \ . \tag{18}$$

Assuming that the exchange-correlation energies for the electron and the hole are equal, which is known to be a good approximation, for Ge (4:2) (18) gives T_e = 0.6 $\times T_h$ implying that $\phi_e > \phi_h$; $i.e.$, the binding energy of an electron is greater than that of hole. On this basis, at finite temperatures due to excess thermionic emission of holes, the "EHD in Ge (4:2) will be negatively charged." The same reasoning for Ge (1:2) leads to T_e = 1.6 T_h implying that $\phi_e < \phi_h$ and the "EHD in Ge (1:2) will be positively charged".

Under very large $<111>$ uniaxial stress, however, the valence band in Ge is considerably modified and one has to consider only one anisotropic hole band with m_{dh} = 0.088.[4,5] For Ge (1:1), $v_e = v_h = 1$, we obtain simply $\frac{T_e}{T_h} = \frac{m_{dh}}{m_{de}}$ = 0.4 implying that $\phi_e > \phi_h$; $i.e.$, the binding energy of electrons is greater than that of holes. Hence at finite temperatures "the EHD in Ge (1:1) will be negatively charged" due to excess evaportion of holes resulting from their lighter density of states mass. We thus predict that under $< 111 >$ uniaxial stress the charge on the EHD in Ge will change sign twice, the sign of charge on the EHD in Ge (4:2) being negative, changing to positive in Ge (1:2) (stress $\approx$ 3 kg/mm^2) and once again changing to negative in Ge (1:1) (under very large $<111>$ stress). So far we are not aware of any experimental observations of the sign of the charge on EHD in Ge (1:1).

We would further like to add that when exchange-correlation gradient correction is added to Rice's[6] calculation of the charge on the EHD in Ge (4:2), the value of charge becomes +1e from −18e (as calculated by Rice) on a drop of 5μ radius. It suffices to say here that to date all the theoretical calculations,[6,13] including our own presented in Section I, give the sign of the charge on the EHD in Ge in disagreement with the experiment of Pokrovsky and Svistunova.[8]

We are now in the process of doing a self-consistent calculation, as described in Section II, in which the condition of local charge neutrality is relaxed.[16] We hope that such a calculation will shed light on the question of charge on the EHD in Ge.

In view of the recent theoretical interest in the surface energy of the EHL and the sign and magnitude of charge on the EHD in Ge under < 111 > uniaxial stress, there is a great need for accurate experiments on the surface properties of the EHL in Ge under < 111 > stress.

ACKNOWLEDGMENT

We are grateful to Dr. N. D. Lang for providing us with a copy of the computer program to solve the Kohn-Sham equation for a metal surface without which we would not have dared to start calculations of surface properties of an inhomogeneous multicomponent plasma.

REFERENCES

1) J. R. Haynes: Phys. Rev. Letters 17 (1966) 860. Ya. Pokrovskii: Phys. Stat. sol. (a) 11 (1972) 385. C. Benoît à la Guillaume and M. Voos: Phys. Rev. B7 (1973) 1723, and references therein.

2) C. Benoît à la Guillaume and M. Voos: Solid State Commun. 12 (1973) 1257. G. A. Thomas, T. G. Phillips, T. M. Rice, and J. C. Hensel: Phys. Rev. Letters 31 (1973) 386. G. A. Thomas, T. M. Rice, and J. C. Hensel: Phys. Rev. Letters 33 (1974) 219. J. C. McGroddy, M. Voos, and O. Christensen: Solid State Commun. 13 (1973) 1801.

3) T. K. Lo, B. J. Feldman, and C. D. Jeffries: Phys. Rev. Letters 31 (1973) 224.

4) M. Inoue and E. Hanamura: Phys. Soc. Japan 34 (1973) 652, *ibid* 35 (1973) 643. W. F. Brinkman, T. M. Rice, P. W. Anderson, and S. T. Chui: Phys. Rev. Letters 28 (1972) 961. W. F. Brinkman and T. M.

Rice: Phys. Rev. B7 (1973) 1508. M. Combescot and P. Nozières: J. Phys. C: Proc. Phys. Soc. London 5 (1972) 2369. M. Combescot Phys. Rev. Letters 32 (1974) 15. P. Vashishta, P. Bhattacharyya, and K. S. Singwi: Phys. Rev. Letters 30 (1973) 1248. P. Bhattacharyya, V. Massida, K. S. Singwi, and P. Vashishta: Phys. Rev. B10 (1974) 5127.

5) P. Vashishta, S. G. Das, and K. S. Singwi: Phys. Rev. Letters 33 (1974) 911.

6) L. M. Sander, H. B. Shore, and L. J. Sham: Phys. Rev. Letters 31 (1973) 533. T. M. Rice: Phys. Rev. B9 (1974) 1540. T. L. Reinecke and S. C. Ying: Solid State Commun. 14 (1974) 381.

7) R. N. Silver: Phys. Rev. B11 (1975) 1569. T. L. Reinecke and S. C. Ying: Phys. Rev. Letters 35 (1975) 311.

8) A. S. Alekseev, T. A. Astemirov, V. S. Bagaev, T. I. Galkina, N. A. Penin, N. N. Sybeldin, V. A. Tsvetkov: *Proceedings of the XIIth International Conference on the Physics of Semiconductors, Stuttgart, Germany,* 1974, edited by M. H. Pilkuhn (Teubner, Stuttgart, Germany, 1974), p.91; Ya. Pokrovsky and K. I. Svistunova: *ibid,* p.71.

9) B. Etienne, C. Benoiît à la Guillaume, and M. Voos: Phys. Rev. Letters 35 (1975) 536.

10) P. Hohenberg and W. Kohn: Phys. Rev. 136 (1964) B864. W. Kohn and L. J. Sham: Phys. Rev. 140 (1965) A1133.

11) The notation Ge $(\nu_e : \nu_h)$, where ν_e and ν_h are, respectively, the number of electron and hole bands, is due to C. Kittel (private communication). J. P. Wolfe, R. S. Markiewicz, C. Kittel and C. D. Jeffries: Phys. Rev. Letters 34 (1975) 275.

12) R. M. Westervelt, J. L. Staehli and E. E. Heller: to be published; C. D. Jeffries: private communication.

13) P. Vashishta, R. K. Kalia, and K. S. Singwi: to be published.

14) N. D. Lang and W. Kohn: Phys. Rev. B1 (1970) 4555.

15) P. Vashishta and R. K. Kalia: to be published.

16) P. Vashishta and R. K. Kalia: the calculations are in progress
for the charge on EHD in Ge and Si.

THERMODYNAMICAL APPROACH TO THE HIGHLY EXCITED
STATES OF SEMICONDUCTORS

Tomitaro Nagashima, Takeshi Watanabe
and Chuji Horie
Department of Applied Physics
Tohoku University,
Sendai, Japan

ABSTRACT

In order to understand a general trend of the photoconductivity
in a highly excited state of semiconductors, the determination of
carrier density against the total excitation has been carried out in a
relatively low density region where electrons, holes and excitons coexist.   The assumption of chemical equilibrium among electrons, holes
and excitons enables one to determine the density of charge carriers.
The variation of binding energy of an exciton with the excitation
density is taken into account.   Carrier density appears to increase
abruptly at a certain critical region of total excitation density.

I.   INTRODUCTION

It has been observed that when the density of electron-hole pairs
in germanium is increased by illumination of an intense laser light,
the photoconductivity shows an abrupt increase at some pair density.[1-2]
This phenomenon has been analysed in association with the presence of
EHD.[3-4]   In the case of tellurium an anomaly in the photoconductivity
has also been observed, but has been interpreted as being due to the
change in the degree of degeneracy of carriers.[5-6]

In the relatively low excitation densities of about 10^{15} cm^{-3} or
below in both Ge and Te, there is no quantitative analysis of the photoconductivity, where the presence of bound excitons is taken into account.
In order to understand the general trend of photoconductivity at low
and intermediate densities of excitation (which will be called the L-
and I-region, respectively), we present a thermodynamical investigation
of determining the densities of electrons, holes and excitons as functions of a given excitation density.   We assume that the system is in

chemical equilibrium. It is shown that the presence of excitons gives rise to significant effects on the screening of the Coulomb force between carriers in the L-region, through the change in the relative densities of carriers and excitons. In the I-region it turns out that the total excitation contributes to the screening. As a result the increase of excitation density proves to increase the carrier density abruptly at a critical density between the L- and I-regions.

In Sec. II we treat the L-region, where particles are considered to be non-degenerate. Whereas the exciton binding energy depends on the carrier density through the screening parameter of a Debye-Hückel type, the chemical equilibrium condition among particles gives the carrier density as a function of the excitation density for a given exciton binding energy. A self-consistent calculation is carried out of the carrier density against the total excitation density.

When the total excitation density is increased, the system is expected to reach the I-region where carriers can be considered degenerate. This region is treated in Sec. III. A model Hamiltonian for the system is constructed by means of the transformation employed by Girardeau.[7] Then, the exciton contribution to the screening and to the self-energy of particles is taken into account diagrammatically besides the usual RPA for the electron and/or hole correlations to obtain the chemical potentials of electrons, holes and excitons. The chemical equilibrium condition among particles is solved numerically for a given temperature and a total excitation density.

Discussion will be made in Sec. IV in connection with an EHD and the exciton condensation.

II.   LOW EXCITATION DENSITY REGION (L-REGION)

It is assumed that there exists a chemical equilibrium among electrons, holes and excitons. Then, the chemical potentials of electrons, holes and excitons, which are denoted by μ_e, μ_h and μ_{ex}, respectively, satisfy the relation

$$\mu_e + \mu_h = \mu_{ex}.$$ (1)

Denoting the number densities of electrons, holes and excitons by n_e, n_h and n_{ex}, respectively, and assuming that n_e is equal to n_h, the total excitation density n is given by

$$n = n_e + n_{ex}.$$ (2)

In the L-region we can assume the Boltzmann distribution for each kind of particle. Combining eqs. (1) and (2), we obtain an equation relating $n_e (= n_h)$ with n for a given temperature and a given exciton binding energy. On the other hand, the exciton binding energy B depends on n_e through the screening parameter $\varkappa$. For the B-$\varkappa$ relation, we use the result of a variational calculation by taking the hydrogen-like trial wave function. The screening parameter is given by

$$\varkappa = (\frac{8\pi n_e e^2}{\varepsilon_o kT})^{1/2}$$ (3)

where ε_o is the lattice dielectric constant.

A self-consistent calculation is carried out of the carrier density and the exciton binding energy, by assuming for simplicity that an electron and a hole have an equal effective mass m.

Since we neglect the exciton contribution to the screening as compared to that of electrons and holes, we are confined to the region where the following two inequalities are satisfied:

$$\frac{n_e}{n_{ex}} > 3(\frac{kT}{B})^2 ,$$ (4)

$$B(n_e, T) > kT.$$ (5)

These inequalities follow from a simple argument using a generalized dielectric constant and the f-sum rule. Since the left-hand side of the inequality (4) decreases faster than the right-hand side as the temperature decreases, this inequality puts a lower limit on temperature.

On the other hand, the inequality (5) puts a lower limit on the magnitude of $B(n_e, T)$ at given temperatures.

The n_e $vs.$ n curves obtained are shown in Fig. 1 together with the

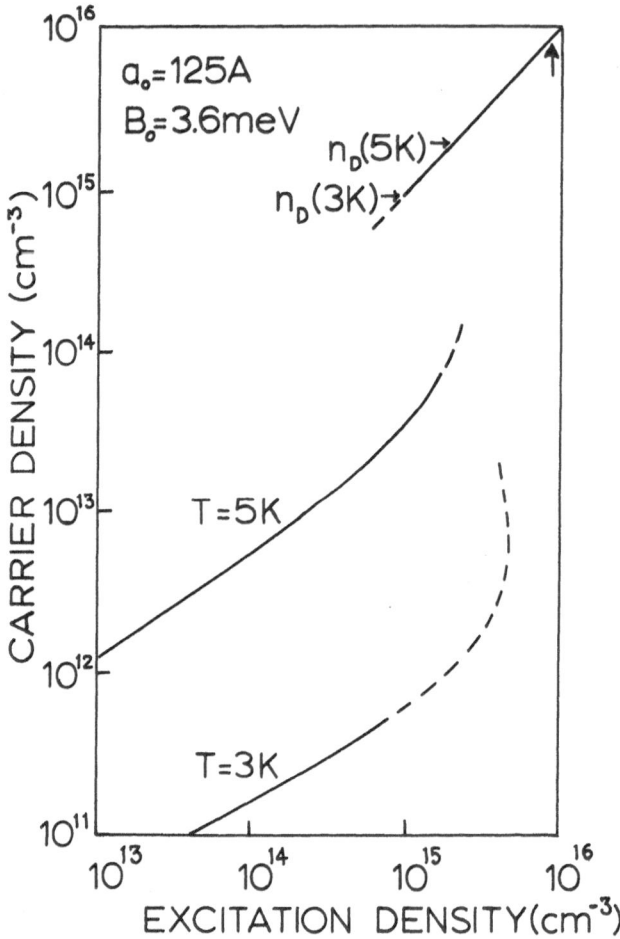

Fig.1. Plot of carrier density n_e against total excitation density n. The lower curves (upper curve) correspond to the lower (intermediate) excitation density region. Parameters denoted by B_0 and a_0 are, respectively, the exciton binding energy and radius in the absence of screening effect. The arrow in the upper-right corner indicates the point where the exciton binding energy vanishes. For the density of carrier $n_e > n_D$, the carriers can be considered degenerate.

results for the I-region. The parameters chosen are shown in the figure. If we were to extend the lower temperature curve to the region where the above criteria are not satisfied, we would observe an S-shape behavior (dotted portion) which suggests an instability.

III.   INTERMEDIATE EXCITATION DENSITY REGION (I-REGION)

When the excitation density is increased, the system is expected to reach the I-region where the carriers are degenerate and excitons are non-degenerate. This is due to an enormous reduction of the exciton binding energy caused by the increase of screening effect.[8]

The model Hamiltonian for the system is constructed by means of the transformation which replaces the exciton operators by boson operators,[7] and is written as follows:

$$H = H_o + H_c + H_{int},$$ (6)

$$H_o = \sum_k \varepsilon_e(k) c_k^+ c_k + \sum_k \varepsilon_h(k) d_{-k}^+ d_{-k} + \sum_k E_{ex}(k) a_k^+ a_k,$$ (7)

$$H_c = \frac{1}{2} \sum_q v(q) \rho(q) \rho(-q),$$ (8)

$$H_{int} = \sum_q \sum_k \{ g(q, k) c_{k+q}^+ d_{-k}^+ a_q + h.c. \},$$

$$g = \tilde{E}_{ex}(q) f(q, k), \quad \tilde{E}_{ex}(q) = E_{ex}(q) - E_G$$ (9)

where c_k, d_{-k} and a_k are the annihilation operators of electrons, holes and excitons, respectively. H_c represents the Coulomb interaction between carriers, and H_{int} represents the carrier-exciton interaction. E_G represents the energy gap between the conduction and valence bands. $\tilde{E}_{ex}(q)$ is the exciton energy relative to the bottom of the conduction band and $f(q, k)$ is the exciton wave function. Here we have assumed for simplicity that the exciton lies in the lowest levels. In the derivation of the Hamiltonian, we have retained terms up to the order of (n_{ex}/n).

In order to consider the effect of carrier-exciton interactions, we perform a perturbational calculation by means of the Green's function method.   As for the contribution to the screening of the Coulomb potential, we calculate the polarization diagrams represented by (a) and (b) of Fig. 2 in the static and low temperature limit.   As for the

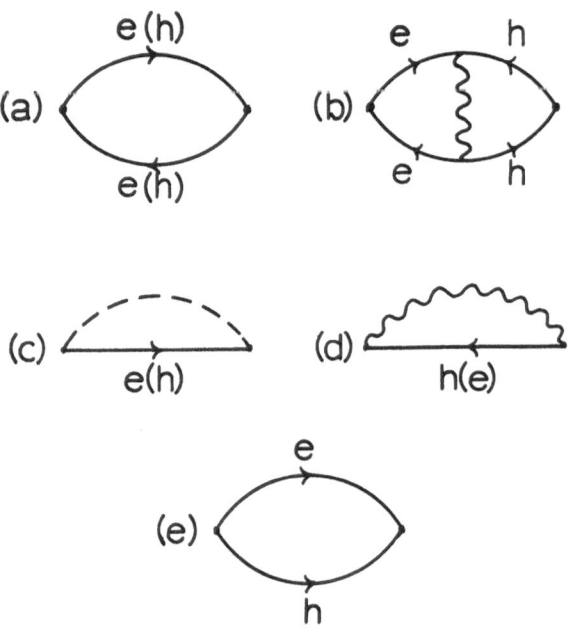

Fig.2.   (a) and (b) are polarization diagrams where wavy line represents the exciton propagator.   (c) and (d) represent the electron (or hole) self-energies $\Sigma^c_{e(h)}$ and $\Sigma^{ex}_{e(h)}$, respectively.   Dashed line represents the screened Coulomb interaction.   (e) is the self-energy for excitons.

self-energy Σ_e (Σ_h) of an electron (a hole) we take diagrams represented by (c) and (d) of Fig. 2.   For the self-energy Π of an exciton we evaluate the diagram (e) in Fig. 2.

In terms of the self-energies the chemical potentials of the respective components are defined by

$$\mu_e = \frac{p_o^2}{2m_e} + \mathrm{Re}\, \Sigma_e\, (p_o,\ 0) + E_G,$$

$$\mu_h = \frac{p_o^2}{2m_h} + \mathrm{Re}\Sigma_h(p_o,\ 0),$$

$$\mu_{ex} = \mu_{ex}^{(o)} + \mathrm{Re}\, \Pi(0,\ 0) - B + E_G,$$

$$\mu_{ex}^{(o)} = kT\ell n\, (n_{ex}\lambda_{ex}^3),$$

where $p_o = (3\pi^2 n_e)^{1/3}$, and λ_{ex} is the thermal de Broglie wavelength of an exciton.

The exciton binding energy B is calculated by the method mentioned in Sec. II. As for $\varkappa$, we use the Thomas-Fermi screening parameter reduced by a factor $\delta\, (<1)$. The factor δ accounts for the fact that the carriers are not fully degenerate in the I-region.

The chemical equilibrium condition is solved numerically to give n_e for a given temperature and for a total excitation density. We have again assumed that $m_e = m_h$. The result is plotted in Fig. 1. It should be noted that in the I-region the exciton concentration n_{ex} is found to be extremely small so that n_e is almost equal to n. It appears from the figure that the carrier density increases abruptly at a critical density between the L- and I-regions.

IV.   DISCUSSION AND CONCLUSION

The calculation is carried out of the chemical potential μ of electron-hole pairs which are in chemical equilibrium with excitons. The result is shown in Fig. 3. The arrow in the figure indicates the excitation density at which the exciton binding energy vanishes. Because of a large negative value of the self-energy the chemical potential is greatly reduced in the I-region. On the other hand, in

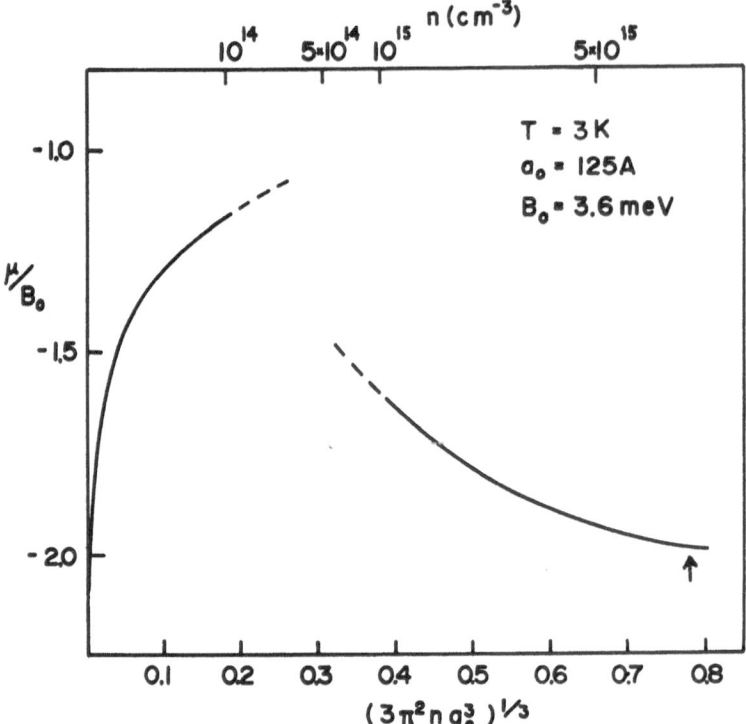

Fig.3. Chemical potentials in the low and intermediate
density regions. The arrow indicates the excitation density at which the exciton binding
energy vanishes.

the L-region the entropy term in the free energy is sufficiently large

to guarantee the chemical equilibrium between carriers and excitons.

Since the appearance of the negative gradient of μ vs. n curve

(T = 3K) in the I-region suggests the occurrence of a gas-liquid

transition as discussed by Mahler[9] and by Combescot[10] in association

with the possibility of EHD, we have to reconsider how the carrier

density behaves after the EHD condensate is formed.   However, even if

the temperature is sufficiently high so that there exists no EHD con-

densate, it seems that there still exists a sharp increase of n_e at some

critical density between the L- and I-regions.

It should be noticed that at sufficiently low temperatures there

appears a critical region between the L- and I-regions, in which the approximation adopted in this paper is no longer valid.   In this

region, the carriers are neither degenerate nor non-degenerate, and

excitons become degenerate and might undergo the exciton condensation.[9]

However, even if the exciton condensation occurs, the exciton binding

energy continues to decrease as the total excitation increases.[8]   As

a result, the exciton density begins to decrease with the increase of

the total excitation and excitons again become non-degenerate while the

carriers become degenerate as assumed in Sec. III.   The critical region

is not treated in the present paper and is a subject of further investi-

gation.

In conclusion, the thermodynamical investigation of the system

composed of electrons, holes and excitons has shown that there is an

abrupt increase of carrier density when a total excitation density n

increases from the L-region to the I-region, and that the n-dependence

of the chemical potential of an electron-hole pair or an exciton is

different in those two regions.

REFERENCES

1)   A. A. Rogachev: *Proc. 9th Int. Conf. Phys. Semiconductors, Moscow*
      1968 ed. S. M. Ryvkin (Nauka, Leningrad, 1968) p.409.

2)   M. N. Gurnee, M. Glicksman and P. W. Yu: Solid State Commun. 11
      (1972) 11.

3)   A. Nakamura and K. Morigaki: Solid State Commun. 14 (1974) 41.

4)   J. C. Hensel, T. G. Phillips and T. M. Rice: Phys. Rev. Letters
      30 (1973) 227.

5)   Y. Nishina, T. Nakanomyo and T. Fukase: *Proc. 10th Int. Conf. Phys.*
      Semiconductors, Cambridge 1970 ed. S. P. Keller *et al.*   (NBS, Springfield,
      Va., 1970) p.493.

6)   Y. Miura and C. Horie: J. Phys. Soc. Japan 33 (1972) 1522.

7)   M. D. Girardeau: Phys. Rev. Letters 27 (1971) 1416.

8)   T. Nagashima and C. Horie: J. Phys. Soc. Japan 37 (1974) 614.

9)   G. Mahler: Phys. Rev. B11 (1975) 4050.

10) M. Combescot: Phys. Rev. Letters 32 (1974) 15.

AUGER RECOMBINATION IN THE ELECTRON-HOLE DROPS IN Si AND Ge

Klaus Betzler

Universität Osnabrück,
FB 4, 45 Osnabrück,
Postfach 4469, Fed. Rep.
Germany

ABSTRACT

Luminescence measurements on Si and Ge at temperatures of about 1.5 K are presented, which indicate that Auger recombination is the main recombination process inside the electron-hole drops (EHD) in both materials.   In silicon a broad spectrum near 2 Eg due to Auger-excited hot electrons could be detected.   From its intensity, an Auger lifetime can be derived which corresponds to the total EHD lifetime. In germanium, the evaluation of magnetooscillation in the luminescence intensity yields a quantum efficiency of only 25% and leads to the con-clusion that 75% of the carriers inside the EHD recombine in Auger processes.

Recently, Auger recombination (AR) in the EHD was taken into con-sideration as important nonradiative process by several authors.[1-3] While in germanium AR was only thought as important for the gen-eration of free electrons and holes from the EHD,[1] in Si and Si-Ge alloys AR was considered as possibly EHD-lifetime-determining.[2,3] In the present paper luminescence measurements on Si and Ge are pre-sented which can only be explained by the assumption that AR is the most important recombination process in the EHD in both materials.

For the case of silicon, the experiments were performed on highly pure samples cooled to about 2 K and excited by a pulsed GaAs laser (peak power 5 W, pulse length 1 μsec, duty factor 1%).   Signal detec-tion after a 0.75 m grating monochromator was carried out using an S 11 photomultiplier tube and special digital Boxcar technique.[4]

The near-2-Eg spectrum of silicon measured under these conditions is shown in Fig. 1.   Besides the relatively intense two-electron transitions line at 2.27 eV,[5,6] a weak, slightly structured spectrum

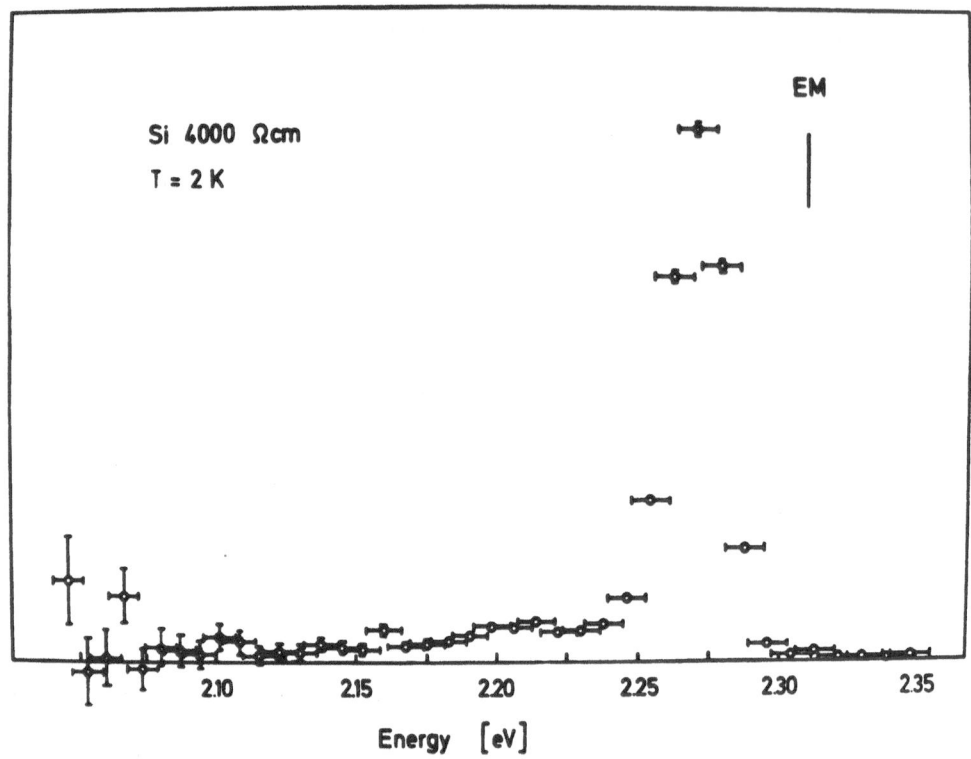

Fig.1.   The near-2-Eg luminescence spectrum of silicon. The relatively intense
         line at 2.27 eV is caused by two-electron band-to-band transitions.

at energies lower than 2 Eg arises.   A better measurement of this

region is shown in Fig. 2, the experimental points are indicated with

their error bars.   The process responsible for this spectrum is the

following:[7]   An electron is excited to a higher conduction band state

by means of a phonon-assisted Auger-process in the EHD.   During the

relaxation from this state, there is the possibility for the electron

to recombine radiatively with a condensed hole.   This recombination

causes the measured spectrum.

     The main energy relaxation process for hot electrons in non-polar

semiconductors - such as Si and Ge - well above the band edge is

optical and acoustical deformation potential scattering.[8]   For these

two processes the relaxation rates have been calculated by Conwell.[9]

Taking her formulas and the set of deformation potential constants from

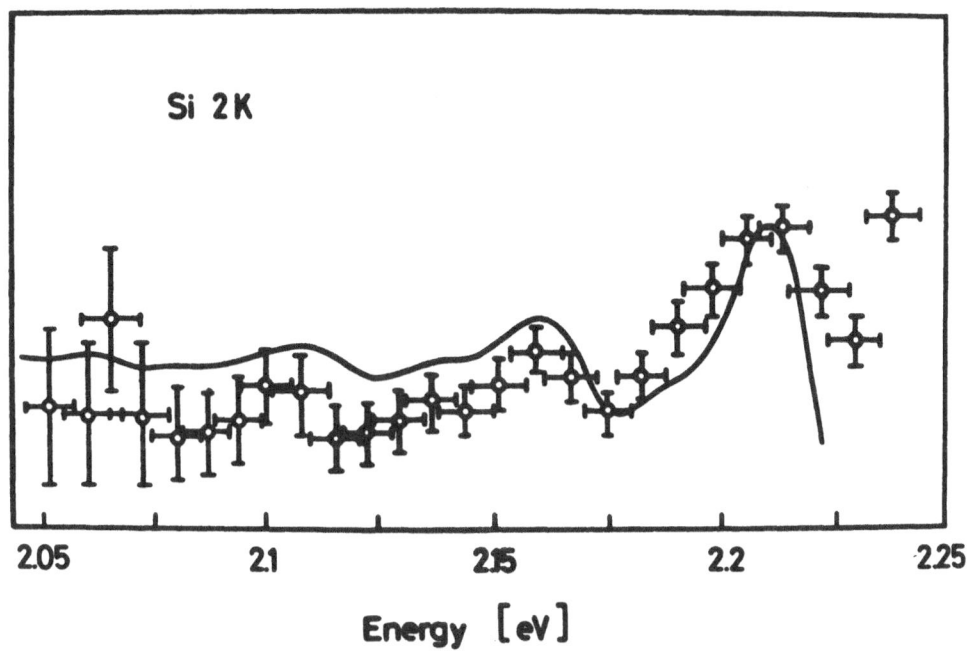

Fig.2. Luminescence spectrum below 2 Eg with better resolution. The full line gives the theoretically expected shape.

Jørgensen *et al.* [10] we have calculated the theoretically expected shape of the near-2-Eg spectrum originating from the described process. This is given by the full line in Fig. 2.

There have been 2 parameters fitted: One is the ratio of optical to acoustical relaxation rates which for a good fit could be chosen only 20% from the calculated value (this may indicate the accuracy of the deformation potential calculation). The other is the absolute intensity of the spectrum, the fit of which, taking into account the radiative recombination coefficient[11] and the reabsorption,[12] yields the AR coefficient for the electron-electron-hole (eeh) process in the EHD in silicon. This value is found to be $C_e = 1.5 \times 10^{-31}$ cm^6 sec^{-1}, which is slightly enhanced compared to the high temperature value.[13] The enhancement will be even more if one takes into account that the AR coefficient should increase with the temperature because of the particip- ipation of an acoustical phonon.

The value of C_e results in an Auger lifetime for the eeh-process of $\tau_{eeh} \approx 500$ nsec; the ehh process should be a little bit stronger in Si,[13] so that the lifetime for both processes in AR is approximately 200 nsec. This is very close to the total lifetime of EHD in Si, which was found to be about 150 nsec.[14]

On germanium, measurements of the magnetooscillations in the luminescence intensity have been performed. To the highly pure samples, which again were excited by a GaAs laser, a magnetic field in (100) direction could be applied in Faraday configuration. For signal detection a Ge photodiode (risetime $\lesssim 1\,\mu$sec) and Boxcar technique was used.

The oscillations found in the luminescence intensity of the EHD-LA line of 709 meV are shown in Fig. 3. They change continuously in their amplitude with increasing delay time between excitation and signal detection, but the extrema positions remain fixed.

The oscillations are connected with oscillations of the carrier density inside the EHD[16] according to the following formula.[17]

$$\frac{\Delta i(H,t)}{i_o(t)} = \frac{\Delta n(H)}{n_o} (1 - t(B \cdot n_o - 2 \cdot C \cdot n_o^2)) \tag{1}$$

where i_o, n_o are intensity and carrier density at zero field, Δi and Δn are the oscillatory parts, B and C are the recombination coefficients for radiative recombination (RR) and AR, respectively, and t is the time between excitation and signal detection. For the oscillations of the time integrated intensity $I(H)=I_o+\Delta I(H)= \int_0^\infty i(t)\,dt$ we can derive:[17]

$$\frac{\Delta I(H)}{I_o} = \frac{\Delta n(H)}{n_o} \cdot \frac{A - C \cdot n_o^2}{A+B \cdot n_o +C \cdot n_o^2} \tag{2}$$

where A is the coefficient for impurity-induced nonradiative recombination.

From the curves in Fig. 3 those for t = 0 and for the integrated

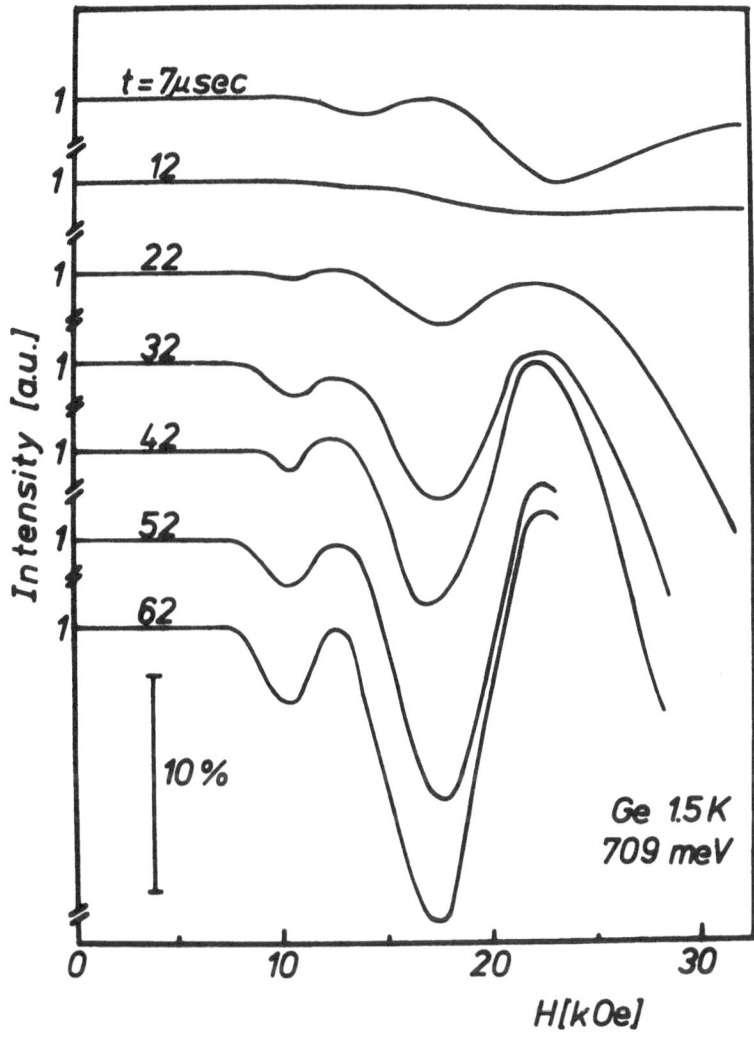

Fig.3. Intensity oscillations of the LA-phonon assisted
EHD line in Ge for different times t between
excitation and detection. All curves are normalized to their intensity at H = 0.

intensity have been derived (Fig. 4). The curve for i(H,0) directly

gives the variation of the equilibrium density inside the EHD as a

function of the magnetic field (see eq. (1)). A comparison of the

two curves gives the following condition for A and C: $A/C \cdot n_o^2 < 0.1$.

This indicates that in the EHD impurity induced nonradiative processes

can be neglected compared to AR. Assuming $A \approx 0$, the quantum efficiency

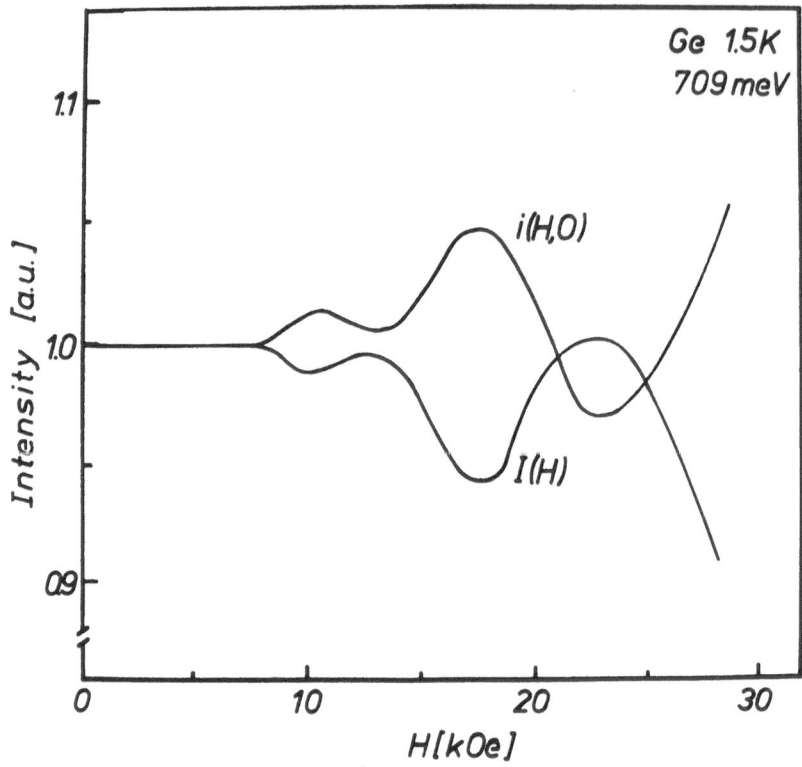

Fig.4. Oscillations of the luminescence intensity
at t = 0 and of the time integrated
luminescence intensity (calculated from
the measurements in Fig. 3).

from the EHD can be derived to be Q ≈ 25%.   This value is less by a
factor of 2 and 3, respectively, than former estimations.[18,19]

    With a total lifetime of τ = 36 μsec,[20] the transition coeffcients
for RR and AR in the EHD can be calculated to be B = 3×10^{-14} cm^3 sec^{-1}
and C = 4×10^{-31} cm^6 sec^{-1}.   The corresponding high temperature values
are[21,11] 1.1×10^{-14} cm^3 sec^{-1} and[22] 2×10^{-31} cm^6 sec^{-1}.   From a comparison we can derive enhancement factors due to electron–hole correla-
tion in the case of RR and to combined electron–electron or hole–hole
and electron–hole correlation in the case of AR of ρ_B = 3 and $\rho_c \gtrsim$ 2.
These values are very close to the theoretical ones which may be derived

from the values calculated by Vashishta *et al.*[23]

In Table 1 the results concerning the lifetime for RR and AR in the EHD are listed. For comparison also the lifetimes for two-

Table 1: Lifetimes for different recombination processes in the EHD in Si and Ge

	τ_{tot}	τ_{rad}	τ_{Auger}	τ_{2-2}
Si	150 nsec [a]	30 μsec [b]	500 nsec (eeh only)	10 sec [c]
Ge	36 μsec [d,e]	150 μsec	50 μsec	3×10^6 sec [c]

a   See ref. 14, b high temperature value from ref. 11
c   ref. 15, d ref. 20, e ref. 17

electron band-to-band transitions have been added, a fourth order process, which also has been found in both materials[6,15] but does not play an important role for the total lifetime. A comparison of the times shows the importance of Auger recombination in the EHD in both materials. Only in Ge radiative recombination may be nearly as strong as AR.

ACKNOWLEDGEMENTS

The author wishes to thank R. Conradt for valuable discussions and B.G. Zhurkin, A. L. Karuzskii, and B. M. Balter for help and discussion concerning the Ge experiments.

REFERENCES

1)   See for instance, T. Ohyama, T. Sanada, K. Fujii, E. Otsuka: *Proc. XIIth Int. Conf. Phys. Semiconductors, Stuttgart* (1974) ed. M. H. Pilkuhn (Teubner, Stuttgart 1974) p.66

2)   J. Barrau, J. C. Brabant, M. Brousseau, J. Collet, M. Heckmann, H. Maareff: Solid State Commun. 16 (1975) 1079.

3) C. Benoit à la Guillaume, M. Voos, Y. Petroff: Phys. Rev. B10 (1974) 4995.

4) K. Betzler, T. Weller, R. Conradt: Phys. Rev. B6 (1972) 1394.

5) K. Betzler, R. Conradt: Phys. Rev. Letters 28 (1972) 1562.

6) K. Betzler, R. Conradt: *Proc. XIth Int. Conf. Phys. Semiconductors, Warsaw* (1972) (Elsevier Pub. Co., Amsterdam, 1972) p.684.

7) K. Betzler: Solid State Commun. 15 (1974) 1837.

8) E. M. Conwell: *High Field Transport in Semiconductors*; Solid State Physics Suppl. Vol.9 (Academic Press, N. Y., 1967)

9) E. M. Conwell: Phys. Rev. 135 A (1964) 1138.

10) M. H. Jørgensen, N. O. Gram, N. I. Meyer: Solid State Commun. 10 (1972) 337.

11) W. Gerlach, H. Schlangenotto, H. Maeder: Phys. Stat. sol. (a) 13 (1972) 277.

12) W. C. Dash, R. Newman: Phys. Rev. 99 (1955) 1151.

13) J. D. Beck, R. Conradt: Solid State Commun. 13 (1973) 93.

14) J. D. Cuthbert: Phys. Rev. B1 (1970) 1552.

15) W. Zeh, K. Betzler, R. Conradt: Solid State Commun. 14 (1974) 967.

16) L. V. Keldysh, A. P. Silin: Fiz. Tverd. Tela 15 (1973) 1532.

17) K. Betzler, B. G. Zhurkin, A. L. Karuzskii, B. M. Balter: P.N. Lebedev Physical Institute Preprint No. 71, Moscow 1975.

18) C. Benoit à la Guillaume, M. Voos, F. Salvan: Phys. Rev. Letters 27 (1971) 1214.

19) Ya. E. Pokrovskii, K. I. Svistunova: Fiz. Tehk. Poluprov. 4 (1970) 491.

20) R. M. Westervelt, T. K. Lo, J. L. Staehli, C. D. Jeffries: Phys. Rev. Letters 32 (1974) 1051.

21) W. Michaelis, M. H. Pilkuhn: Phys. Stat. sol. 3b (1969) 311.

22) R. Conradt, J. Aengenheister: Solid State Commun. 10 (1972) 321.

23) P. Vashishta, P. Bhattacharya, K. S. Singwi: Phys. Rev. B10 (1974) 5108.

TRANSPORT PROPERTIES OF HIGH DENSITY ELECTRON-HOLE PLASMAS AT LOW TEMPERATURES [†]

M. Glicksman, M. N. Gurnee[*] and J. R. Meyer

Division of Engineering, Brown University,
Providence, R. I., U.S.A.

ABSTRACT

Measurements of the photoconductivity of germanium single crystals as a function of electron-hole pair density show the effects of carrier-carrier scattering on the mobility, for sample temperatures in the range 48 - 300 K, and for densities up to 2×10^{18} cm^{-3}. Calculations including the effects of carrier-carrier scattering but neglecting conduction band anisotropy appear to underestimate the effect of scattering on the mobilities, especially at the lower temperatures. At 2 K, the photoconductivity shows evidence for the formation of electron-hole drops at intermediate excitation levels.

Although there has been considerable interest in the transport properties of high densities of electrons and holes in semiconductors, there has been limited experimental study which can be compared critically with theory. Experimental studies in germanium,[1-5] silicon[6] and tellurium[7] have been reported, and some inconclusive comparison[8] with theory for the tellurium experiments has been made. Among the other theoretical treatments available, that of Appel[9,10] appears suitable (with modifications) for testing against available data, or newly-generated experimental results. Of particular interest is the regime of electron-hole pair densities sufficiently high that the electron-hole interactions have an appreciable influence on the conductivity.

The work reported here deals with germanium, in basically two different temperature regimes. At temperatures above 10 K, there is no evidence of liquid drop condensation,[11] so that the experiments should deal with a high density electron-hole fluid. In experiments below 6 K,

---

[†] Supported in part by The National Science Foundation

[*] Present address: Honeywell, Inc., Minneapolis, Minnesota.

the germanium can contain electrons, holes, excitons and liquid condensate, so that the analysis is much more complex. In this paper, the high temperatures chosen for study range from 48 K to 300 K, and the low temperature studies were carried out in the range 1.6 to 2 K.

Samples of germanium of two different kinds were used in the experiments. Group A were single crystals, n-type, with approximately 10^{13} cm^{-3} impurity concentration; group B were single crystals, initially p-type, with initial impurity concentrations less than 10^{11} cm^{-3}. All samples were of dimensions $6 \times 3 \times t$ mm^3 in size, with the thicknesses t ranging from 30 to 525 μm. The large area surfaces were optically polished and etched. The thickness was tapered at about 1° to reduce multiple reflections of the penetrating laser radiation used in the diagnostics.

The large surfaces were each illuminated simultaneously over the whole area by pulsed optical radiation, to produce electron-hole pairs in a region within about 1 μm of the surface of the sample. In one set of experiments, Xe flash lamps were used as the source of radiation, with their output filtered through water cells to eliminate the infrared. The pulse length was about 3 μsec. In the other set of experiments, the source was a 1.06 μm Nd: glass laser, Q-switched to provide a 25 nsec pulse. Calibrated optical filters varied the intensity falling on the samples.

In addition to this primary source of electron-hole pairs, a 3.39 μm CW He-Ne laser was used to monitor and measure the density of holes produced, since the holes absorb[12,13] at this wavelength, making a transition to the split-off valence band in the process. A fast-response InSb photoconductive detector allowed accurate observation of the time-dependence of this absorption (and thus of the hole density). This enabled independent measurements of carrier density to be made and compared with those estimated from photoconductivity; at the higher temperatures, the 3.39 μm absorption was studied over the whole range

of excitation, and the absorption cross section could thus be checked.
We found cross-section values in the range $1.6 - 2 \times 10^{-16}$ cm^2/hole,
which agree fairly well with earlier measurements;[14] we used the values
1.75, 1.75, 1.88 and 1.71×10^{-16} cm^2/hole at 48, 72, 124 and 298 K respectively. At the low temperatures there is some uncertainty[13] in
the proper value of the cross section and our measurements were unable
to provide an independent test of the value used. We used[12] 1.5×10^{-16}
cm^2.

A four-point probe configuration was used to measure the photoconductance. The current contacts covered the sample ends; each consisted
of three n^+ and three p^+ alloyed dots, alternately arranged and soldered
together to provide ohmic contacts under all conditions of sample injection. The voltage probes on the sides of the crystals were also
alloyed contacts. The conductance was measured as a function of applied
electric field, and was found to be independent of field below about 1
V/cm at all temperatures and densities. Above 1 V/cm at 2 K and below,
increases[4] of photoconductance with electric field were observed, which
can be related to the heating of the carriers and the ionization of any
excitons present.[15,16] At the high electric fields, the level of conductivity produced would imply that any drops present are also evaporated, although no evidence for this was reported in the work of Nakamura
and Morigaki.[5] This is discussed in more detail below.

Conductivity Above 10 K

The photoconductivity was calculated from the conductance measured,
assuming a homogeneous filling of the germanium by the electron-hole
plasma. This assumption seemed well verified by the fact that samples
of varying thicknesses (30 up to 400 μm) yielded the same values of
photoconductivity at the same average carrier density. Measurements of
the conductance were made at times of the order of 2-4 μsec after the
laser pulse, during a period where the density was varying very little
with time (lifetimes were 10 μsec and longer). Measurements made with

Xe flash lamps and with the 1.06 μm laser yielded results which agreed
very well with each other; there was also no difference observed at 72
K and higher between A-type and B-type samples.   Figure 1 shows the

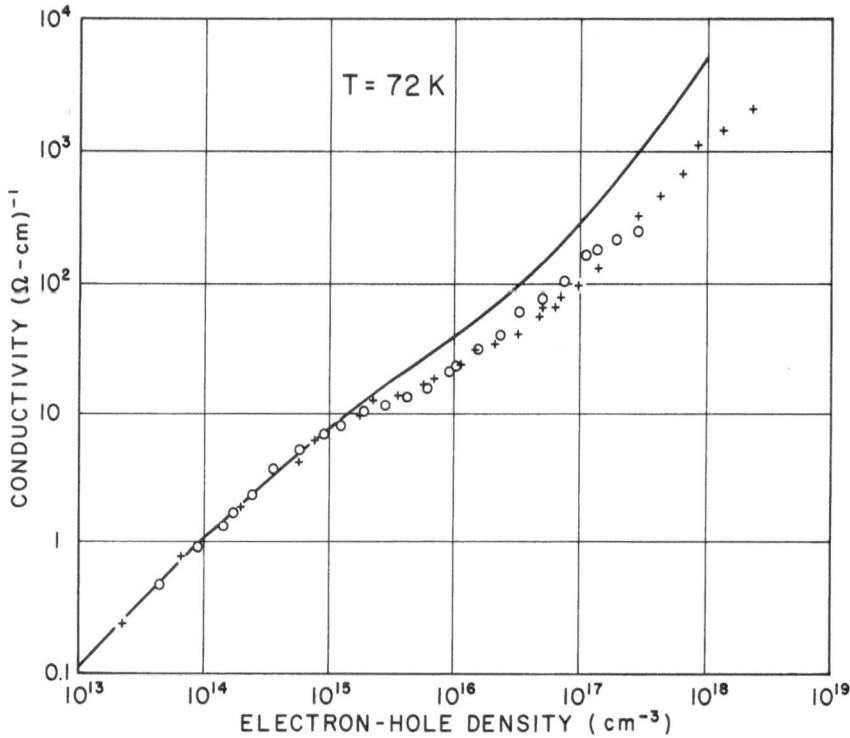

Fig.1.   The photoconductivity of germanium as a function of electron-hole
         concentration at 72 K.   Samples of thicknesses ranging from 30 to
         400 μm were used.   Curve is theoretical.

results of our measurements with several samples at 72 K.

The curve plotted in Fig. 1 is calculated using the expressions[10]
of Appel, which in zero-order of the variational solution are:

$$\sigma^{(0)} = \frac{\sigma_e^{(0)} + \sigma_h^{(0)}}{1 + (\sigma_h^{(0)} + \sigma_e^{(0)}) \; C \; J_o}$$

$$C = \frac{2^{5/2} \pi^{1/2} e^2 (m_e m_h)^{1/2}}{3 \; K_o^2 (kT)^{3/2} (m_e + m_h)^{1/2}}$$

$$J_o(y) = \int_o^\infty x \exp(-x^2) \left[\ln(1 + 2x^2/y) - (1 + y/2x^2)^{-1} \right] dx$$

$$y = \hbar^2/4\lambda_D^2 (m_e m_h/m_e + m_h) \, kT$$

$$\lambda_D^2 = K_o \, kT/4\pi e^2 (n + p)$$

We included both light and heavy holes in the calculation, and took it to first order in the expansion, using[17] m_e = 0.12, m_{1h} = 0.043 and m_{hh} = 0.347; K_o is the germanium dielectric constant, 16. The anisotropy of the conduction band ellipsoids was not included; preliminary calculations which include this anisotropy give reduced conductivities, in better agreement with experiment. The behavior of the photoconductivity is qualitatively as expected: as the density of electron-hole pairs is increased, the mobilities decrease due to electron-hole scattering; further increases in density increase the screening of the Coulomb interaction and decrease the effectiveness of the carrier-carrier scattering.

The same behavior occurs at all of the temperatures studied, with the calculated conductivity being somewhat higher than the measured values at the highest densities. The experimental data are summarized in the form of a plot of the reduced mobilities (the sum of electron and hole mobilities, divided by their value at very low densities) as a function of electron-hole density in Fig. 2. We see that the density of carriers at which the mobilities drop sharply decreases with decreasing temperature.

Conductivity at 2 K

These experiments were carried out with samples of types A and B, and Xe flash lamp excitation. Preliminary results were published earlier.[4] A summary of data taken with type A samples is given in Fig. 3, which is a plot of conductivity as a function of electron-hole density produced. In these measurements, it was not possible to detect the 3.39

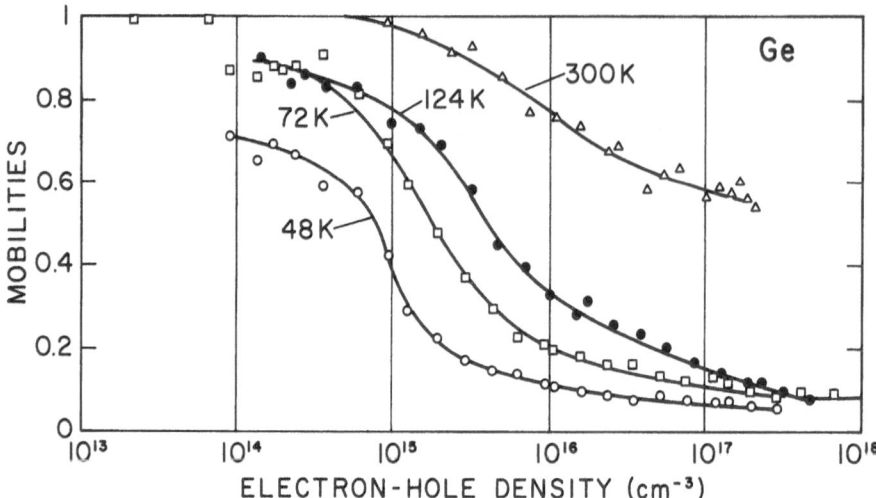

Fig.2. The mobilities of electrons and holes as a function of electron–hole
density. The mobilities are normalized to 1 at low density. The
curves are drawn to fit the data.

μm absorption over the whole range of excitation; in general this was

possible for densities above about 2×10^{16} cm^{-3} only. The abscissa

then represents the relative numbers of electron–hole pairs <u>produced</u>

for densities lower than this value, since the number of absorbing

holes is not known.

As noted earlier, at the highest densities, the data indicate a

dependence of conductivity on carrier density somewhat stronger than

linear. It should be noted, however, that there is heating of the

sample for injection levels above 10^{17} cm^{-3}, so that the actual lattice

and carrier temperatures can be appreciably greater[4] than 2 K. As the

excitation level is decreased, the conductivity drops sharply, and this

is interpreted as due to the formation of electron–hole liquid drops in

the sample, now not sufficiently strongly excited to fill the whole

sample with the liquid (density about 2×10^{17} cm^{-3}). With further

intensity decrease we find the onset of a conductivity "plateau" which

appears to have a dependence on sample thickness and some dependence on

the purity of the sample. Figure 4 shows the plateau region data for

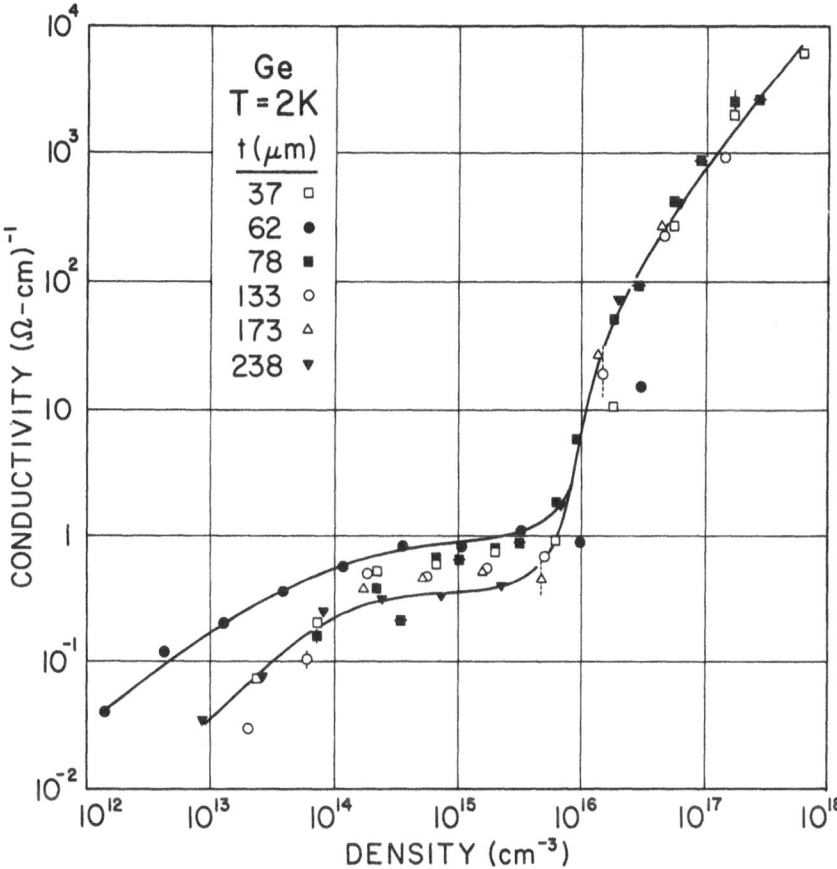

Fig.3. The photoconductivity of germanium at about 2 K (temperatures were in the range 1.8 to 2.0 K), as a function of injection density. Samples were of type A. Curves are drawn to fit the data. The values for the carrier density were calculated from the 3.39 μm absorption using the cross section 1.5×10^{-16} cm^2 (ref. 4 and 12).

samples of type B; the values of the conductivity are very much the same as those in type A germanium samples, but the region of excitation level over which the plateau is seen is greater than $1:10^3$, as compared to about $1:10^2$ shown in Fig. 3. This may be due to differences in the nucleation conditions for droplets in the two kinds of samples; it could also be due to the difference in carrier mobilities and the stronger effect of electron-hole scattering in the purer samples, with their lower concentration of ionized and neutral impurities. A clearer interpretation of this region should await conductivity measurements on

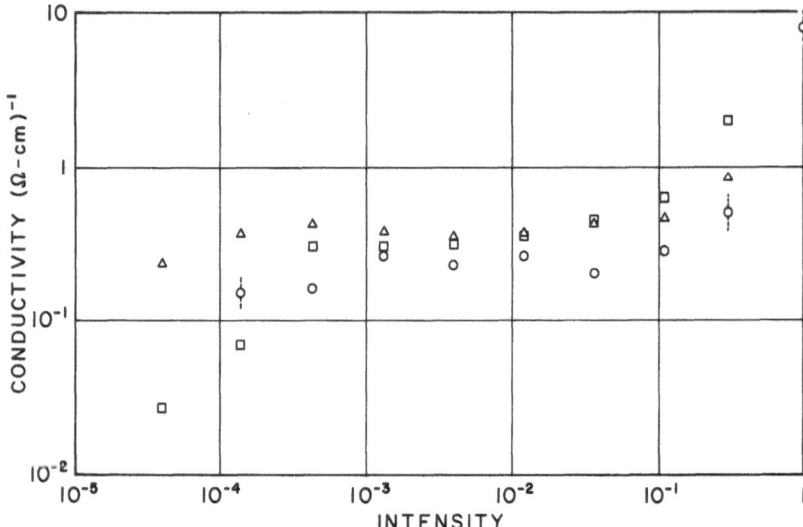

Fig.4. The photoconductivity of type B samples of germanium as a function
of injecting light intensity, in the "plateau" region, at 1.9 K.

these samples covering the whole temperature range, so that the second

suggestion may be evaluated.

It should also be noted that our measurements of the electric-

field dependence of the conductivity indicate an inappreciable number

of excitons present in the lowest region of excitation. If we refer to

Fig. 3, measurements of the I-V characteristic showed essentially linear

dependences up to fields in excess of 10 V/cm, underline except in the region we

have called the "plateau". In that region, a rise in conductivity was

observed between 2 and 4 V/cm, followed at higher fields by ohmic

behavior. These results lead to the conclusion that, in this region,

there are carriers which do not contribute to the conductivity unless

the average energy of the carriers is increased to about the energy

required to ionize excitons,[15] or presumably to evaporate the elec-

tron-hole drops. Since this is not observed for lower excitation

levels, it is believed this is evidence for the negligible fraction of

the excitation, under our conditions, present in excitons in this region.

Figure 5 plots the range of values of the plateau "conductivity"

observed as a function of thickness. The curve is calculated for a

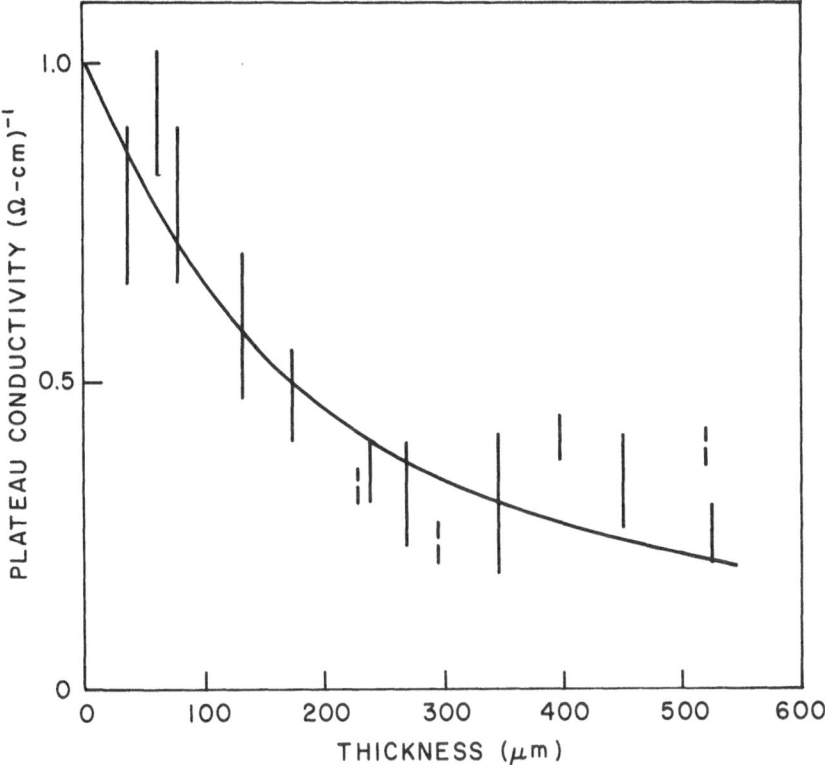

Fig.5. Values of the "plateau" conductivity as a function of sample thickness, for germanium crystals at about 2 K. Curve is calculated for an ambipolar diffusion length of 110 μm.

steady-state distribution of electrons and holes present, with an ambipolar diffusion length of about 110 μm, under the assumption that the conductivity does not vary with density. This would lead us to conclude that the variation of plateau conductivities with thickness is due to the inhomogeneous electron-hole plasma distribution. The density of electrons and holes is here then directly related to the presence of liquid drops, and its value is determined by the evaporation of electrons and holes from the drops.

We thank Sheldon Mittleman for his assistance in the experimental effort.

Notes Added in Proof:

According to our recent infrared measurements of absorption (unpublished), the value of 1.5×10^{-16} cm^2 for the absorption cross section quoted from ref. 12 should be replaced by 4.5×10^{-17} cm^2/hole. Hence the scale of the abscissa for the density in Fig. 3 should be multiplied by the factor 3.3.

REFERENCES

1) L. W. Davies: Nature 194 (1962) 762.

2) V. M. Asnin and A. A. Rogachev: ZhETF Pis. Red. 7 (1968) 464 (JETP Lett. 7 (1968) 360).

3) V. M. Asnin and A. A. Rogachev: ZhETF Pis. Red. 14 (1971) 494 (JETP Lett. 14 (1971) 338).

4) M. N. Gurnee, M. Glicksman and P. W. Yu: Solid State Commun. 11 (1972) 11.

5) A. Nakamura and K. Morigaki: Solid State Commun. 14 (1974) 41.

6) A. A. Patrin, M. Ryvkin, V. M. Salmanov and I. D. Yaroshetskii: Fiz. Tekhn. Poluprovod. 3 (1969) 449 (Sov. Phys.-Semicond. 3 (1969) 383).

7) Y. Nishina, T. Nakanomyo and T. Fukase: *Proc. Xth Int. Conf. Physics of Semicond.*, ed. S. P. Keller, J. C. Hensel and F. Stern (Nat. Tech. Info. Serv., Springfield, Va., 1970) p.493.

8) Y. Miura and C. Horie: J. Phys. Soc. Japan 33 (1972) 1522.

9) J. Appel: Phys Rev. 122 (1961) 1760.

10) J. Appel: Phys. Rev. 125 (1962) 1815.

11) G. A. Thomas, T. M. Rice and J. C. Hensel: Phys. Rev. Lett. 33 (1974) 219.

12) W. Kaiser, R. J. Collins and H. Y. Fan: Phys. Rev. 91 (1953) 1380.

13) Ya. E. Pokrovskii and K. I. Svistunova: Fiz. Tverd. Tela 13 (1971) 2788 (Sov. Phys.-Solid State 13 (1972) 2334).

14) W. E. Pinson and R. Bray: Phys. Rev. <u>136</u> (1964) A1449.

15) V. M. Asnin, A. A. Rogachev and S. M. Ryvkin: Fiz. Tekhn.
Poluprovod. <u>1</u> (1967) 1740 (Sov. Phys.-Semicond. <u>1</u> (1968) 1445).

16) T. Yao, K. Inagaki and S. Maekawa: Solid State Commun. <u>13</u> (1973)
533.

17) M. Neuberger: *Group IV Semiconducting Materials* (IFI/Plenum, New York,
1971) p.18-19.

ON THE SHAPE OF THE DROPLET IN UNIAXIALLY STRESSED Ge

M. Morimoto, K. Shindo, and A. Morita

Department of Physics,
Tohoku University,
Sendai, Japan

ABSTRACT

Density functional formalism developed by Hoenberg, Kohn, and Sham is applied to the calculation of the surface tension of electron-hole droplets in uniaxially stressed Ge. It is shown that the shape of droplets in highly excited Ge under a uniaxial stress of about 0.3 k bar along the [111] direction is oblate spheroid because of the anisotropy of electron mass parameter. The ratio of the minor axis to the major one is estimated to be at most 1.9.

Determination of the shape of droplet by the light scattering experiment is discussed.

I. INTRODUCTION

Combescot and Nozieres,[1] Brinkman and Rice,[2] and Vashishta et al.[3] have estimated the binding energy per electron-hole pair in droplets of stressed Ge and Si by using the theory of uniform electron-hole liquid; in which the surface energy was not considered explicitly. Theoretical investigation on the surface property of droplets has been done so far for unstressed Ge.

The minima in the conduction band of Ge consist of four ellipsoidal valleys. Although each valley is anisotropic, the symmetry of the whole valleys is cubic. Therefore the mass anisotropy of each valley does not influence on the surface property of droplets such as surface tension in unstressed Ge.

Benoit à la Guillaume and Voos[6] observed electron-hole drops in uniaxially stressed Ge. In their experiment Ge is uniaxially stressed with the pressure of about 0.3 k bar in the [111] direction, so that conduction electrons in the droplets belong only to [111] valley, while holes in the droplets still consist of the heavy and light ones.

The aim of this report is to make clear theoretically the effect of the mass anisotropy on the surface tension of droplets in properly stressed Ge.

In what follows we employ units of $\hbar = m_0 = 1$ (where m_0 is electron mass in vacuum) but electron charge e is conserved.

II. DENSITY FUNCTIONAL FORMALISM AND SCALE TRANSFORMATION

Let us consider a droplet consisting of fixed number N of electron hole pairs in stressed Ge.

According to Benoit à la Guillaume and Voos's experiment[6] we deal with only one electron valley with anisotropic masses $(m_{e/\!/}, m_{e\perp})$ and one isotropic hole band for simplicity. We use the number density functional formalism developed by Hoenberg, Kohn and Sham.[7,10] The ground state energy of the droplet is given as a functional of the electron density $n^e(\vec{r})$ and hole density $n^h(\vec{r})$ by

$$E_{NG}(n^e(\vec{r}), n^h(\vec{r}))$$

$$= \int d\vec{r} \left[\varepsilon_{NG}(n^e(\vec{r}), n^h(\vec{r})) \right.$$

$$+ A_{/\!/}(n^e, n^h)(\vec{\nabla}_{/\!/} n^e)^2 + A_{\perp}(n^e, n^h)(\vec{\nabla}_{\perp} n^e)^2$$

$$+ B_{/\!/}(n^e, n^h)(\vec{\nabla}_{/\!/} n^h)^2 + B_{\perp}(n^e, n^h)(\vec{\nabla}_{\perp} n^h)^2$$

$$+ C_{/\!/}(n^e, n^h)(\vec{\nabla}_{/\!/} n^e)(\vec{\nabla}_{/\!/} n^h) + C_{\perp}(n^e, n^h)(\vec{\nabla}_{\perp} n^e)\cdot(\vec{\nabla}_{\perp} n^h)$$

$$\left. + 0(\nabla^4) + \cdots \right]$$

$$+ \frac{e^2}{2\varepsilon_0} \int\int d\vec{r} d\vec{r}\,' [(n^h(\vec{r}) - n^e(\vec{r}))(n^h(\vec{r}\,') - n^e(\vec{r}\,'))/|\vec{r} - \vec{r}\,'|\,]. \qquad (2.1)$$

The gradient dependent terms come from variation of the carrier densities and the coefficients $A_{/\!/}$, $A_{\perp}$, $B_{/\!/}$, $B_{\perp}$, $C_{/\!/}$ and $C_{\perp}$ are connected with the proper polarization parts of uniform electron-hole liquid.[7] Hereafter we take into account up to second order in the gradient expansion. The notations $/\!/$ and $\perp$ mean the directions parallel and perpendicular to [111] respectively.

We make use of R.P.A. in calculation of these coefficients $A_\parallel$, $A_\perp$, $B_\parallel$, and so forth and furthermore assume local neutrality;

$$n(\vec{r}) \equiv n^e(\vec{r}) = n^h(\vec{r}) .$$
(2.2)

Then after some mathematical manipulations

$$E_{NG} = \int d\vec{r}\{ \ \varepsilon_{NG}(n,n) + \frac{1}{72n} \ [(\ \frac{1}{m_{e\parallel}} + \frac{1}{m_h} \ )(\vec{\nabla}_\parallel \ n)^2$$

$$+ (\ \frac{1}{m_{e\perp}} + \frac{1}{m_h} \ )(\vec{\nabla}_\perp \ n)^2 \ ] \ \} \quad \cdots \ ,$$
(2.3)

where $m_{e\parallel}$ $(=1.58)$ and $m_{e\perp}$ $(=0.082)$ are the masses of electron parallel and perpendicular to the $[111]$ direction, respectively and m_h is the hole mass. It should be noted that the gradient terms in eq.(2.3) come from kinetic energies and do not depend upon electron charge e.

The coefficients of the second term in eq.(2.3) is smaller than that of the third term. The second term determines the surface energy $S_\parallel$ of the surface whose normal vector is parallel to the $[111]$ direc tion and the third term determines the surface energy $S_\perp$ of the surface whose normal vector is perpendicular to the $[111]$ direction.

Because $S_\parallel$ is smaller than $S_\perp (S_\parallel < S_\perp)$ we can expect that the stable shape of the droplet is not spherical but oblate-spheroidal. In order to determine the shape quantitatively, we perform scale transfor- mation of coordinate. The new coordinate $\vec{\tilde{r}}$ is defined by

$$\vec{\tilde{r}}_\parallel \equiv (\ \frac{M_\parallel}{M_\perp} \ )^{1/3}\vec{r}_\parallel \ , \qquad \vec{\tilde{r}}_\perp \equiv (\ \frac{M_\perp}{M_\parallel} \ )^{1/6}\vec{r}_\perp \ ,$$
(2.4)

where

$$\frac{1}{M_\parallel} = \frac{1}{m_{e\parallel}} + \frac{1}{m_h} \quad , \qquad \frac{1}{M_\perp} = \frac{1}{m_{e\perp}} + \frac{1}{m_h} \ .$$
(2.5)

The density in the scaled coordinate $\tilde{n}(\vec{\tilde{r}})$ should be defined to equal to $n(\vec{r})$ since the volume element is conserved in transformation to the new coordinate.

The gradient terms appearing in energy functional become spherical in this scaled coordinate $\vec{\tilde{r}}$ as

$$E_{NG} = \int d\vec{r} \; [\varepsilon_{NG} + \frac{1}{72M^{2/3}{}_M{}^{1/3}\tilde{n}(\vec{r})} \; | \vec{\triangledown}\tilde{n}(\vec{r}) |^2].$$ (2.6)

If the density distribution $\tilde{n}_G(\vec{r})$ minimizing the density functional E_{NG} is unique, this distribution should be a spherical function of $\vec{r}$ from symmetrical consideration. This means that the shape of the droplet is oblate-spheroidal in the ordinary coordinate.

The ratio of the minor axis parallel to $[111]$, $R_{/\!/}$, to the major one perpendicular to $[111]$, $R_\perp$ is given by

$$\frac{R_\perp}{R_{/\!/}} = \left(\frac{\frac{1}{m_{e_\perp}} + \frac{1}{m_h}}{\frac{1}{m_{e_{/\!/}}} + \frac{1}{m_h}} \right)^{1/2}$$ (2.7)

from the definition of the scaled coordinate $\vec{r}$. The mass m_h is determined as the average value of those of heavy and light ones, m_{hh} and m_{lh};[11]

$$\frac{1}{m_h} = \frac{\alpha_{hh}}{m_{hh}} + \frac{\alpha_{lh}}{m_{lh}} \quad , \qquad \alpha_{hh} + \alpha_{lh} = 1,$$ (2.8)

where α_{hh} and α_{lh} are the rates of the densities of heavy hole and light hole. Assuming that the hole bands are degenerate at Γ point and using the values $m_{hh}=0.347$, $m_{lh}=0.042$ we get $m_h=m_{hh}/1.3$ and from (2.7) the value of $R_\perp/R_{/\!/}$ is 1.9.

III. EFFECT OF EXCHANGE ENERGY AND CORRELATION ENERGY

When we go beyond R.P.A., taking into account the contribution to the properpolarization parts of electron-hole liquid from exchange and correlation energy roughly,[9,10] we may estimate the ratio of the axes to be larger than 1.2.[11] After all we may expect the ratio of the axes is between 1.2 and 1.9.

IV. LIGHT SCATTERING EXPERIMENT

The warping of the droplet discussed so far can be measured in a

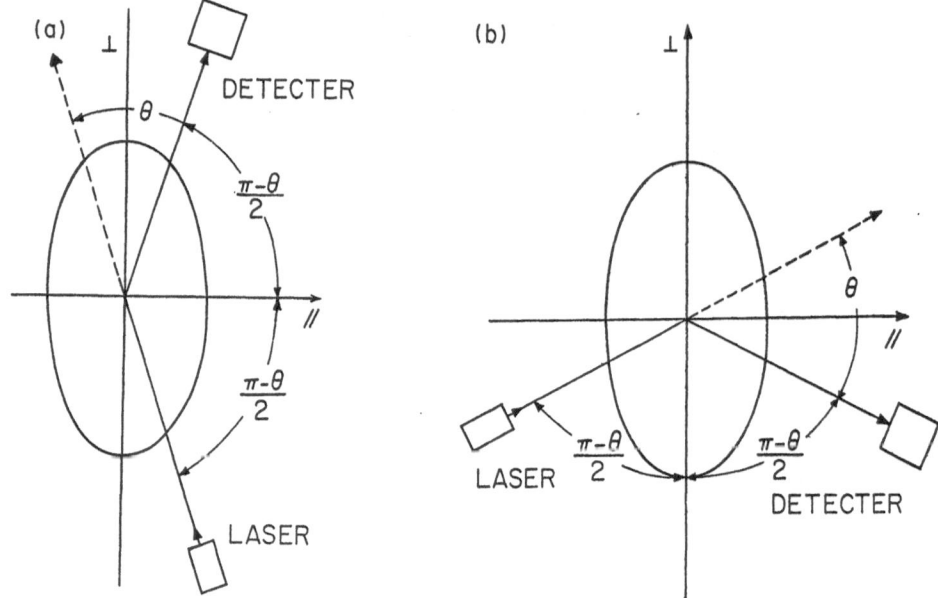

Fig.1. The symbols ∥ and ⊥ denote the direction parallel and perpendicular to the [111] direction. A droplet is represented as an ellipsoid.

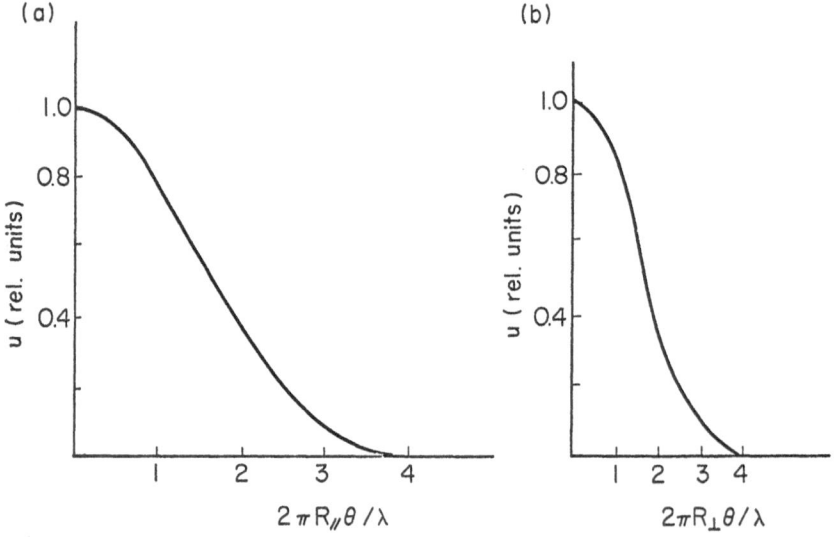

Fig.2. The angle distribution of the scattered light.
Fig.2a corresponds to Fig.1a and Fig.2b to Fig.1b.
The unit of the abscissa in Fig.2a is $\lambda/2\pi R_{\parallel}$ and that in Fig.2b $\lambda/2\pi R_{\perp}$,
Note that $1/R_{\parallel}$ is larger than $1/R_{\perp}$.

light scattering experiment. The interesting geometries in the case of deformed droplet are sketched in Fig.1a and Fig.1b. A droplet is represented as an ellipsoid. The He-Ne laser and the detector of scattered light should be moved relatively to the sample crystal so as to keep the bisect between the incident beam and scattered light in the [111] direction or the direction perpendicular to [111], respectively.[11]

The angular distribution of the scattered light for the cases (a) and (b) are shown in Fig.2a and Fig.2b. We can observe that the axis ratio $R_\perp/R_{/\!/}$ is given by the ratio of half widths in Fig.2a and Fig.2b.

V. CONCLUSION

In conclusion, the shape of the droplet in properly stressed Ge is oblate spheroid and the ratio of minor axis to major one is at most 1.9.

REFERENCES

1) M. Combescot and P. Nozieres: J. Phys. C5 (1972) 2369.

2) W. Brinkman and T. M. Rice: Phys. Rev. B7 (1973) 1568.

3) P. Vashishta, P. Bhattacharyya, V. Massida, K. S. Singwi and P. Vashishta: Phys. Rev. B10 (1974) 5127.

4) T. M. Rice: Phys. Rev. B9 (1974) 1540.

5) L. M. Sander, H. B. Shore and L. J. Sham: Phys. Rev. Letters 31 (1973) 533.

6) C. Benoit à la Guillaume and M. Voos: Phys. Rev. B5 (1972) 3079.

7) P. Hoenberg and L. J. Sham: Phys. Rev. 136 (1965) B864.

8) W. Kohn and L. J. Sham: Phys. Rev. 140 (1965) A133.

9) L. J. Sham: *Computational Methods in Band Theory* ed. by P. M. Marais, J. F. Jank and A. R. Williams (Plenum, New York, 1971) p.458.

10) S. K. Ma and K. A. Bruckner: Phys. Rev. 165 (1968) 18.

11) M. Morimoto, K. Shindo and A. Morita: J. Phys. Soc. Japan 41 (1976) 91.

12) Y. Pokrovskii: ZhETF Pis. Red. 9 (1969) 435 [Soviet Physics -

JETP Letters *2* (1969) 261.]

13) H. C. Van de Hulst: *Light Scattering by Small Particles* (Wiley, N.Y., 1957) p.93.

14) L. D. Landau and E. M. Lifshitz: *Statistical Physics* (Pergamon, N.Y., 1958) p.458.

TWO-DIMENSIONAL ELECTRON-HOLE METALLIC LIQUIDS

Yoshio Kuramoto[*] and Hiroshi Kamimura

Department of Physics,
Faculty of Science,
University of Tokyo,
Tokyo, Japan

ABSTRACT

The ground state energy of two-dimensional electron-hole liquids
is calculated in the generalized random phase approximation. It is
shown that the cohesive energy per electron-hole pair exceeds the
binding energy of a two-dimensional exciton in the case of a system
with two conduction band minima and a single valence band maximum with
a heavier effective mass. A possibility of producing electron-hole
"pancakes" in layer-type semiconductors is pointed out. The many-body
correction of the effective masses and dispersion relations of collective modes are also derived. Finally the condition for occurrence of
excitonic phases is discussed.

I. INTRODUCTION

Electron-hole metallic liquids in semiconductors such as Ge and Si
have been investigated in detail. Theoretical analyses[1] have shown
that the nature of band structures of these semiconductors plays an
important role in stabilizing the liquid phase. Recently remarkable
progress has also been made in study of semiconductors with large
anisotropy. It would be interesting to see how strong anisotropy in
energy bands influences the stability of the electron-hole metallic
liquid (EHL). From this standpoint we will study a system of two-
dimensional (2D) electron-hole metallic liquids in this paper.[2] Such
a system will be realized in layer-type semiconductors with both conduction and valence bands being two-dimensional. We note that the
observation of 2D indirect excitons has been reported in a layer-type
semiconductor GaSe.[3] The binding energy of a 2D free exciton is four
times larger than that of a 3D one. In this paper we show that the
cohesive energy of the 2D EHL is more than four times larger than that

of the 3D one in the case of a system with a suitable many-valley

structure in energy bands.   This means that the EHL in the 2D system

is energetically more stable than that in the 3D one.   We also calculate

excitation spectra of quasi-particles and collective modes.   Further

we discuss a possibility of the formation of electron-hole "pancakes"

in highly excited layer-type semiconductors.   Effects of van der Waals

interaction between layers is also examined.   Finally the condition

for the occurrence of excitonic phases is discussed.

II.   GROUND STATE ENERGY

We consider a 2D model of a layer-type semiconductor in which

electrons in a conduction band and holes in a valence band with

effective masses m_e and m_h respectively are confined in a single layer.

The coulomb interaction in an anisotropic medium with dielectric tensor

components $\in_\parallel$ and $\in_\perp$ is given by

$$V(\rho, z) = \frac{e^2}{\sqrt{\in_\parallel \in_\perp}} \left(\rho^2 + \frac{\in_\perp}{\in_\parallel} z^2\right)^{-\frac{1}{2}}, \quad \rho = (x, y) . \tag{1}$$

The ground state energy per electron-hole pair of the metallic state is

calculated as a function of the interparticle spacing r_s defined by

$\pi(r_s a_B)^2 = S/N_p$, where a_B is the exciton Bohr radius $\sqrt{\in_\parallel \in_\perp} \hbar^2/\mu e^2$, N_p/S

the density of electron-hole pairs.   First we consider a simple system

in which each of conduction and valence bands has a single valley.

Then the ground state energy ε_{tot} is given by

$$\varepsilon_{tot} = r_s^{-2} - 2.401 \, r_s^{-1} + \varepsilon_{corr} , \tag{2}$$

where energies are measured in unit of the effective Rydberg defined as

$e^2/(2\sqrt{\in_\parallel \in_\perp} \, a_B)$.   The correlation energy ε_{corr} depends on the effective

mass ratio $\sigma \equiv m_e/m_h$ of electrons to holes and is calculated in the

generalized RPA.   The calculated results of ε_{tot} are shown in Fig. 1,[2]

where those of the 3D system are also shown for comparison.   The results

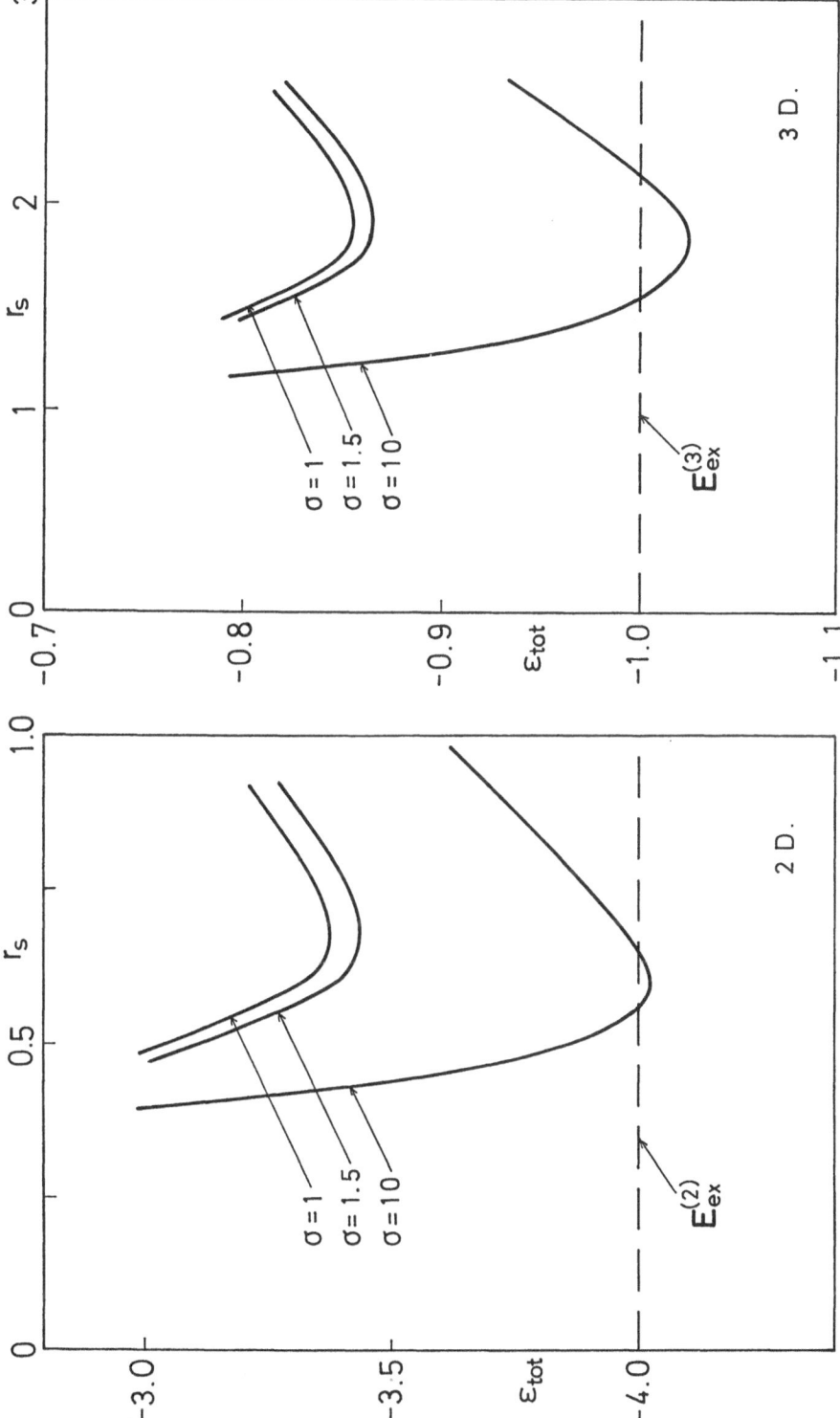

Fig.1. Ground state energies per electron-hole pair in the 2D system (left) and the 3D one (right) without many-valley structures in energy bands. Dashed lines indicate the binding energies of free excitons.

indicate that for each value of σ the minimum of ε_{tot} in the 2D system is about four times as deep as that in the 3D system, and that the values of r_s at the minimum of ε_{tot} are about three times smaller than that in the 3D system. These circumstances may correspond to the difference in binding energies and radii of free excitons in the 2D and the 3D systems.

Next we discuss effects of many-valley structures on the ground state energy of the 2D EHL. As an example we consider a case where a conduction band has two minima and a valence band has a single maximum. The kinetic energy of electrons is reduced by the existence of many valley structure and therefore the EHL is more stabilized.[2] The calculated results are shown in Fig. 2 together with these in the corresponding 3D system. In the case of $\sigma = 1$ the cohesive energy in the 2D system is again four times as large as that in the 3D one. In contrast with the previous case of a single extremum for each band, however, ε_{tot} in the 2D system depends more strongly on σ. As a result, the cohesive energy in a 2D system becomes more than four times larger than that in a 3D system as σ becomes smaller than unity. We therefore conclude that the effect of many-valley structures on the cohesive energy of the EHL is more prominent in the 2D system when the conduction band has a many-valley structure with the effective mass being lighter than that of the valence band.

III. EXCITATION SPECTRA

We now investigate the excitation spectra of the 2D EHL. First the many-body correction of the effective masses is derived in the generalized RPA and compared with that of the 3D system. We consider the case of a single-valley with $m_e = m_h \equiv m$. The effective mass m^* of an electron-like quasi-particle (or a hole-like quasi-particle) is calculated by the equation

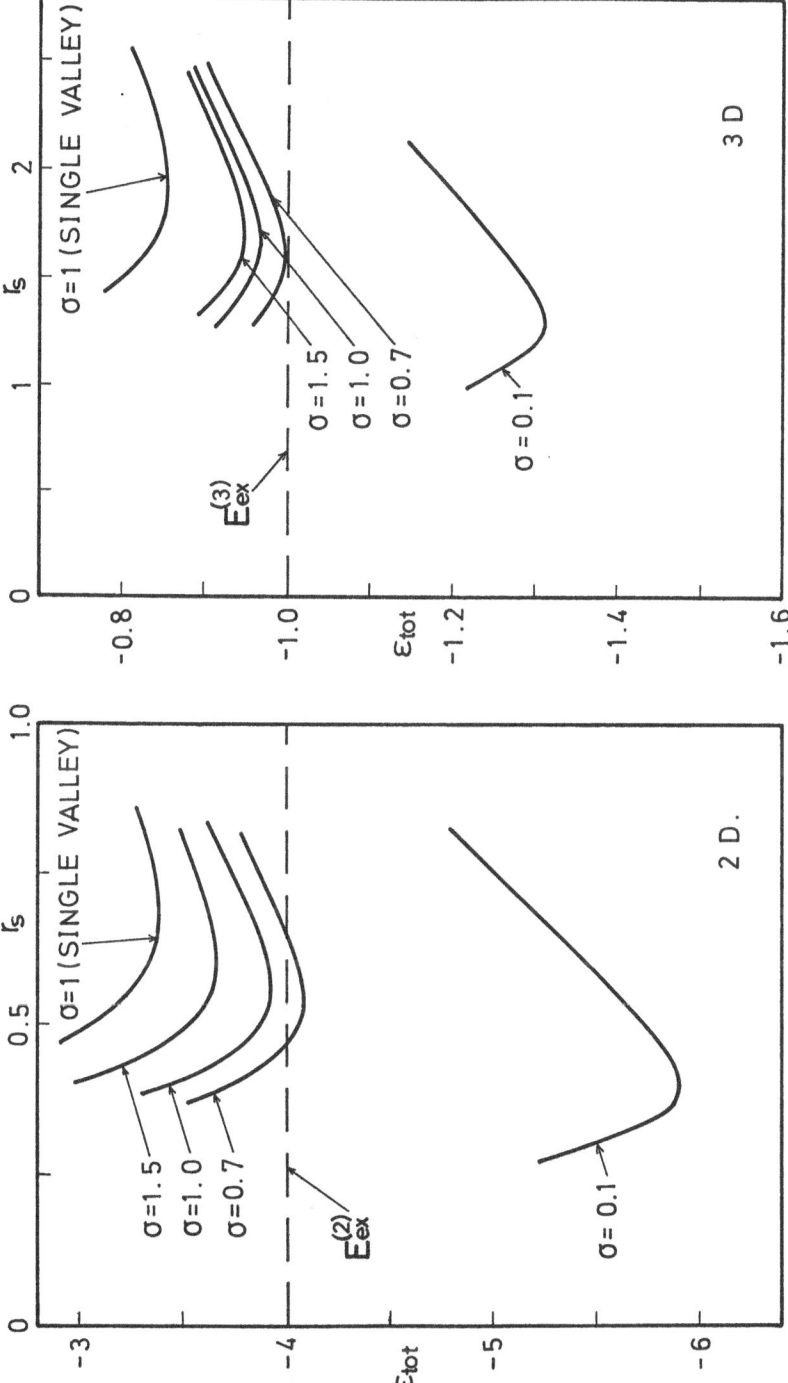

Fig.2. Ground state energies in the case of two minima in the conduction band and a single maximum in the valence band.

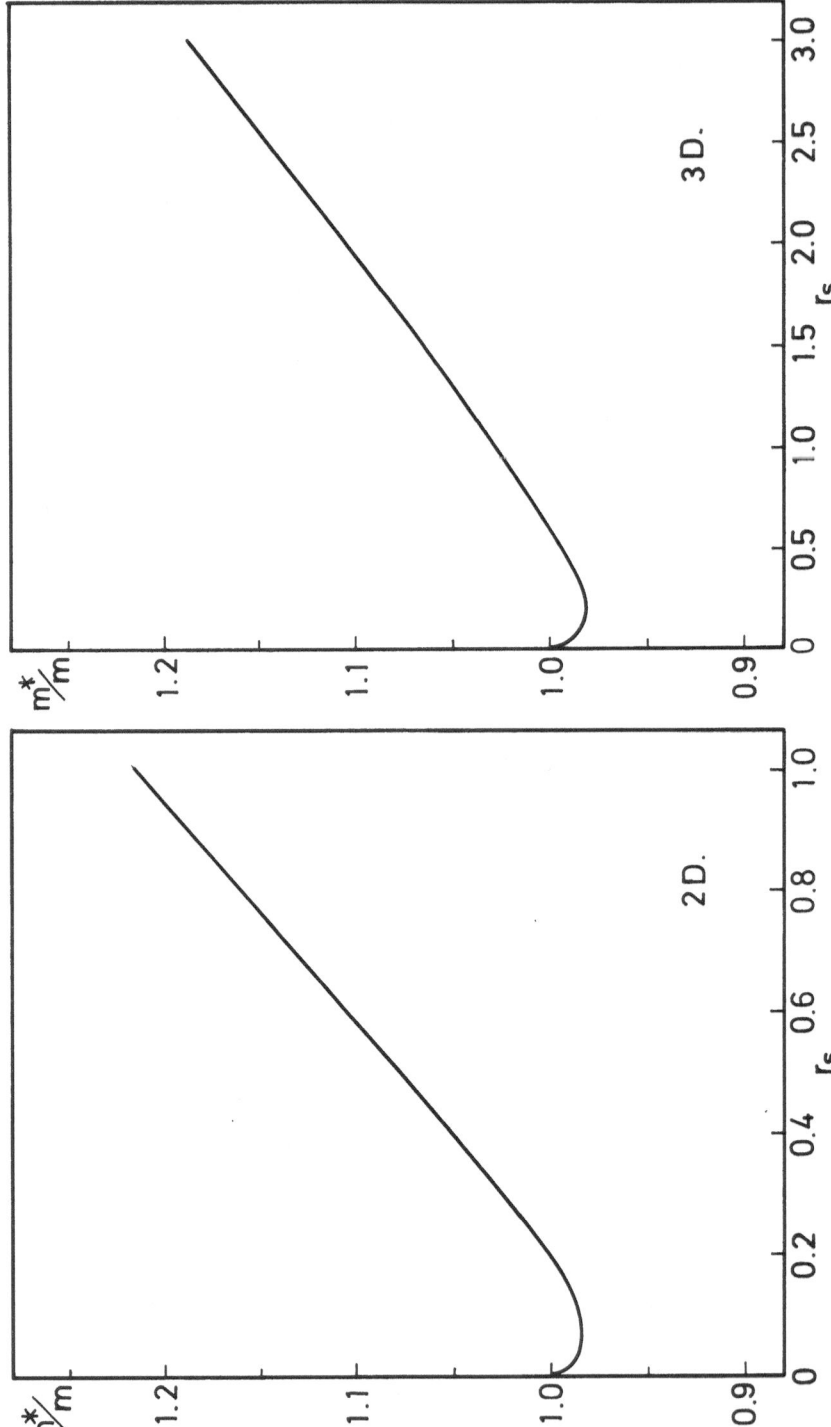

Fig.3. Effective masses of quasi-particles in the 2D system (left) and in the 3D one (right).

$$\frac{1}{m^*} = \frac{1}{m} + \frac{1}{\hbar^2 k_F} \left[\frac{d}{dk} \Sigma \left(k, \omega = \frac{\hbar k^2}{2m} \right) \right]_{k=k_F} , \tag{3}$$

where k_F is the Fermi wave number and $\Sigma \left(k, \omega \right)$ the self-energy of an electron (or of a hole) in the generalized RPA. The computational procedure is similar to that in the case of the 3D electron liquid.[4] We show the calculated results of m* in Fig. 3 together with those of the 3D EHL. The magnitude of m* in the 2D system is nearly equal to that in the 3D one when r_s in the former system becomes three times smaller than that in the latter one. We note that the effective mass in each system is enhanced by about 10% at the most stable density of the EHL.

Next we discuss collective modes.[2] One of them is a 2D plasmon mode and its dispersion relation is given by

$$\omega_p(q) = \left(\frac{2\pi e^2 N_p q}{\sqrt{\epsilon_\parallel \epsilon_\perp} \mu s} \right)^{\frac{1}{2}} , \tag{4}$$

in the small wave number region with neglect of small polariton effects.[5] The $\sqrt{q}$ dependence is explained by the absence of the depolarizing field against the uniform charge displacement in a 2D system. If the holes are quite heavier than electrons (or the other way round), an acoustic mode also becomes a well defined collective mode. Since the screened short-range interaction is responsible for this mode, the dispersion relation in the 2D system is rather close to that in the 3D system.[2]

IV. DISCUSSION

On the basis of the results obtained in the previous sections we now discuss a possible situation which will occur when a layer-type semiconductor is highly excited by a laser or by other means and 2D electrons and holes are densely generated.   If the semiconductor has a

suitable many-valley structure as discussed in Sec. II, the EHL state is

more stable than a free exciton state and possibly than a state of

free excitonic molecules.   In such a case we can expect that clusters

of the metallic state will appear at low temperatures with appropriate

averaged densities of electrons and holes.   Since the metallic state

thus produced in a 2D layer is supposed to take a pancake shape so as

to minimize the boundary energy, we call it an electron–hole pancake

(EHP).   It is likely to occur that some EHP's in nearby layers

aggregate balancing the gain of mutual polarization energy with the

cost of entropy of individual motion.   The additional contribution of

the mutual polarization to the cohesive energy is estimated to be 3%

when the spacing of interacting two layers is half the exciton Bohr

radius.   The interaction between EHP's in different layers is of van

der Waals nature and is quite different from the interaction within a

layer.   Therefore we expect that a possible aggregate of EHP's would

also be pancake–shaped and be highly anisotropic.   This anisotropy may

be observed in light scattering experiments.[6]   If the EHL is composed

of N layers, the frequency of the plasmon is $\sqrt{N}$ times higher than that

given by eq. (4).[2]   In the limit of N = ∞ , it becomes equal to the

3D plasmon frequency which is given by

$$\omega_{p,\infty} = \left(\frac{4\pi e^2 N_p}{\mu \in_\perp cS} \right)^{\frac{1}{2}} \tag{5}$$

where c is the spacing of the successive layers.   Therefore the degree

of aggregation along the axis perpendicular to layers is reflected on

the behavior of the collective mode in an EHL.

Finally we briefly discuss the condition for the occurrence of

excitonic phases.   As has been shown in the present paper, the EHL

phase becomes more stable for more complicated band structures.   This

means that the occurrence of excitonic phases becomes more severe for

more complicated band structures.   From this standpoint Kamimura and

Kuramoto[7] have recently investigated a 3D system with a simple band structure of two conduction band minima and a single valence band maximum in order to find the most generous condition for the occurrence of excitonic phases. In this case they have concluded that the excitonic phase occurs only when the effective mass of an electron is 0.7 times heavier than that of a hole. On this ground we suggest that the first order semiconductor-semimetal transition bypassing excitonic phases is likely to occur in many real indirect gap semiconductors when one reduces the energy gap continuously.

REFERENCES

*) Present Address: Dept. of Applied Physics, Faculty of Engineering, Tohoku University, Sendai, Japan 980.

1) M. Combescot and P. Nozières: J. Phys. C5 (1972) 2369;
   W. F. Brinkman and T. M. Rice: Phys. Rev. B7 (1973) 1508.

2) Y. Kuramoto and H. Kamimura: J. Phys. Soc. Japan 37 (1974) 716.

3) H. Kamimura, K. Nakao and Y. Nishina: Phys. Rev. Letters 22 (1969) 1379.

4) T. M. Rice: Ann. Phys. 31 (1965) 100.

5) F. Stern: Phys. Rev. Letters 18 (1967) 546.

6) Ya. Pokrovskii: Phys. Stat. solidi (a) 11 (1972) 385.

7) H. Kamimura and Y. Kuramoto: Prog. theor. Phys. Suppl. No.57 (1975) 138.

MOTION OF ELECTRON-HOLE DROPS IN Ge AT LOW EXCITON DENSITIES

J. M. Hvam

Institute of Physics,
Odense University
DK-5000 Odense,
Denmark

ABSTRACT

The expansion of the electron-hole drop (EHD) cloud after a short
and local excitation by a pulsed dye laser is investigated by EHD
detection in a metal-semiconductor contact.   In regions with low free
exciton (FE) densities a free flight of the drops is observed, suggesting a collision time of EHD $\tau_c \geq 25\mu sec$.   It is proposed that the
EHD are ejected out of the excitation region by a strong FE density
gradient.   Velocities in the range 200 - 2000 cm/s are observed in-
dependently of an applied electric field, indicating the EHD to be
essentially uncharged.

I.   INTRODUCTION

In order to reveal the dynamics of the exciton condensation occurring in Ge at low temperatures it is of importance to know the detailed
conditions for nucleation and growth of the electron-hole drops (EHD)
formed.   The excitation by an intense laser beam being usually very
local, these conditions depend strongly on the possible motion of EHD.
Thus giant drops have been formed recently by applying an inhomogeneous
stress field forcing the drops towards the point of maximum stress.[1]
In the absence of external fields, on the other hand, experiments[2,3]
show a sharp boundary of the EHD cloud from a local excitation, indicating negligible drop diffusion.   Hence, the spatial distribution
of EHD previously reported[4-6] is governed mainly by free exciton
diffusion in connection with a finite growth rate of the liquid
drops.[6,7]

In the present work is discussed some experiments revealing real
EHD motion in a region with a low FE density.   The analysis of the

results suggests a revised view on diffusive motion of EHD as well as on their drift mobility.

II.   EXPERIMENTAL PROCEDURE

In the present experiment Ge samples (2.7 Ωcm, n-type), at 2 K, were excited by a pulsed dye laser emitting at 6200 Å and with a peak power of 50 W.   Pulse width and repetition rate were 10 nsec and 20 Hz, respectively.   The experimental geometry is shown in Fig.1.   The

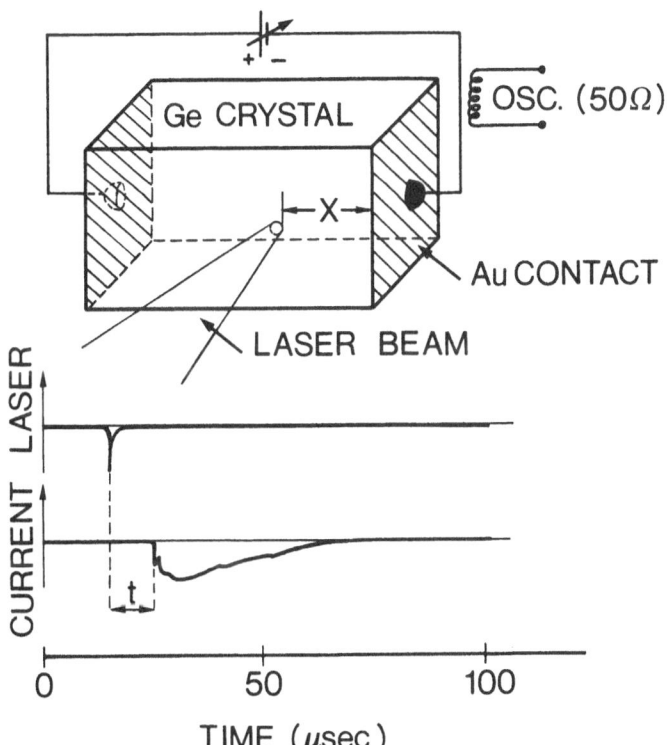

Fig.1.   Experimental geometry and oscilloscope traces of laser light
pulse (upper trace) and typical photocurrent signal (lower
trace).   The delay time t depends only on the distance x
between laser spot and negative contact.

samples ($0.6 \times 0.6 \times 0.6$ mm^3) were supplied with Au contacts symmetrically on two opposite faces, and a bias voltage of 0-5 V was applied across the sample.   The transient photocurrent was monitored via a current transformer and displayed on the oscilloscope.   The laser beam was sharply focussed on the surface of the sample and it could be moved

along the sample in a direction perpendicular to the contact planes.
The laser pulse and a typical photocurrent signal is also displayed in
Fig.1. For lower bias voltages (about 0.2 V was usually applied) the
instantaneous photocurrent was negligible owing to the fast formation
of excitons. However, with some delay relative to the laser pulse a
current pulse appeared. This current pulse was identified as a pile-
up of current spikes originating from the break-up of EHD arriving at
the negatively biased contact, acting as a reverse biased Schottky
barrier diode.[8] The delay of the photocurrent onset is thus a direct
measure of the time elapsed before the first EHD reach the negative
contact, being dependent solely on the distance between this contact
and the laser spot.

III. RESULTS AND DISCUSSION

In Fig.2 is shown the relation between x, the distance from laser
spot to contact, and t, the delay time for EHD detection, at different
excitation powers, revealing the expansion of the region containing EHD
after a short and local excitation. The maximum range of the EHD
cloud is about 0.4 mm, and is reduced by decreasing the excitation
power. The major part of the EHD cloud expansion occurs within a time
that is short compared with the FE life time. This part can be explained by FE diffusion in connection with a finite growth rate of EHD
at rest[8] as illustrated by the dashed curve in Fig.2. This curve
shows, as a function of time, the calculated linear extension of the
region containing EHD with radius $R \geq 1\mu m$, assuming the EHD to be at
rest. The numerical values employed in the calculations are given in
the caption of Fig.2. As a "fitting" parameter is used the number N
of excitons created per pulse. The curve in Fig.2 is calculated for
$N \approx 7 \times 10^{10}$ (for further details on these calculations, see ref. 8).

A more interesting result of the experiments, shown in Fig.2, is
that the EHD cloud still expands slightly after several exciton life

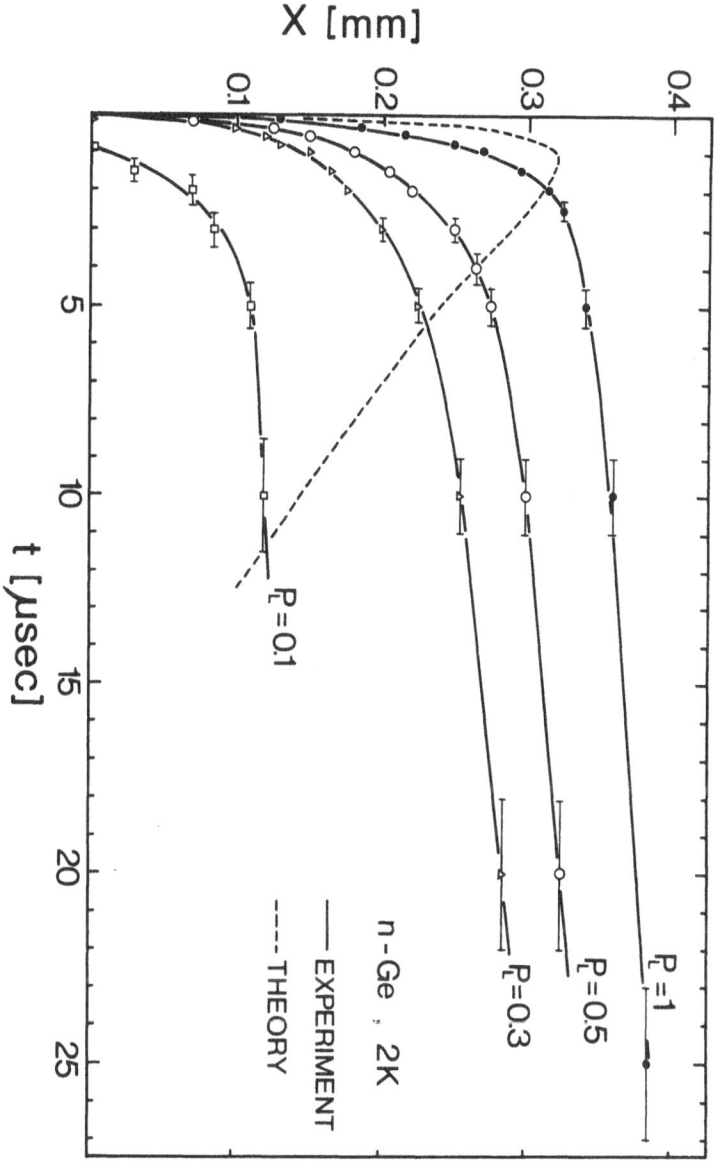

Fig.2. Expansion of EHD cloud as a function of time for different excitation powers P_L. $P_L=1$ corresponds to 0.5µJ/pulse. The dashed curve is a theoretical calculation based on EHD at rest (see ref. 8). The numerical values employed were: $n_c=2 \times 10^7 cm^{-3}$, $n_0=2 \times 10^{13} cm^{-3}$, $<v>=10^6 cm/s$, $D=2000 cm^2/s$, $\tau_{ex}=5µsec$, where n_c is the density of the e-h plasma and n_0 is the saturation value of the FE density at 2K. $<v>$, D, and τ_{ex} are thermal velocity, diffusion coefficient and life time of FE, respectively.

times.   The velocity of expansion is relatively low, about 200 cm/s,
and appears to be constant and independent of initial excitation power.
This expansion, occurring in an environment where the FE density is far
below its saturation value, is a clear evidence of real EHD motion.
The expansion velocity of 200 cm/s sets the lower limit of the EHD
velocities, whereas an upper limit of about 2000 cm/s can be found by
comparing the actual expansion of the EHD cloud with the theory based
on EHD at rest.

3.1. Motion of EHD in a Low FE Density

The observed drop velocities in the range from 200 cm/s to 2000 cm/s
exceed the thermal velocity $<v>_{EHD}$ = 100 cm/s of typical EHD with R =
5μm, eliminating an explanation in terms of drop diffusion.   Moreover,
the experimental curves of Fig.2 indicate a straight-line motion of the
drops during their evaporation in the vacuum, implying a collision
time τ_c of EHD exceeding their total life time.   Hence, we can estimate
τ_c > 25μsec, suggesting that the usual scattering from phonons and impurities is effectively screened by the highly degenerate electron-hole
plasma within the drops.   If a collision-free motion of EHD is accepted
it is obviously meaningless to introduce diffusion coefficient and drift
mobility of the drops in the way it is usually done, and furthermore it
has some interesting consequences in the interpretation of our
experiment.

It has previously been suggested that EHD possess a net charge owing
to a difference in the work-function for electrons and holes.[9,10)]
In the presence of an electric field a net charge on the EHD will cause
an acceleration of the drops.   In a time t (without collisions) an EHD
with a net charge z × e (e being the elementary charge) and mass M will
obtain a velocity of $V_d = \frac{z \times e \times E}{M} \times t$, where E is the bulk electric field.
With E=1V/cm and t = 25μsec we find V_d=380 × z cm/s for a drop containing
10^8 e-h pairs (R=5μm).   A net charge should thus be revealed in our
experiment with a resolution of the order of a single elementary charge.

However, the EHD motion experimentally observed was unaffected by an increase of the electric field up to values causing dissociation of EHD an FE in the bulk. The latter was observed as a decrease in the delayed photocurrent simultaneous with the appearance of an instantaneous photocurrent. We are therefore led to conclude that the EHD in our experiment are essentially uncharged, or at least that a possible net charge is screened effectively within a small distance from the drop surface.

3.2 Influence of FE Gas

It should be pointed out that the experiment and the above considerations refer to a situation with EHD moving in an ambient of negligible FE density ("vacuum"). In regions, where the FE density n is appreciable, the drops no longer move freely, but are influenced by the surrounding FE gas in two major respects.[7]

1. An FE gas, of density n, exerts a friction on the EHD motion implying a velocity relaxation time $\tau_v = 2Rn_c/3n<v>$, where the symbols have the same meaning as in Fig.2. For a drop with R=5μm moving in an FE density near saturation, $n \approx n_o \approx 2 \times 10^{13}$ cm^{-3}, we find $\tau_v \approx 3\mu$sec. In this case a diffusion coefficient can be defined $D_{EHD} \approx \tau_v kT/M \approx 0.01$ cm^2/s. It is, however, negligible small owing to the gross drop mass.

2. If the FE density is not uniform but exhibits an appreciable gradient, which is usually the case in the vicinity of a local surface excitation, the condensing excitons provide the EHD with a net momentum in the direction of decreasing FE density, tending to eject the EHD out of the highly excited region. In steady state this force is balanced by the above friction yielding a drop velocity $\bar{V}_d = -\frac{1}{9} R<v> \cdot (\nabla n/n)$. Under typical conditions $(\nabla n/n = 10$ cm^{-1}, R= 5 μm) the steady state drop velocity is $V_d = 500$ cm/s and is reached within a few microseconds.

In the present experiment steady state is never reached. Still, with the intense pulsed excitation applied, the EHD are formed in a strong FE density gradient and are thus accelerated during their growth.

Furthermore, when this driving force diminishes (after about a micro-second) so does the friction from the FE gas, leaving the EHD moving with constant velocity during their entire evaporation. Hence, this model explains qualitatively the observed EHD motion, and quantitatively it predicts velocities of the right order of magnitude.

ACKNOWLEDGEMENTS

Many inspiring and fruitful discussions with Prof. I. Balslev during the course of this work are gratefully acknowledged.

REFERENCES

1) J. P. Wolfe, R. S. Markiewicz, C. Kittel and C. D. Jeffries: Phys. Rev. Lett. 34 (1975) 275.

2) J. C. Hensel and T. G. Phillips: *Proc. XIIth Int. Conf. on Physics of Semiconductors, Stuttgart 1974,* ed. M. H. Pilkuhn (Teubner, Stuttgart, 1974) p.51.

3) M. Voos, K. L. Shaklee and J. M. Worlock: Phys. Rev. Lett. 33 (1974) 1161.

4) Ya. Pokrovskii and K. I. Svistunova: Fiz. Tverd. Tela 13 (1971) 1485 [Sov. Phys. - Solid State 13 (1971) 1241]

5) C. Benoit à la Guillaume, M. Voos, and F. Salvan: Phys. Rev. Lett. 27 (1971) 1214.

6) R. W. Martin: Phys. Stat. sol (b) 61 (1974) 223.

7) I. Balslev and J. M. Hvam: Phys. Stat. sol. (b) 65 (1974) 531.

8) J. M. Hvam and I. Balslev: Phys. Rev. B11 (1975) 5053.

9) T. M. Rice: Phys. Rev. B9 (1974) 1540.

10) Y. E. Pokrovskii and K. I. Svistunova: in ref. 2. p.71.

LUMINESCENCE AND TRANSPORT PROPERTIES OF ELECTRON-HOLE DROPS IN HIGHLY EXCITED GERMANIUM

Arao Nakamura[*] and Kazuo Morigaki

Institute for Solid State Physics,
University of Tokyo,
Roppongi, Tokyo 106, Japan

ABSTRACT

Luminescence and electrical measurements have been performed in pure and As-doped germanium at 1.6 - 4.2 K under intense optical excitation. Under extremely high excitation of pure germanium, it is found that the electron-hole liquid phase is uniformly generated in the whole volume of the sample. From the observation of the giant fluctuating current, the electron-hole drops turn out to be negatively charged as a whole. Even in heavily doped germanium, it is found that the EHD is formed stably in an atmosphere of metallic electrons released from donor impurities.

I. INTRODUCTION

Since Keldysh[1] predicted the existence of the electron-hole drop (EHD) in germanium, the properties of EHD have extensively been investigated by various kinds of means at low temperatures in germanium.[2] In this paper, we present the results of luminescence and electrical measurements of EHD in pure and As-doped germanium. Here, main emphases are laid on the following three problems; one of them is concerned with the density-dependent luminescence properties of EHD in pure germanium. Secondly, we are concerned with the transport properties of EHD, especially, its net charge in pure germanium. Thirdly, an attention is paid on the existence of EHD in heavily doped germanium whose donor concentration exceeds critical concentration for the metal-nonmetal transition. The second section is mainly devoted to the description of the second problem. As regards the first and third problems, only the

---

[*] Present address: Laboratoire de Physique de la Matière Condensée, Ecole Polytechnique, Plateau de Palaiseau - 91120 Palaiseau, France.

essential results are summarized in section II.

II. EXPERIMENTAL RESULTS AND DISCUSSION

A. Density-dependent Luminescence Properties of EHD in Pure Germanium

In a previous paper[3] we have pointed out that when the average density of electron-hole(e-h) pairs, N_p, over the sample exceeds the equilibrium density of e-h pairs contained within an EHD, the electron-hole liquid(EHL) phase occupies the whole volume of the sample and thus a remarkable change with the variation of the average density of e-h pairs could be observed in the luminescence due to EHD. Also we have shown that this expectation actually occurs in the luminescence line due to EHD under an intense optical excitation at low temperatures. Further-more, from an analysis of the line shape, we have obtained the ground state energy of EHL as a function of N_p. Such a change of the lumi-nescence line with N_p can be seen in the time-resolved luminescence spectra of EHL which are taken after an optical excitation pulse is turned off. Figure 1 shows the time-resolved spectra of the LA(large peak) and TO(small peak) phonon-assisted luminescence from EHL at 1.6 K. The magnitude of N_p at the end of the excitation pulse, which is gen-erated by a Nd-doped yttrium aluminum garnet Q-switched laser, with the pulse width of 11 nsec, is 6.4×10^{17} cm^{-3}. The spectra are taken for the delay time ranging from 1 μsec to 60 μsec. As seen from this figure, the peak position is shifted to the low energy side and the line width is decreased when the delay time is increased. However, when the delay time exceeds 40 μsec, the peak position and the line width do not change with time. The increase in the delay time corresponds to the decrease in the average density of non-equilibrium e-h pairs in the crystal. Therefore, this result can be interpreted as follows; after the laser pulse is turned off, the e-h pair density of EHL, which occupies the whole volume of the sample, decreases, so the line width gradually decreases with time and also the peak position is shifted

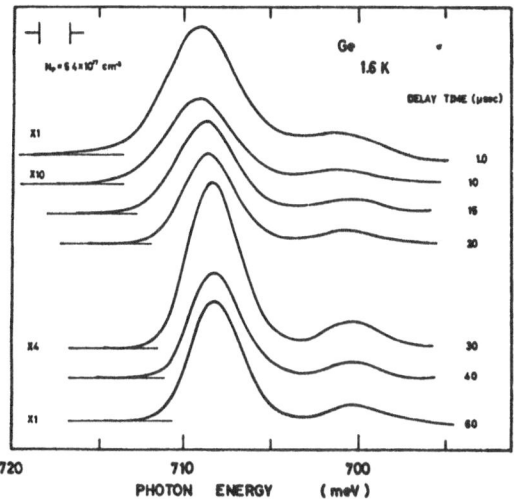

Fig.1. Time-resolved spectra of EHL at 1.6 K in Ge with the
delay time of 1.0 μsec to 60 μsec. The magnitude of
N_p is 6.4×10^{17} cm^{-3} at the instant of the end of
the excitation pulse.

towards the low energy side. After the e-h pair density of EHL reaches
the equilibrium density of EHD, the EHL phase exists as drops in the
crystal, so that the peak position and the line width do not change
with time.

In germanium, in addition to the main recombination mechanism
which involves an LA-phonon, there is also a TA-phonon assisted process
which is forbidden for an e-h pair at the band extrema. It is expected
that the ratio of the integrated intensities of the TA lines and LA
lines as a function of N_p is dependent of the e-h pair density of EHL
phase with power of 2/3.[4] As shown in Fig.2, the ratio of the integrated intensities of the TA lines and LA lines as a function of N_p is
critically increased at N_p = 2.6×10^{17} cm^{-3}. This result also suggests
that the EHL phase occupies the whole volume of the sample when N_p
exceeds its critical density ($\simeq 2.6 \times 10^{17}$ cm^{-3}), being consistent with
the observations mentioned before.

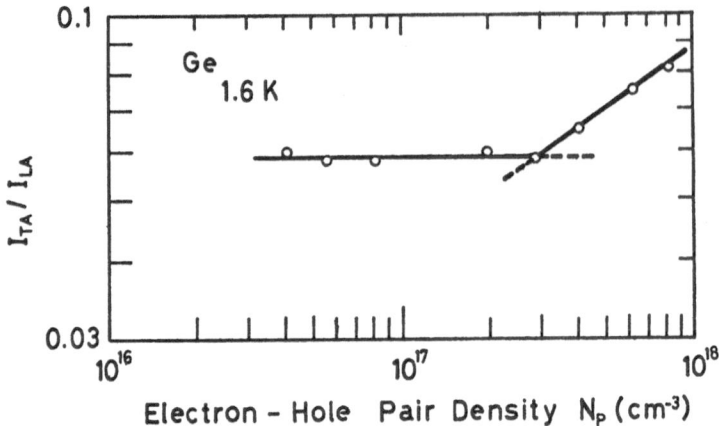

Fig.2.   The ratio of the integrated intensities of the TA lines
and LA lines as a function of N_p.

B.   Transport Properties of EHD

In what follows, we summarize previous results[5] of the transient
photoconductivity measurements in pure germanium.   We have observed an
anomalous profile of the transient decay curve of the photocurrent, as
shown in Fig.3.   The sample is excited by a pulsed dye laser with the

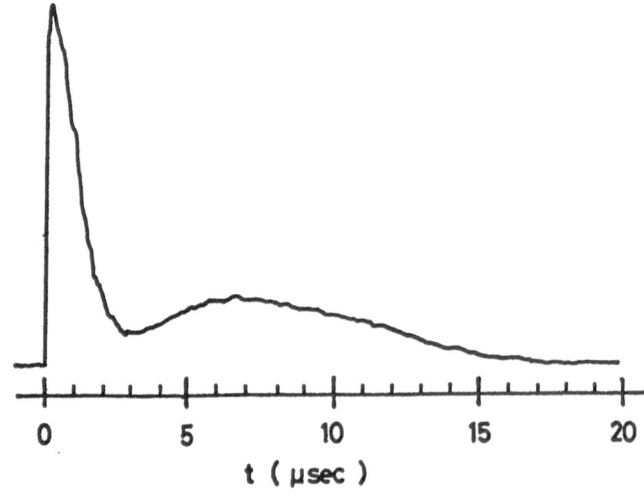

Fig.3.   Decay profile of the photocurrent at $N_p = 2 \times 10^{15}$ cm^{-3} and 4.2 K.
The magnitude of applied electric field is 1.1 V/cm.

pulse width of 5 nsec and the wavelength of 630 nm. The first peak
which appears after being delayed by the response time (~ 0.5 µsec) of
the apparatus is due to free electrons created by the band to band
excitation, while the second peak arises from the contribution of free
electrons ejected from EHD as a result of the Auger recombination of
EHD. The N_p-dependence of the photoconductivity given by the second
peak shows that the Auger recombination rate of EHD rapidly increases
with N_p above $N_p \simeq 10^{15}$ cm^{-3}.

As was pointed out previously, the spike-like giant fluctuating
photocurrent has been observed as superposing in the second peak, when
the temperature is below 2.6 K. One pulse of the fluctuating photocurrent gives the charge of about 4×10^{-10} coulomb. Since this value
is consistent with the results of p-n junction experiments,[6,7] such a
spike-like giant fluctuating photocurrent is interpreted as being due
to dissociation of an EHD containing N~2×10^9 electron-hole pairs. In
order to understand the possible mechanism of such a giant fluctuating
photocurrent observed in the sample with ohmic electrodes, we have performed the following measurement; samples of pure germanium (residual
donor concentration $\leqslant 10^{13}$ cm^{-3}) were used and the beam of the laser
(Nd:YAG Q-switched laser with the width of 11 nsec) was focused with a
spot of diameter of 0.6 mm into a face of the sample with the electrode
whose separation was 4 mm. The delay times for the appearance of
the first giant fluctuating pulse were measured as a function of
the distance between the electrode and the laser spot. This distance was changed by moving the lens along the direction normal to
the laser beam. From this measurement, we have found that the delay
time at which a photocurrent pulse first appears in the decay curve
changes when the laser spot is moved towards the electrode. Figure 4
shows the delay time of the first pulse as a function of the relative
distance between the electrode and the laser spot at both polarities of
applied voltages. In this experimental condition, the EHD's are

generated in the vicinity of the laser spot, so that the delay time
corresponds to the drift time of an EHD towards the electrode.   As
shown in Fig.4, it is decreased when the laser spot is shifted towards

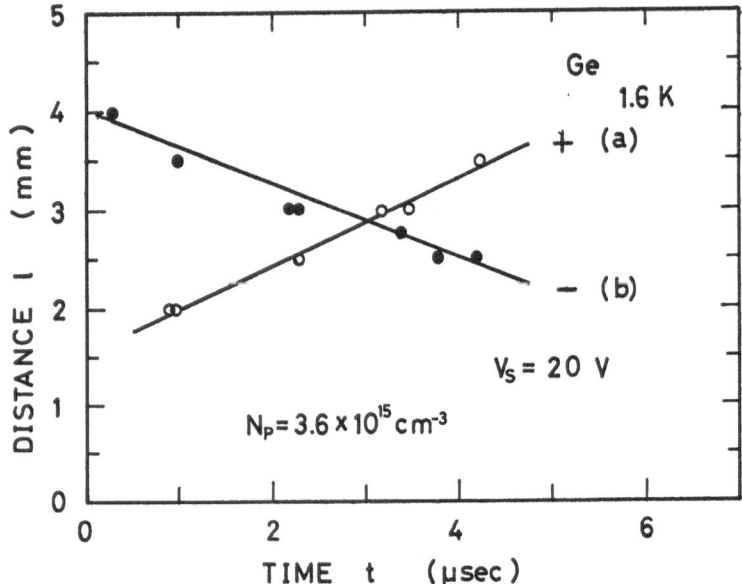

Fig.4.   The time of the first giant fluctuating pulse as a function
of the relative distance between the electrode and the small
spot of the laser light. (a): the cathode on the upside of
the ordinate; (b): the anode on the upside of the ordinate.
The applied voltage is 20 V.

the anode.   This result allows us to conclude that an EHD is negatively

charged as a whole, consistently with the experimental result of

Pokrovskii and Svistunova[8] and also with the theoretical result of

Rice.[9]

    In what follows, let us consider an equation of motion of an EHD

regarding it as a classical particle. We assume that the diffusion of

EHD which arises from its density gradient is negligible compared with

the drift motion by applied electric fields.   Assuming that the electric

field is uniform over the sample, we can write the equation of motion of

EHD as follows:

$$M\left(\dot{v} + \frac{v}{\tau_p}\right) = Ze\,\frac{V_s}{d} \quad , \tag{1}$$

where M, v, τ_p, and Ze are the mass, the drift velocity, the collision time, and the net charge of EHD, respectively, and V_s and d are the applied voltage and the distance between electrodes, respectively. If the drift time of EHD between the laser spot and the anode is long enough compared with the collision time, the motion of EHD becomes a constant velocity motion. Thus, the following equation of motion of EHD is obtained from eq.(1),

$$v = \frac{Ze\tau_p}{M} \cdot \frac{V_s}{d} . \qquad (2)$$

The slope of the straight line of the relative distance versus the delay time of the first pulse in Fig.4 gives the drift velocity of EHD. Thus, the drift velocity obtained from such a procedure should depend linearly on the applied voltage if eq.(2) holds. We have observed that this is the case in our experimental condition (V_s is between 18 V and 26 V). From eq.(2), we can derive the value of the net charge of EHD. The mass of EHD is estimated by using the magnitudes of the effective masses of electrons and holes and the number of electron-hole pairs contained in an EHD, which is known from the charge giving one photo-current pulse. Since the experimental value of τ_p is not available at present, we take τ_p as 5×10^{-7} sec, which is by Bagaev et al.[10] Then, inserting these values of the parameters and also the experimental values of V_s, d and v into eq.(2), we can derive the magnitude of the net charge as a function of N_p as follows; Ze = -18000 e at N_p = 4.3 $\times 10^{14}$ cm^{-3} and Ze = -2300 e at N_p = 3.6×10^{15} cm^{-3}, and the net charge is linearly dependent of N_p in the range of N_p = 4.3×10^{14} cm^{-3} to 3.6 $\times 10^{15}$ cm^{-3}.

These values are large compared with the previous experimental result[8] and theoretical one.[9] However, we consider that this dis-crepancy is due to the large radius of EHD and the high electric fields applied to the sample in this experiment; in the range of applied voltages of this experiment, excitons and residual impurities in the crystal are fully ionized (in this sample, the threshold field for the impact ionization of impurity and exciton is about 10 V/cm and 3 V/cm, respectively), so that the EHD exists in a gas of free carriers rather than in an exciton gas. Since the ionized carriers outside the EHD form a screening cloud around a charged EHD, it is expected that the EHD has a large value of net charge, depending on the density of ionized carriers. If we follow the formula of Rice[9] which contains the temperature, the radius of EHD and the density of free carriers as the parameters, we can estimate a net charge of EHD in case of the presence of free carriers outside the EHD. For $N_p = 4.3 \times 10^{14}$ cm^{-3}, we can obtain an almost consistent value of net charge with the experimental one, using the density of free carriers of 10^{13} cm^{-3}, and the radius of EHD of 13 µm.

C. EHD in Heavily Doped Germanium

As was reported previously,[11] the luminescence measurement allows us to conclude that EHD stably exists even in metallic samples of As-doped germanium whose donor concentrations exceed the critical concentration for the metal-nonmetal transition, that is, $N_D = 4.7 \times 10^{17}$ cm^{-3}. This conclusion is consistent with recent theoretical results.[12,13]

In heavily As-doped germanium, the peak position of the EHD luminescence is shifted towards the low energy side with the donor concentration in comparison with the case of pure germanium. This shift is interpreted as being due to the interaction of electrons with donor impurities. The line width of the EHD luminescence depending on the donor concentration which has been observed in the samples of the metallic concentration region ($N_D \gtrsim 4.7 \times 10^{17}$ cm^{-3}) is mainly determined by the sum of the Fermi energies of electrons and holes. An additional broadening observed in the intermediate concentration region for the impurity conduction (1×10^{17} cm$^{-3} \lesssim N_D \lesssim 4.7 \times 10^{17}$ cm^{-3}) could be accounted for by the fluctuation of impurity potentials.   The luminescence measurement of EHD as a function of the donor concentration will be useful to study the nature of the metal-nonmetal transition in doped germanium.

REFERENCES

1) L. V. Keldysh: *Proc. Int. Conf. Physics of Semiconductors, Moscow* (1968) ed. S. M. Ryvkin (Nauka, Leningrad, 1968) p.1303.

2) Ya. E. Pokrovskii: Phys. Status solidi (a) $\underline{11}$ (1972) 385, and numerous references contained therein.

3) A. Nakamura and K. Morigaki: Solid State Comm. $\underline{14}$ (1974) 1212.

4) C. Benoît à la Guillaume, M. Voos and F. Salvan: Phys. Rev. $\underline{B5}$ (1972) 3079.

5) A. Nakamura and K. Morigaki: Solid State Comm. $\underline{14}$ (1974) 41.

6) V. M. Asnin, A. A. Rogachev and N. I. Sablina: ZhETF Pis. Red. $\underline{11}$ (1970) 162. [Soviet Phys. JETP Lett. $\underline{11}$ (1970) 99.]

7) C. Benoît à la Guillaume, M. Voos, F. Salvan, J. M. Laurant and A. Bonnot: C. R. Acad. Sc. Paris $\underline{t272}$ (1971) 236.

8) Ya. E. Pokrovskii and K. I. Svistunova: ZhETF Pis. Red. $\underline{19}$ (1974) 92. [Soviet Phys. JETP Lett. $\underline{19}$ (1974) 56.]

9) T. M. Rice: Phys. Rev. $\underline{B9}$ (1974) 1540.

10) V. S. Bagaev, T. I. Galkina, O. V. Gogolin and L. V. Keldysh: ZhETF Pis. Red. $\underline{10}$ (1969) 309. [Soviet Phys. JETP Lett. $\underline{10}$ (1969) 195.]

11) A. Nakamura and K. Morigaki: *Proc. Int. Conf. Physics of Semiconductors, Stuttgart* (1974) ed. M. H. Pilkuhn, (Teubner, Stuttgart, 1974) p.86.

12) G. Mahler and J. L. Birman: Phys. Rev. $\underline{B12}$ (1975) 3221.

13) B. Bergersen, P. Jena and A. J. Berlinsky: J. Phys. C, Solid State Phys., $\underline{8}$ (1975) 1377.

CYCLOTRON RESONANCE STUDY OF DIFFUSION PROBLEMS
IN HIGHLY EXCITED GERMANIUM

T. Sanada, T. Ohyama and E. Otsuka

Department of Physics,
College of General Education,
Osaka University,
Toyonaka, Osaka 560, Japan

ABSTRACT

Diffusion coefficients of excitons and electron-hole drops in pure germanium are measured. Use is made of 35 GHz time-resolved cyclotron resonance method. The diffusion coefficient of excitons at 4.2 K turns out to be ~ 1000 cm^2/s. For electron-hole drops it is lower than ~ 500 cm^2/s at 1.6 K.

One of the attractive topics in high excitation physics in solid will be the diffusion problem of electron-hole drops. A new method for obtaining the diffusion coefficient, not only of the electron-hole drops but of excitons in germanium, is introduced here. It takes advantage of the technique of the time-resolved cyclotron resonance at 35 GHz.

The essential part of the apparatus is given in Fig. 1. A rectangular germanium crystal, having the dimensions of $1.1 \times 1.2 \times 9.5$ mm^3, is placed through holes drilled on the broader faces of the waveguide. Excitation is made either with a xenon flash-tube in combination with a water-liquid filter, or with a Q-switched Nd: YAG laser (1.06 μm). In order to cut off undesirable excitation lights, the crystal is covered with silver paint and with tin-foil, the only exposed part being the top face. With our excitation lights, the carrier generation region is restricted within 10^{-3} cm from the surface. This thickness is much smaller than the distance between the illuminated surface and the first observation edge —— denoted x_1 in Fig. 1 —— of the inside of the waveguide, which is about 0.3 cm. A very steep gradient of the density of the photo-produced particles is expected to exist within the crystal

XENON + WATER FILTER
OR
Q-SWITCHED YAG LASER

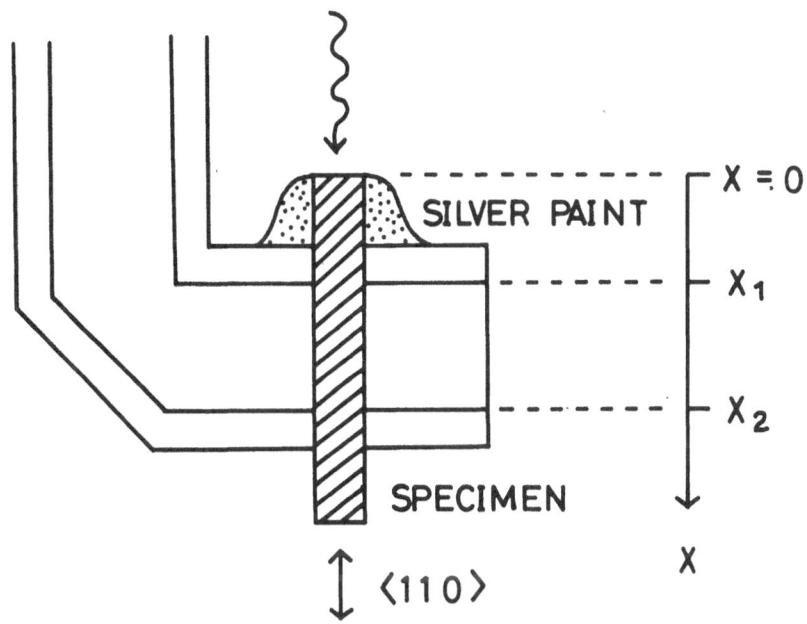

Fig.1. Scheme of the essential part of the experimental set-up.

at each excitation pulse.

With our sample shape, the problem can nearly be treated in one dimension. The diffusion equation is written as

$$\frac{\partial n(x,t)}{\partial t} = D \frac{\partial^2}{\partial x^2} n(x,t) - \frac{1}{\tau_r} n(x,t) \ ; \tag{1}$$

where $n(x,t)$, D and τ_r are, density, diffusion coefficient and lifetime of particles, respectively. The solution of (1) is

$$n(x,t) = \frac{N_0}{\sqrt{4\pi Dt}} \exp\left(-\frac{x^2}{4Dt} - \frac{t}{\tau_r}\right) \ ; \tag{2}$$

where N_0 is the number of particles produced at $x = 0$ when $t = 0$.

Here the illuminated face is taken at x = 0. The total number of
particles lying between x_1 and x_2 is given by

$$N(t) = \int_{x_1}^{x_2} n(x,t)dx. \tag{3}$$

Figure 2a shows the linewidth of electron cyclotron resonance
at 4.2 K transformed to the inverse relaxation time as a function of

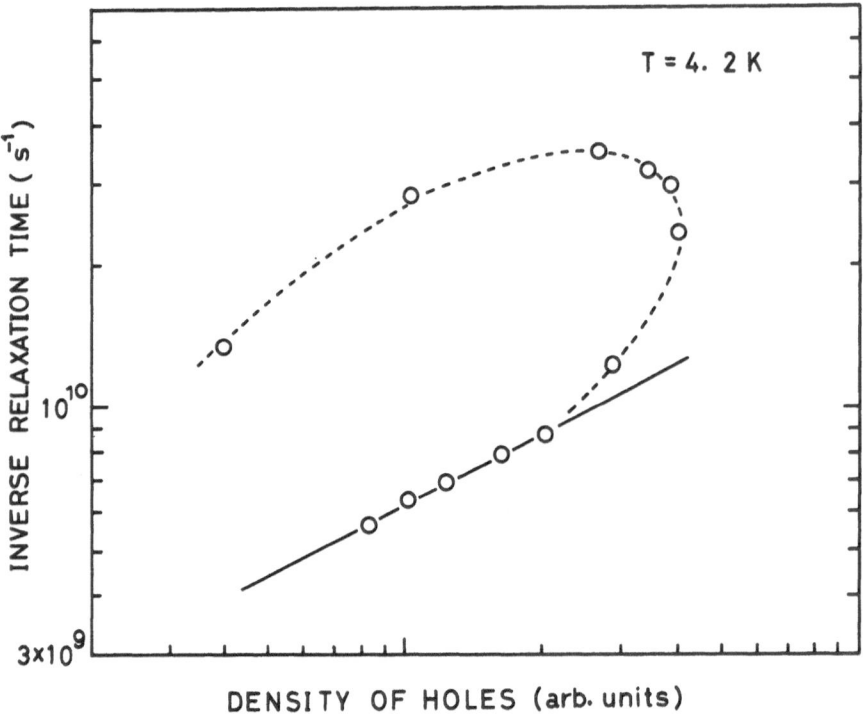

Fig.2a. Carrier density dependence of the inverse relaxation time at 4.2 K.
The gradient of the solid line is 0.5, indicating the electron-hole
interaction. [1)]

the relative density of holes. A xenon flash-tube was used for excitation here. The contribution of the electron-phonon interaction to the
linewidth is subtracted beforehand. Circles in this figure correspond
to different delay-times, though not shown explicitly. For larger
delay-times, the linewidth shows proportionality to the square-root of
the density of holes. This proportionality shows the electron-hole

interaction[1] to be dominant. For smaller delay-times there exists
another interaction contributing to the linewidth. It is the electron-
exciton interaction, which we treated before.[2] We assume the above-
mentioned proportionality for the electron-hole interaction also for
smaller delay-times and subtract this contribution from the linewidth.
The remainder of the linewidth then yields the density of excitons.
Figure 2b shows the density of excitons against the delay-time. The

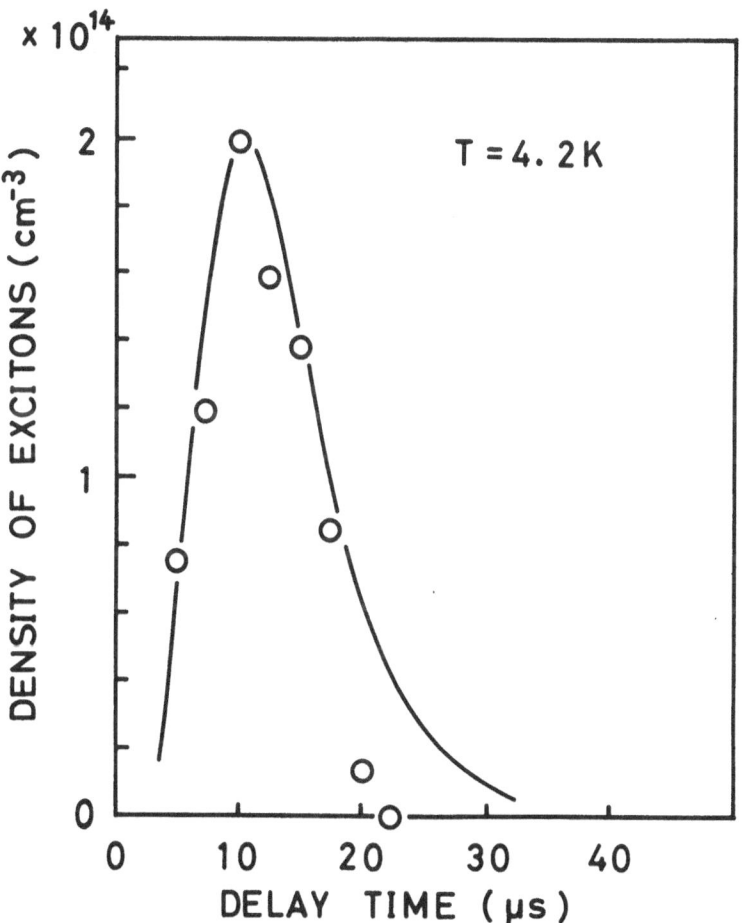

Fig.2b. Diffusion profile of the density of excitons at 4.2 K.
The solid line gives the calculation with $D = 1000$ cm^2/s
and $\tau_r = 4$ μs.

peak value corresponds to a density of 2×10^{14} cm^{-3} when estimated from
the theory by Matsuda *et al.*.[3]   Fitting the peak position of the

calculated curve to that of the experimental observation, we find the diffusion coefficient of excitons to be ~ 1000 cm^2/s, using $\tau_r = 4$ μs for excitons.[*]

At 1.6 K, the diffusion profile of electrons consists of two parts as shown in Fig. 3a. What is occurring inside the crystal may be as

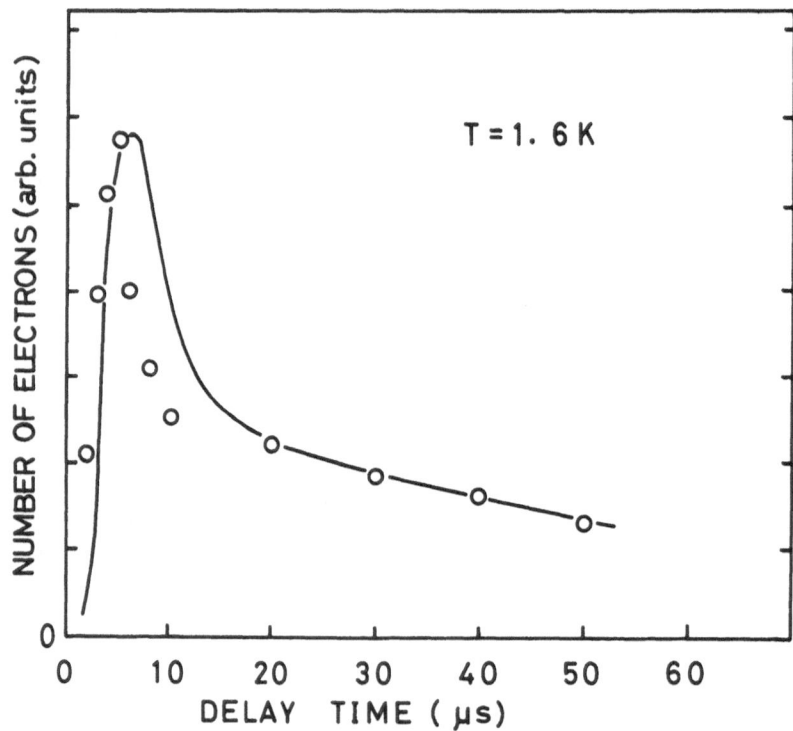

Fig.3a. Diffusion profile of electrons at 1.6 K illuminated by a xenon flash-tube.

follows. Particles created at the surface, and not associated in the drop-formation, may run inward first and will show a rapid rise-and-fall immediately after the photo-excitation. For mild excitation, drops are born only at the surface, where the density of particles is high enough to produce drops. The drops start moving and throw free particles

---

[*] The observed value of electron lifetime at this temperature is 8 μs. It is easily shown that the lifetime of excitons is just a half of that of electrons under some appropriate conditions.[4]

around themselves. The evaporated particles run after the primarily
created ones and give rise to the slowly decaying response. If one
takes this picture into account, the number of particles observed should
be expressed as

$$N(t) = \exp{(-t/\tau_0)} \int_{x_1}^{x_2} dx \int_0^t dt' \frac{N_0}{\sqrt{4\pi D t'}}$$

$$\times \exp{\left\{ \frac{-(x-\sqrt{D_0(t-t')})^2}{4Dt'} - (\frac{1}{\tau_r} - \frac{1}{\tau_0})t' \right\}} \tag{4}$$

in place of eq.(3); where τ_0 is the decay-time of supply from drops and
D_0 is the diffusion coefficient of drops. We put $\tau_r=2$ μs, the value
observed at this temperature. The quantity D for electrons is estimated
to be ∼ 1800 cm^2/s from the value obtained at 4.2 K. By parameter fitting the diffusion coefficient of drops is found to be less than 500
cm^2/s.

A stronger excitation is achieved with the help of a Q-switched Nd:
YAG laser. When the power of microwaves is increased, sharp spikes are
observed in the photoresponse signal (Fig. 3b). This feature no doubt
indicates the effective bombardment on the drops by microwave-acceler-
ated carriers. By tracing the drop-indicating photoresponse signal, or
the location of the spikes, with the help of eq.(3), the diffusion coefficient of drops is found to be considerably larger than 500 cm^2/s, the
upper-limiting value for the relatively low excitation case.

It seems somewhat strange that D_0 depends on the degree of excitation and that drops can have a large diffusion coefficient despite their
huge masses. The high value of D_0 may be accounted for by the following
reason: when excitation is very strong, the high particle density
region and hence the formation of a drop will not be restricted close
to the illuminated surface. In other words, the jammed-up particles
produce drops while travelling inward. The observed diffusion coefficient of drops will then be hard to distinguish from that of free

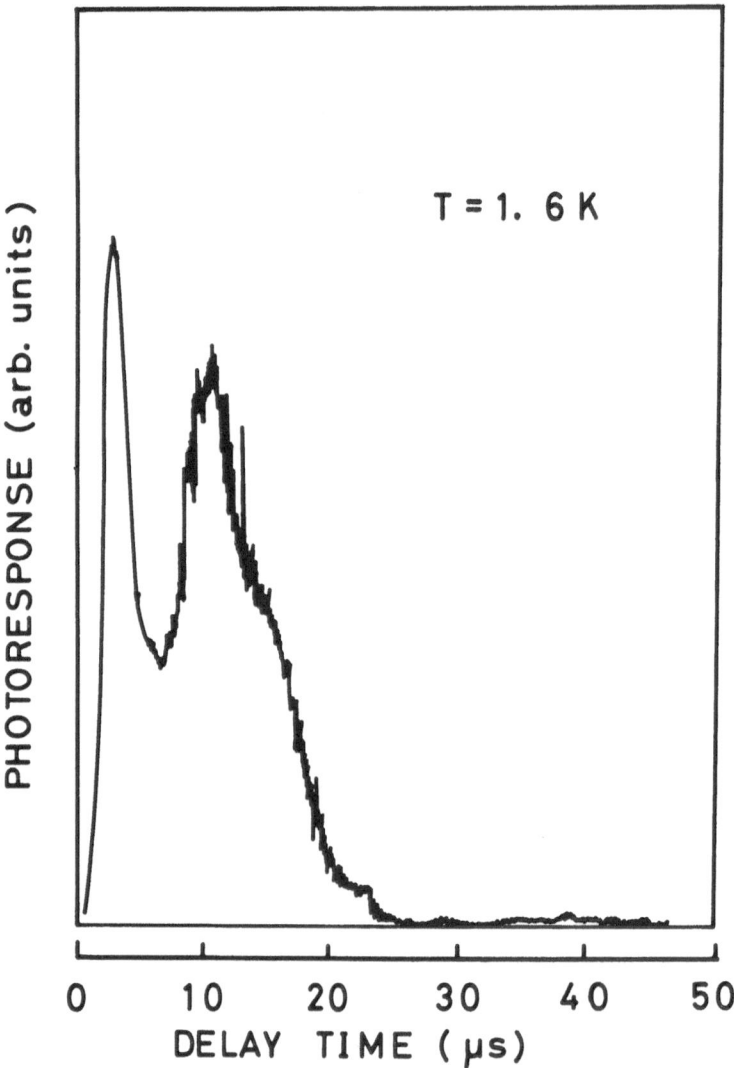

Fig.3b. Photoresponse signal with much more intense excitation
lights by a Q-switched Nd-YAG laser and with a stronger
power of microwaves.

particles.

It is true that too much emphasis cannot be laid on the absolute
value of the diffusion coefficient for the present. But the analysis
introduced here, though full of simplification, will offer some help
for solving the diffusion problem in highly excited materials.

REFERENCES

1) H. Kawamura, H. Saji, M. Fukai and I. Imai: J. Phys. Soc. Japan 19 (1964) 288.

2) T. Ohyama, T. Sanada, T. Yoshihara, K. Murase and E. Otsuka: Phys. Rev. Letters 27 (1971) 33.

3) K. Matsuda, M. Hirooka and S. Sunakawa: Prog. theor. Phys. 54 (1975) 79.

4) E. L. Nolle: Sov. Phys. Semiconductors 2 (1969) 1397.

NUCLEATION PHENOMENA IN ELECTRON-HOLE
DROP CONDENSATION IN ULTRA-PURE Ge[†]

R. M. Westervelt, J. L. Staehli,[*] E. E. Haller, and C. D. Jeffries

University of California,
Berkeley,
California 94720 USA

ABSTRACT

By careful measurements and appropriate theory, we are able to
observe and explain quantitatively for the first time major aspects of
electron-hole drop nucleation phenomena in ultra-pure Ge. The free
exciton-drop system above 1.3 K is shown to be always in a metastable
state, i.e. dependent upon the history of optical excitation. We quantitatively explain the observed luminescence hysteresis and measure the
drop surface tension, $\sigma = 2.6 \times 10^{-4}$ erg cm^2 at 2 K. The metastability
lifetime is experimentally found to be $\sim 8 \times 10^6$ sec. The gas-liquid
up-going and down-going threshold curves are measured and explained
using an exciton condensation energy $\phi \cong 2$ meV. The theory also predicts the drop radius and drop concentration as a function of temperature and excitation history.

I. NUCLEATION PHENOMENA: EXPERIMENTAL

Optical hysteresis in the luminescence of electron-hole drops
(EHD)[1] in Ge was first reported by Lo *et al.*[2] The relative intensities I_{709} of the drop luminescence and I_{714} of the free exciton (FE) luminescence were found to depend on the history of optical excitation.
This was interpreted as a gas-liquid supersaturation phenomenon providing direct evidence for the existence of the surface energy of EHD. In
this paper we present fuller experimental and theoretical details of

---

[†] Supported in part by the U.S. Energy Research and Development Administration

[*] Fellow of the Schweizerischer Nationalfonds

hysteresis and other nucleation phenomena in ultra-pure Ge.[3,4] The experiments are done as follows: Single crystals of polished and etched ultra-pure Ge of size $3 \times 4 \times 9$ mm^3 are mounted without stress and without metallic contacts over a 2×8 mm^2 hole in a copper plate, and immersed in liquid He in an optical dewar. One surface is uniformly illuminated with the light from a highly stabilized electronically controlled incandescent lamp, using a beam splitter to accurately monitor the absolute power flux P absorbed by the Ge surface. Both surface excitation ($0.5 < \lambda < 1.4$ µm, $\lambda_{ave} = 0.9$ µm) and volume excitation ($\lambda \simeq 1.52$ µm) are used. The luminescence is collected from the opposite crystal face, analysed by a spectrometer and recorded by a sensitive Ge detector using lock-in techniques. We are able to observe direct thresholds down to 1.3 K where P is only a few microwatts/mm^2. A variety of crystal samples were used, both n and p type, with net impurity density in the range 3×10^9 to 4×10^{12} cm^{-3}, some dislocation-free with vacancy clusters, and some with dislocations and no vacancies. The crystals were grown in the $\langle 100 \rangle$ or $\langle 111 \rangle$ direction in H$_2$ gas.

Figure 1 shows the observed luminescence I_{709} and I_{714} as functions of excitation power flux P for both "up-going" and "down-going" excitation. The pronounced hysteresis near threshold is always observed and is accurately reproducible when the data are taken in the following pattern: First, P $\rightarrow$ 0 for 5 sec, then the power is switched smoothly and monotonically to a low value for 50 sec at which I_{709} and I_{714} are recorded. Then P $\rightarrow$ 0 for 5 sec, and is then switched to a slightly higher value, $etc.$, thereby obtaining the points on the up-going excitation curve. At the up-going threshold, $P_+ \simeq 4.4$ mW cm^{-2}, I_{709} rises sharply while I_{714} breaks downward as expected, because the e-h pairs produced by increased excitation go largely into the liquid phase once drop nucleation has begun. In Fig. 1 the down-going excitation data are taken in a similar pattern: P $\rightarrow$ 0 for 5 sec, then P

→ 100 mW cm^{-2}, *i.e.* far above threshold, for 5 sec, then P → 3.4 mW cm^{-2} for 50 sec, at which I$_{709}$ and I$_{714}$ are recorded; then P → 0, P → 100 mW cm^{-2}, P → 3.2 mW cm^{-2}, *etc.* A very sharp down-going threshold at P$_-$ ≃ 1.4 mW cm^{-2} is quite evident, as well as a corresponding break in I$_{714}$. The fact that the up-going threshold P$_+$ is considerably larger than the down-going one P$_-$ has a simple explanation: It is necessary to supersaturate the exciton gas density to initiate nuclea-tion because the sur-

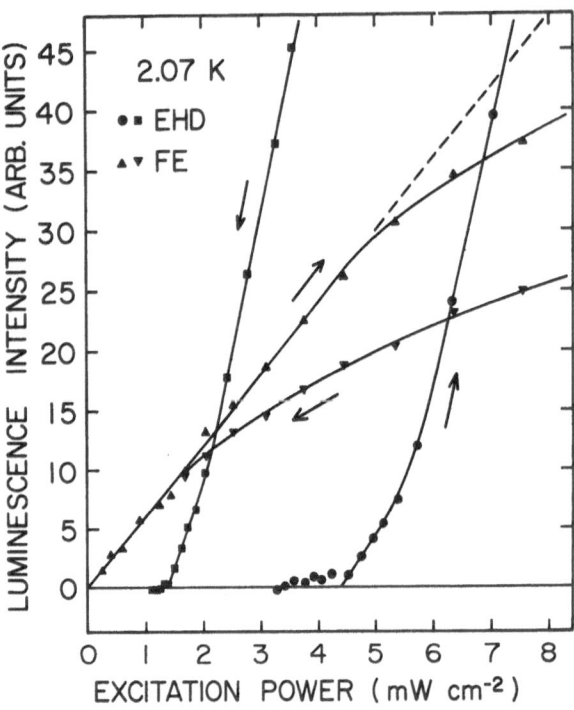

Fig.1. Luminescence intensity of free excitons (FE) in ultra-pure Ge at 714 meV and of electron-hole drops (EHD) at 709 meV, as a function of calibrated excitation power absorbed in the crystal at λ_{ave} ≃ 0.9 μm. The crystal has a net impurity concentration N_A ≃ 5×10^{10} cm^{-3}, and dislocation density ~10^2 cm^{-2}. The data points are taken by increasing (up arrows) and by decreasing (down arrows) the excitation, and display a large hysteresis owing to the droplet surface tension.

face tension of embryonic droplets tends to reduce them to zero size. A measurement of P$_+$ and P$_-$ is the most direct way to experimentally determine the surface tension, as discussed below. The hysteresis is an intrinsic and invariant feature of drop nucleation in ultra-pure Ge, observed in all samples between 1.3 < T < 4.2 K, for both surface and volume excitation. The persistence of the hysteresis shows that the free exciton-drop system is in a long-lived metastable state, not in thermodynamic equilibrium. Failure to observe hysteresis may be due to insufficient detector sensitivity, excess impurities, or a number of mechanisms which destroy drops, such as unstable excitation, sample strains, and electrical contacts. If the lifetime of drops against destruction is very short, the observed state of the system is determined by these external mechanisms.

How long is the metastability lifetime in pure crystals? We have measured this, Fig. 2, by switching from P = 0 to a power P = 25 mW cm^{-2} on an up-going excitation curve, then recording $I_{709}(t)$ while P is accurately held constant; a very slow increase in drop luminescence is observed. Similarly, in the down-going case we reduced the excitation from P = 140 mW cm^{-2} to P = 25 mW cm^{-2}, the same level, then recorded $I_{709}(t)$ while holding P constant; a slow

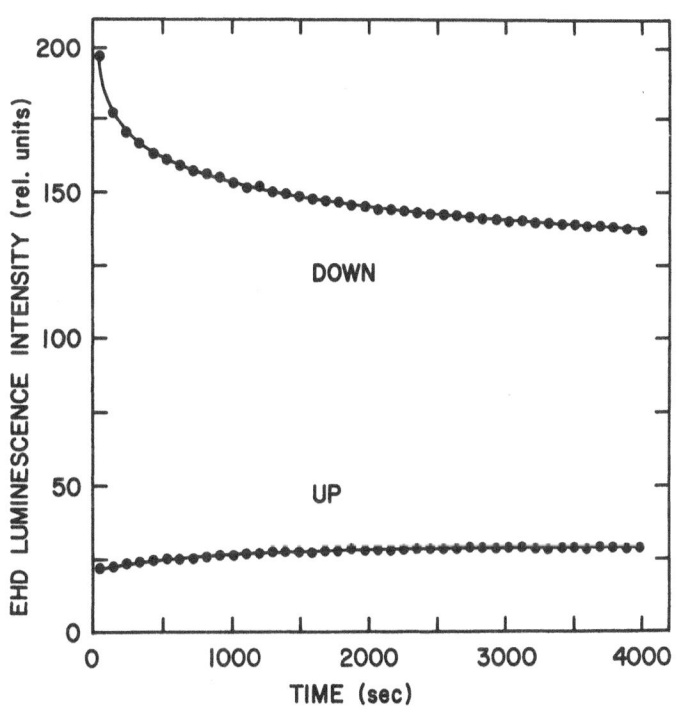

Fig.2. Slow time behavior of drop luminescence following an up-going and a down-going step in excitation (see text) at T = 2.01 K in ultra-pure Ge, net $N_D \simeq 3 \times 10^9$ cm^{-3}, dislocation density ~10^4 cm^{-2}. These data clearly show that the exciton-drop system is in a long-lived metastable state, dependent on the excitation history.

decrease is observed.   As detailed below, the luminescence varies logarithmically with time, in excellent agreement with our theory— this allows us to predict that, barring external destruction of drops, the two curves will cross, *i.e.* the hysteresis would be naturally quenched, after a time of $\gtrsim 8 \times 10^6$ sec.   Thus the exciton gas and drops, even though excited states of the Ge crystal, can be conveniently conceptualized as stable components of a quasi-thermodynamic system —as long as the excitation is uniform and stably maintained.

The magnitude of the hysteresis, *i.e.* the ratio P_+/P_-, is observed to depend on the temperature, Fig. 3, where it is evident that the hysteresis vanishes at temperatures below 1.3 K, a fact quantitatively explained by the theory below. As shown in Fig. 3, we observe sharp thresholds at all temperatures down to the lower limit of our apparatus, $T \simeq 1.25$ K.   At temperatures above the λ point of liquid He, the hysteresis is found to be smaller and more difficult to measure, but non-vanishing.   By measuring the up-going and down-going thresholds at various temperatures, we obtain the data of Fig. 4 showing that the surface energy and finite drop size give two

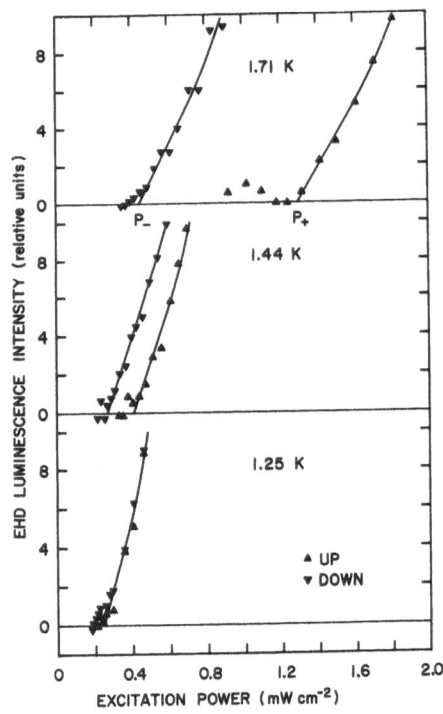

Fig.3. Drop luminescence for up-going and down-going excitation showing that below $T \simeq 1.3$ K the hysteresis vanishes.   The sample is the same as Fig. 2. The ratio of the up-going threshold power P_+ to the down-going threshold power P_- experimentally determines the surface tension.

branches of the
gas-liquid co-
existence curve.
Below 2 K both
branches roll
upward, away
from the ex-
ponential de-
pendence on 1/T
usually assumed.
Previous thresh-
old experiments
were performed
in such a way
that only the up-
going curve was
measured.

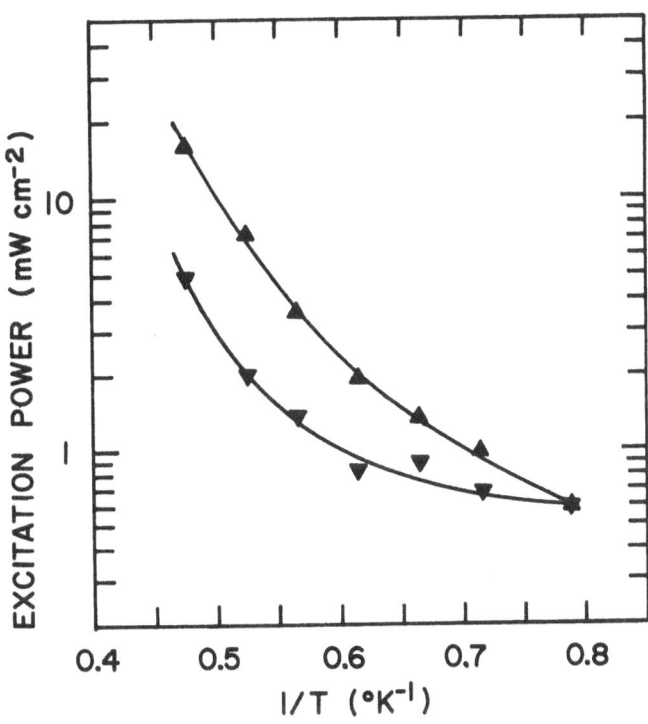

Fig.4. Observed up-going ($\triangle$) and down-going ($\triangledown$) thresholds versus reciprocal temperature for ultra-
pure Ge, net $N_D \simeq 6 \times 10^{10}$ cm^{-3}, dislocation
density ~10^4 cm^{-2}. The power is not absolutely
calibrated (in contrast to Figs. 1 and 3).

Figures 1 to
4 can be considered
to represent in-
trinsic nucleation

phenomena of EHD, to be explained through an adequate theoretical model

possessing these features: 1) quantitative prediction of hysteresis

(Figs. 1 and 3) providing a procedure for data reduction to obtain the

drop surface tension; 2) quantitative explanation of coexistence curves

(Fig. 4) to be used to extract the exciton condensation energy ϕ; 3)

quantitative explanation of metastable time behavior (Fig. 2); 4) pre-

diction of drop radius near both up-going and down-going thresholds, as

a function of temperature; 5) prediction of drop radius and concentra-

tion, and free-exciton density in time evolution for arbitrary excita-

tion P(t); 6) applicability to drop formation both on and free of nucleation centers.   Such a model has been developed.[3,4]

II.   OUTLINE OF NUCLEATION THEORY

The point of departure is the classical theory of vapor condensation in a spatially uniform system.[5]   Fluctuations in the exciton
gas density, or impurities, will nucleate embryonic droplets; those
smaller than a critical radius R_c are unstable because surface tension
quickly reduces them to zero size.   Embryos with $R > R_c$ quickly grow
into observable macroscopic drops.   Lord Kelvin showed that R_c =
$(2\sigma/n_o kTx)$, where σ is the surface tension, n_o is the particle density
in the drop and $x = \ln (n/n^s)$ is the degree of gas supersaturation.
For the exciton gas the saturated density can be found approximately
by setting the chemical potential of an ideal gas equal to the condensation energy:[6]

$$n_s = \gamma \left[\frac{m^*kT}{2\pi\hbar^2} \right]^{3/2} e^{-\phi/kT} \tag{1}$$

The rate of macroscopic drop formation J_+ (drops cm^{-3} sec^{-1}), is determined by the density of drops of critical size, which can be estimated
from the entropy decrease ΔS in forming a critical drop from free
particles, $i.e.$ $J_+ \propto \exp(\Delta S/k) = \exp(-16\pi\sigma^3/3k^3T^3n_o^2x^2)$.   Because J_+ is
extremely sensitive to the degree of supersaturation, this theoretical
model yields sharp thresholds.

However, because the classical theory does not include the effects
of the finite pair lifetime τ_0 within drops, it does not yield a finite
drop size, and cannot be used to calculate the rate of breakup J_- (drops
cm^{-3} sec^{-1}) of macroscopic drops.   It therefore cannot explain central
features of hysteresis and metastability and provides no adequate procedure for measuring the surface energy.   Similarly, an equilibrium
model[7] including τ_0, but neglecting the time required to reach the
steady state, does not predict hysteresis and is inapplicable above

1.3 K. It is in fact necessary to calculate the separate non-equilibrium rates of drop formation J_+ and breakup J_- including the lifetime τ_0. We have done this by extending the classical nucleation theory of Becker and Döring:[8] to their rate equations we add the volume decay rate of drops.

An important parameter is ψ_ν, the difference in free energy between ν e-h pairs condensed in a drop and the same number of free excitons:

$$\psi_\nu = 4\pi R_o^2 \nu^{2/3}\sigma - xkT\nu + f_\nu{}^\tau \tag{2}$$

The first term is the surface free energy. The second is the chemical potential difference, and the third is a new effective free energy due to the finite pair lifetime. This function is plotted in Fig. 5 *versus* the drop radius $R = R_o\nu^{1/3}$, for various values of the supersaturation ratio $S = (n/n^S)$. For ratios larger than a minimum value S_M, below which no drop size is stable, there is a minimum in ψ_ν at the stable radius R_s which can be populated by macroscopic drops, and a maximum ψ_c at the critical radius R_c. For $R > R_s$, ψ_ν turns up sharply due to the finite lifetime term, limiting the drop size to a finite value with a

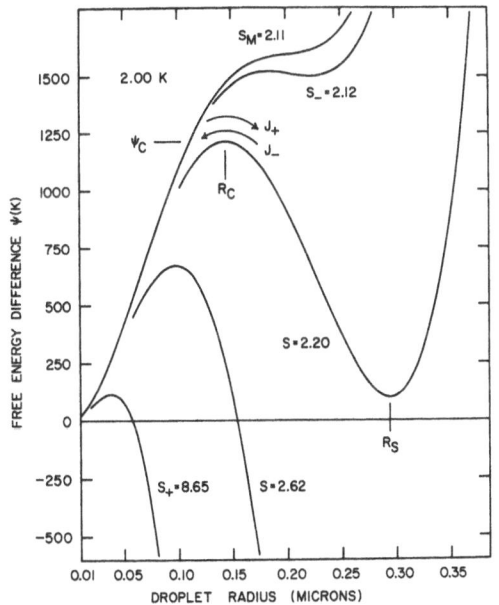

Fig. 5. Effective free energy difference ψ_ν, eq. (2), for EHD in Ge versus droplet radius. The curves are calculated assuming $T = 2.0$ K, $\phi = 2.03$ meV and $\sigma = 2.6 \times 10^{-4}$ erg cm^{-2}. [From the nucleation theory of Westervelt[4]]

sharply peaked distribution. We see that the maximum ψ_c due to the surface energy, acts as a "barrier" to the drop formation rate J_+; an excitation P_+ that can produce a supersaturation ratio S_+ = 8.65 at 2 K is necessary to reduce the barrier sufficiently to allow J_+ to become significant, *i.e.* to "turn on." Similarly the "barrier" $(\psi_c - \psi_s)$ limits the rate of breakup J_- at 2 K until the supersaturation ratio is reduced to S_- = 2.12. A plot of J_+ and J_- *versus* ln S, Fig. 6, shows the sharp dependence on the degree of supersaturation. The minimum in ψ_v for S - 8.65 lies offscale at $R_s \simeq 2.5$ μm, the stable drop size produced at up·· going threshold. If the excitation is increased, a larger number of identical drops are formed. If the excitation is now decreased, the

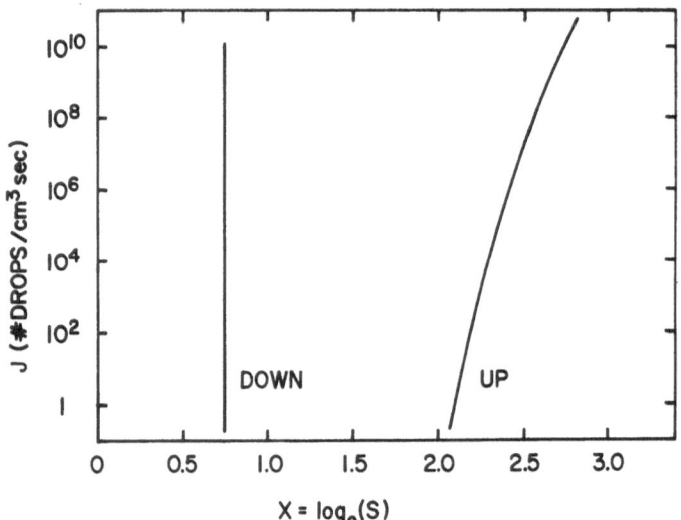

Fig.6. Up-going (J_+) and down-going (J_-) homogeneous nucleation rates for EHD in Ge, eq. 3, versus the degree of exciton gas supersaturation x = $\log_e(n/n^s)$. Calculated from nucleation theory [Westervelt[4]], using the parameters of Fig. 5.

nucleated drops continue to be maintained by condensation of excitons on their surface, but the stable drop size R_s, Fig. 5, decreases steadily until the supersaturation ratio S_- = 2.12 is reached, where the breakup current J_- "turns on", Fig. 6, and the down-going threshold is reached, at $R \simeq 0.22$ μm. The nucleation currents are found to be[4]:

$$J_+ = \beta n v_c^{2/3} h_c / \Delta v_c \; ; \; J_- = J_+ N/h_s \Delta v_s \; . \tag{3}$$

Here $\beta = 4\pi R_o^2 v_x$, where v_x is the FE mean speed, the subscripts c and s refer to the critical and stable sizes v_c and v_s, Δv is the Gaussian width of the appropriate extremum in ψ_v, N is the drop concentration (drops cm^{-3}) in the well about R_s, and h_c and h_s are the densities (cm^{-3}) of drops of size v_c and v_s. It is clear from Fig. 6 that for values of S between S_+ and S_- _both_ J_+ and J_- are vanishingly small and thus N must remain constant. In fact, for $T \gtrsim 1.3$ K we find no value of S for which J_+ and J_- are simultaneously appreciable. This predicts that the system will always be in a metastable state, _i.e._ $J_+ \neq J_-$.

By incorporating eq. 3 into a set of rate equations somewhat like those of Pokrovskii,[1] it is possible to include a treatment of nucleation in the time evolution of the EHD-FE system

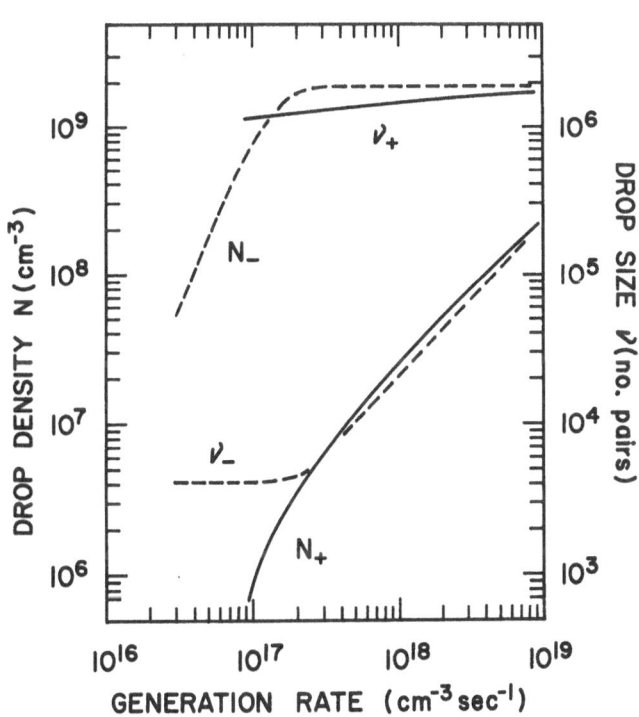

Fig.7. Drop concentration N (cm^{-3}) and number of pairs per drop v versus e-h pair generation rate, computed from exciton-drop rate equations including non-equilibrium homogeneous nucleation currents J_+ and J_-, eq. (3). Results are given for both up-going and down-going exponential step excitation, ($\tau_{on} = 0.1$ sec), and show a marked dependence on excitation history. The computations assume T = 2.1 K, $\sigma = 2.5 \times 10^{-4}$ erg cm^{-2} and $\phi = 2.04$ meV. [Staehli[9]]

subject to an arbitrary exciton generation rate.[9]   Figure 7 shows

the computed solutions for the up-going drop concentration N_+ and

number of pairs per drop ν_+ at 10 sec following excitation switch-on

to the value shown on the abscissa.   It turns out that the solutions

depend markedly on the rate of turn-on of the excitation; the calcula-

tions shown are for a step with an exponential time constant 0.1 sec.

Figure 7 also shows N_- and ν_- at 10 sec following a step down from a

high generation rate of 10^{20} excitons cm^{-3} sec^{-1} to the values on the

abscissa.   Noteworthy is the prediction that for up-going excitation

the drop size remains nearly constant while the drop concentration

increases.   For down-going excitation the drop concentration remains

nearly constant while the drop size decreases until threshold is

approached.   The predicted total EHD up-going luminescence intensity

$I_{709}^+ \propto N_+\nu_+$ and down-going intensity $I_{709}^- \propto N_-\nu_-$ calculated from Fig.

7 are in reasonable agreement with our data.

III.   COMPARISON OF THEORY AND EXPERIMENT

To extract the value of the surface tension σ from the hysteresis

data, the drop luminescence intensity *versus* generation rate has been

computed from the theory; representative results for up-going and

down-going excitation are shown in Fig. 8.   The indicated thresholds

G_+ and G_- are linear continuations of the curves to the axis; they are

well-defined theoretically and correspond exactly to the measured

thresholds P_+ and P_- shown in Fig. 3.   Because the threshold ratio

G_+/G_- is a monotonic function of the surface tension, linearly extra-

polated threshold ratios P_+/P_- from data such as that in Figs. 1 and 3

can be fit to the theory with only σ as an adjustable parameter.   The

only other important parameters are the temperature and the pair

density in the drop n_o, both well known.   Since no measurement of σ

could be considered reliable unless a large number of different crystal

samples were investigated, we have taken threshold data on seven crystals in the temperature range 1.3 < T < 2.1 K. We have shown that dislocation densities in the range 0 to 10^4 cm^{-2} have no discernible effect on P_+/P_- This ratio was found to decrease somewhat at 2.1 K as the net impurity concentration varied from 3×10^9 to 4×10^{12}. Since *a priori* we do not know whether nuclea-

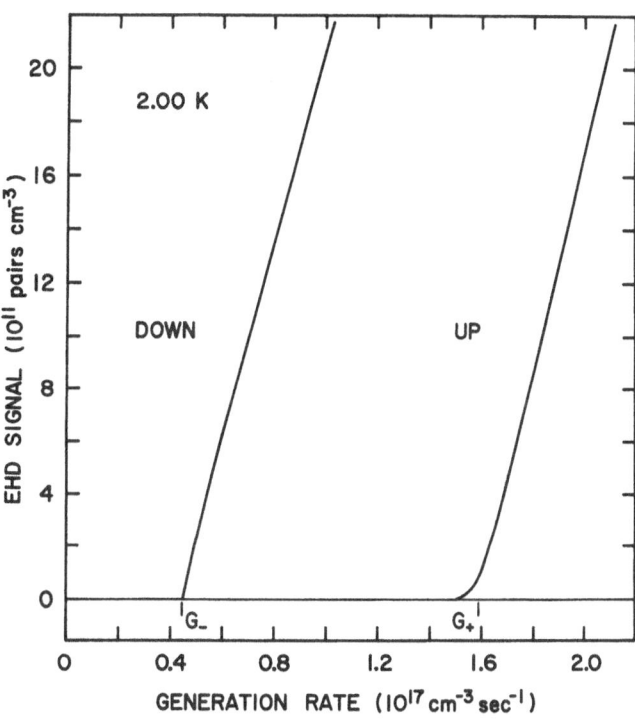

Fig.8. Predicted up-going and down-going drop lumines-cence intensity for pure Ge versus e-h pair generation rate. The scales have been chosen to correspond approximately to those of Fig. 1. This computation assumes $\sigma = 2.4 \times 10^{-4}$ erg cm^{-2} and $\phi = 2.03$ meV. [Westervelt[4]]

tion is homogeneous or inhomogeneous, we reduced all the data for both cases. Assuming homogeneous nucleation, we find $\sigma = (2.4 \pm 0.2) \times 10^{-4}$ erg cm^{-2} at the mean temperature T = 2 K. Assuming inhomogeneous nucleation we find values $\sigma = (2.4 \pm 0.2) \times 10^{-4}$ erg cm^{-2} and $(2.8 \pm 0.3) \times 10^{-4}$ erg cm^{-2} assuming representative values[10] of 4.3 meV and 8.6 meV for the barrier lowering due to neutral impurities. Thus we can conclude that the surface tension at 2 K lies in the range

$$\sigma = (2.4 \text{ to } 2.8) \times 10^{-4} \text{ erg cm}^{-2} \qquad (4)$$

regardless of the role of impurities in our samples.   The surface

tension at T = 0, denoted as the surface energy w, is found using the

approximate expression $w \simeq \sigma \left[1 - (T/T_c)^2\right]^{-1}$ to be

$$w = (2.7 \text{ to } 3.1) \times 10^{-4} \text{ erg cm}^{-2} , \tag{5}$$

which agrees well with the recent theoretical prediction of Vashishta

et al.,[11] who, by including the exchange correlation gradient correc-

tion, find $w \simeq 3.5 \times 10^{-4}$ erg cm^{-2}.   Earlier theoretical estimates

yielded $w \simeq 10^{-4}$ erg cm^{-2}.[12]   To our knowledge our value, eq. 5, is

the first and only accurate measurement of the surface energy.[13]

The detailed theory[4] also predicts the slow time behavior of the

EHD luminescence under the experimental conditions described above for

Fig. 2.   Under up-going excitation the prediction is that, near thresh-

old, $I_{709}^+ = I_o^+ + I_1^+ \ln(Dt)$ where D depends only on fundamental drop

parameters and the excitation level.   Similarly $I_{709}^- = I_o^- - I_1^- \ln(Et)$.

The data of Fig. 2 fit these forms very well, as shown in Fig. 9, in

the range 100 < t < 4000 sec.   If extrapolated, the curves will cross

at t = 8×10^6 sec, which represents a lower limit to the time it takes

to reach the equilibrium state, $J_+ = J_-$.   The fact that the drop

signals obey these expressions for a time exceeding 1 hour suggests

that the droplets in our samples, once formed, are not subject to

significant external destruction mechanisms.   It is likely that the

droplets nucleate or become trapped on impurities and remain virtually

fixed in the crystal, the decaying pairs being replenished by exciton

condensation.   Momentum transferred from an exciton density gradient

is far too small to dislodge the drops.

The theory also predicts the up-going threshold curve, Fig. 10(a),

and the down-going threshold, Fig. 10(b), including the effects of

hysteresis and surface tension.[4]   The coexistence curve for the ideal

exciton gas, eq. 1, is also shown, Fig. 10(c).   The data, Fig. 4, show

the predicted behavior, in particular that the hysteresis vanishes at

283

Fig.9. Semilogarithmic replot of the data of Fig. 2; the solid lines are $I_{709}^- = 235 - 11.9 \times \log_e t(\text{sec})$ and $I_{709}^+ = 11.2 + 2.22 \log_e t(\text{sec})$. These forms are expected theoretically [Westervelt[4)]]. The lines would cross at $t = 8 \times 10^6$ sec.

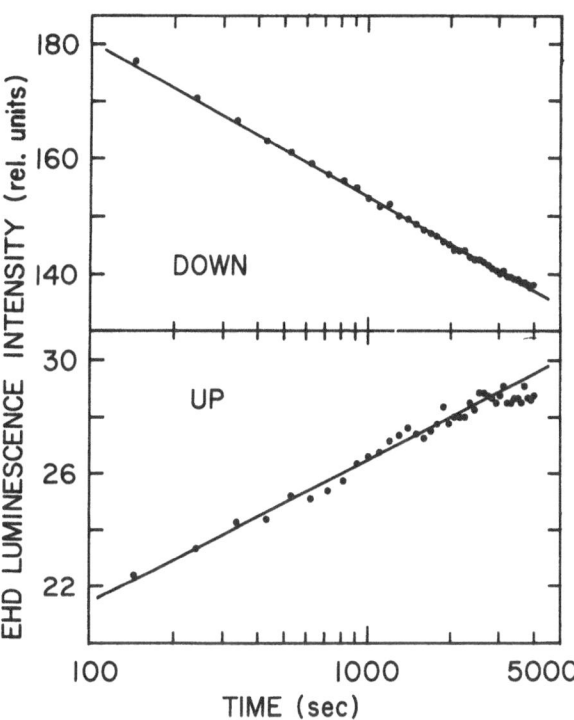

Fig.10. Threshold curves for (a) up-going excitation, and (b) down-going excitation for EHD in pure Ge, from homogeneous non-equilibrium nucleation theory [Westervelt[4)]]. Curve (c) is the saturated density of an ideal exciton gas, eq. (1). Values assumed are $\sigma = 2.7 \times 10^{-4} [1 - (T/6.5)^2]$ erg cm^{-2} and $\phi = (1.97 + 0.016\, T^2)$ meV. The predicted vanishing of the hysteresis at $T \gtrsim 1.3$ K is observed, Fig. 4.

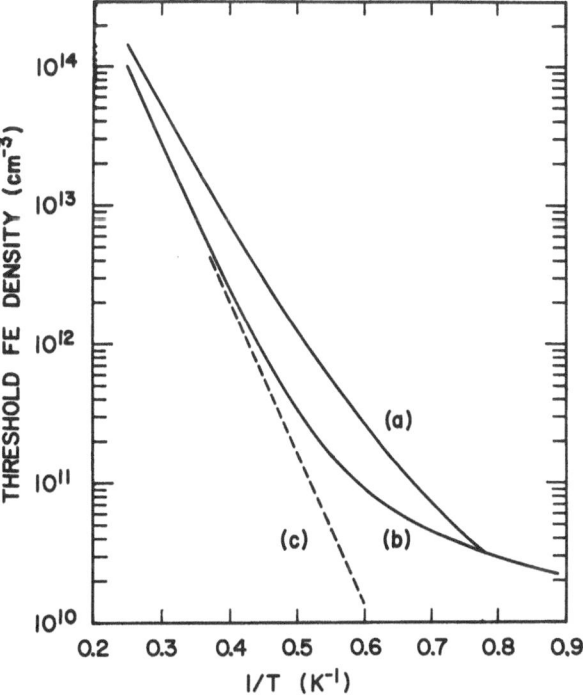

$T \simeq 1.25$ K; the homogeneous nucleation theory predicts this at $T \simeq 1.3$ K, and the inhomogeneous theory at $T \simeq 1.4$ K. What has happened physically below 1.3 K is that the barrier ϕ_c to nucleation has lowered sufficiently that J_+ and J_- are simultaneously large, and the system rapidly reaches thermal equilibrium.

Previous threshold measurements of the condensation energy, typically $\phi \simeq 1.5$ meV, were made by fitting data apparently taken on the up-going curve, Fig. 10(a), to the saturated exciton density, eq. 1. The apparent value deduced in this way from Fig. 10(a) between 2 K and 4 K is $\phi \simeq 1.3$ meV. As pointed out earlier,[3] this seems to explain the apparent discrepancy between threshold and spectroscopic measurements of ϕ. Furthermore, from the absolute value of threshold measurements at 2 K, Fig. 1, calibrated in FE density, we find $\phi = 2.1 \pm 0.2$ meV in good agreement with the spectroscopic value at 2 K, $\phi = 2.03$ meV.[1] (The error in ϕ quoted above corresponds to an uncertainty of a factor ~ 3 in FE

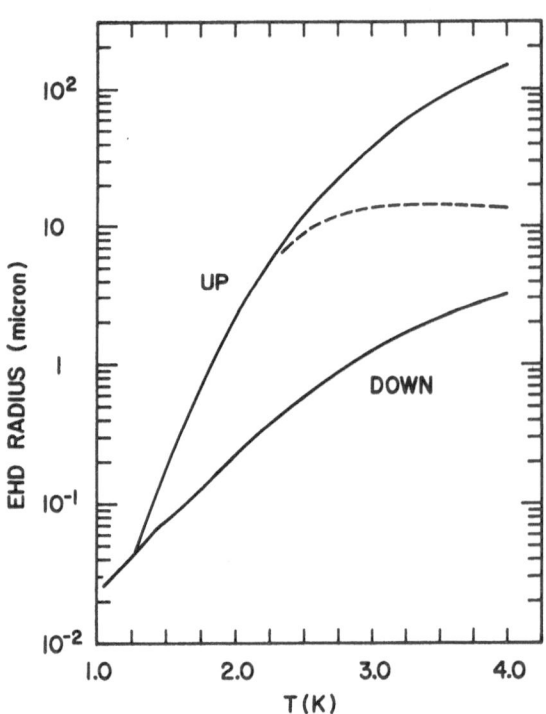

Fig.11. Predicted EHD radius near up-going and down-going thresholds versus temperature in pure Ge from homogeneous non-equilibrium nucleation theory [Westervelt[4]], with the parameters of Fig. 10. The dashed curve shows the reduced up-going radius predicted when local diffusive exciton depletion is also take into account.

density).

The stable drop radius R_s near threshold can be predicted from the theory for both up- and down-going excitation, and these results are shown in Fig. 11 as a function of the temperature. This plot explicitly illustrates the point discussed earlier, that the down-going radius is much smaller than the up-going radius, except below T $\simeq 1.3$ K, where they are equal. The predicted large radii, $R \sim 100$ μm, near the up-going threshold at higher temperatures have never been observed. It is possible to add to the theoretical model the effects of local diffusive exciton density depletion, shown as the dashed line, to account for the reported drop radii $R \sim 10$ μm from light scattering experiments.[14]

In summary, careful measurements on ultra-pure Ge crystals reveal a number of intrinsic nucleation phenomena of EHD arising from the surface energy of drops and their finite pair lifetime. Above T ~ 1.3 K the system is always in a long-lived metastable state which depends on the history of optical excitation; equilibrium nucleation theory is invalid. An adequate theory has been developed that quantitatively explains the observed phenomena and yields an accurate measurement of the surface energy, eq. 5. We acknowledge valuable discussions with C. Kittel, and thank W. L. Hansen for providing Ge crystals.

REFERENCES

1) For a review of electron-hole drops see Ya Pokrovskii: Phys. Stat. solidi (a) 11 (1972) 385; for a recent review see C. D. Jeffries: Science 189 (1975) 955.

2) T. K. Lo, B. J. Feldman and C. D. Jeffries: Phys. Rev. Letters 31 (1973) 224.

3) Some results of this paper have been briefly published by R. M. Westervelt, J. L. Staehli and E. E. Haller: Bull. Am. Phys. Soc. 20 (1975) 471; J. L. Staehli, R. M. Westervelt and E. E. Haller:

Bull. Am. Phys. Soc. 20 (1975) 471.

4) A detailed nucleation theory is given by R. M. Westervelt: Part I, Phys. Stat. sol. (b) 74 (1976) 727; and Part II, Phys. Stat. sol. (b), in press.

5) For a clear review of homogeneous nucleation theory see J. E. McDonald: Am. J. Phys. 30 (1962) 870; also see J. Frenkel: *Kinetic Theory of Liquids* (Oxford Press, Oxford, 1946).

6) See, *e.g.* C. Kittel: *Thermal Physics* (John Wiley, New York, 1969) p.163. For Ge we use the effective mass $m^* = 0.335\ m_0$ and degeneracy $\gamma = 16$, from ref. 2. At temperatures above ~ 3 K, corrections to eq. 1 become significant, A. Frova, G. A. Thomas, R. E. Miller and E. O. Kane: Phys. Rev. Letters 34 (1975) 1572.

7) An equilibrium treatment of the effects of surface tension has been given by R. N. Silver: Phys. Rev. B11 (1975) 1569.

8) R. Becker and W. Döring: Ann. Physik 24 (1935) 719; R. Becker: *Theory of Heat* (Springer-Verlag, New York, 1967), 2nd Edition, p.239.

9) J. L. Staehli: Phys. Stat. sol. (b) 75, issue 2 (June 1, 1976).

10) R. W. Martin: Solid State Comm. 14 (1974) 369.

11) P. Vashishta: Private communication; P. Vashishta, R. Kalia and K. S. Singwi: *Proc. of the Oji Seminar.*

12) L. M. Sander, H. B. Shore and L. J. Sham: Phys. Rev. Letters 31 (1973) 533; H. Büttner and E. Gerlach: J. Phys. C6 (1973) L433; T. M. Rice: Phys. Rev. B9 (1974) 1540; T. L. Reinecke and S. C. Ying: Solid State Comm. 14 (1974) 381.

13) The value $\sigma = 1.6 \times 10^{-4}$ erg cm^{-2} is estimated by V. S. Bagaev, N. N. Sibeldin, and V. A. Tsvetkov: J.E.T.P. Letters 21 (1975) 80, by measuring the temperature dependence of the concentration of drops by light scattering. However, this paper uses an inaccurate equation (their eq. (2)) and classical nucleation theory. The paper of B. Etienne, C. Benoit à la Guillaume and M. Voos: Phys.

Rev. Letters <u>35</u> (1975) 536 uses equilibrium theory to estimate an empirical value of (A/σ), where A is the Richardson-Dushman constant, not known for EHD.

14) A. S. Alekseev, T. A. Astemirov, V. S. Bagaev, T. I. Galkina, N. A. Penin, N. N. Sybeldin, V. A. Tsvetkov: *Proc. of 12th Intern. Conf. on Physics of Semiconductors, Stuttgart* (Teubner, Stuttgart, 1974) p.91.

AN OLD STORY OF NEW CYCLOTRON RESONANCE PEAKS
IN HIGHLY EXCITED GERMANIUM

E. Otsuka, T. Ohyama and T. Sanada

Department of Physics,
College of General Education
Osaka University,
Toyonaka, Osaka 560, Japan

ABSTRACT

Some modification is suggested for the interpretation of the emergence of new strange peaks in time-resolved cyclotron resonance of
germanium under high excitation. The excitonic polaron model suggested
earlier is still kept, but the presence of electron-hole drops as the
source of supply of free excitons and carriers has to be called for.

In earlier papers,[1,2] we reported the observation and the tentative interpretation of new strange peaks of cyclotron resonance in
germanium in terms of the new-type polaron. It is essentially an
electron (or hole) in the boson field of polarized excitons.

The observed phenomenon is the emergence of an extra peak on the
high magnetic field side of the ordinary free carrier resonance. Relative intensities of the ordinary and extra-ordinary peaks as well as
their separation depend on the delay-time after photoexcitation (Fig.
1). The emergence of strange peaks is accompanied with the appearance
of a huge pulse on the tail of the microwave photoconductance signal as
shown in Fig. 2. The trace is taken with the magnetic field being set
at slightly higher than the position of the proper resonance peak for
electron.

Our interpretation of the phenomenon, as mentioned above, is
through the formation of a quasiparticle which we tentatively called
excitonic polaron. There is, however, an alternative interpretation,
suggested by Numata, in terms of dimensional resonance associated with
the spherical electron-hole drop.[3] Numata, using a fixed radius of
the drop and suitable parameters, predicts appearance of extra peaks

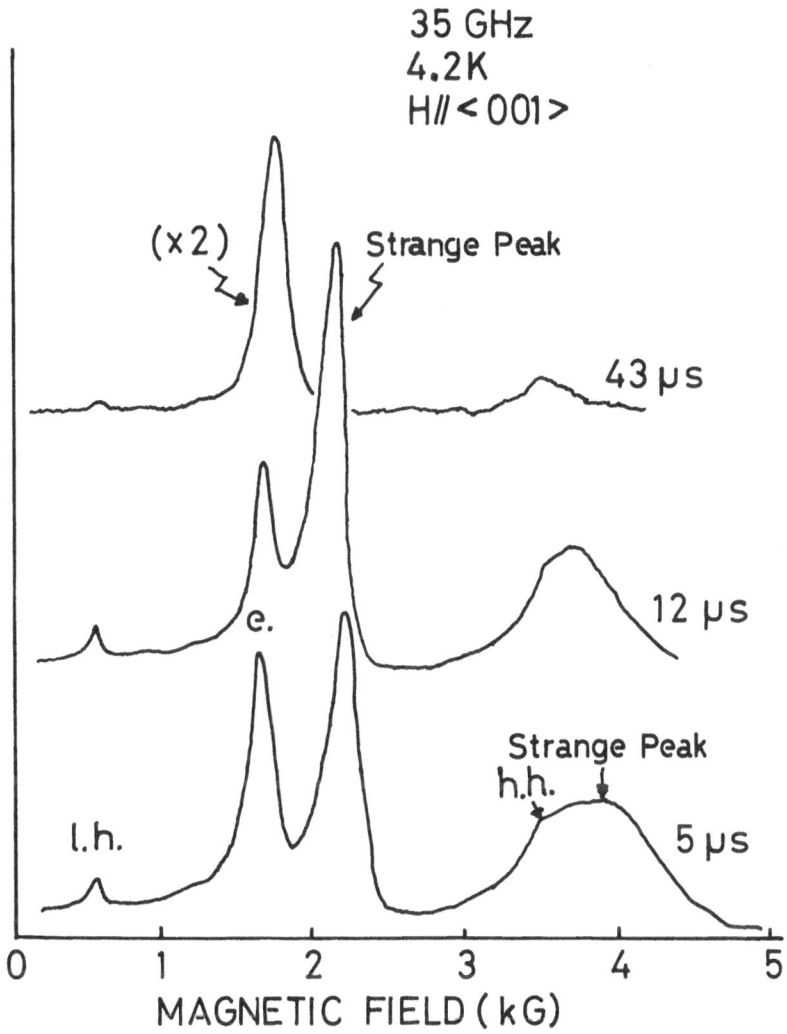

Fig.1. Some traces of time-resolved cyclotron resonance signals, indicating emergence and shift of the strange peak. The delay-times
after photoexcitation are shown on the right.

having close resemblance with our experimental data at a fixed

delay-time. It is true that, so long as one compares the experimental

result with theory at a single time-profile, the agreement is exceeding-

ly good. But the agreement is found only at one suitably chosen section

of the time-resolution. If one watches the phenomenon through a passage

of time, the theory of dimensional resonance confronts a difficulty.

4.2 K

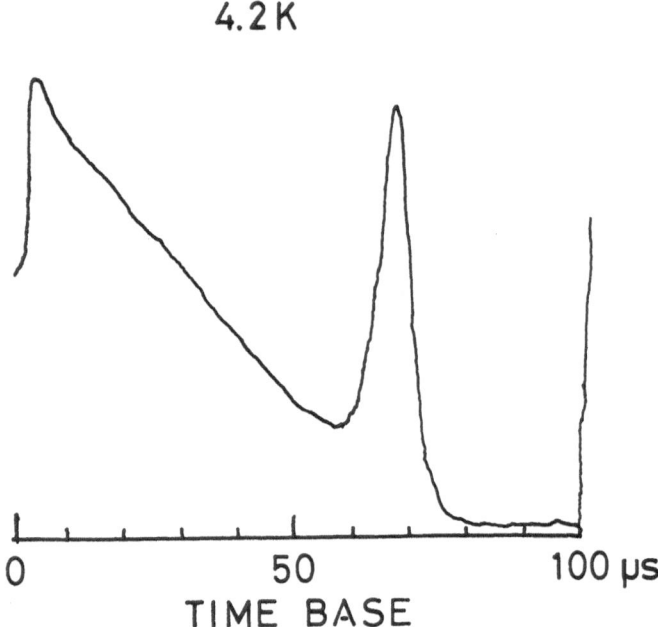

0 50 100 μs

TIME BASE

Fig.2. Microwave photoconductance trace under the application
of a magnetic field. The field is set somewhat above
the proper electron resonance field. It is characterized
by the emergence of a striking overshoot on the tail.

Namely, the theory predicts a decrease in relative size of the strange

peak with decreasing separation.   That is contrary to the experimental

observation.

      The polaron model, on the other hand, has its own difficulty.   It

is hard to find any good reason why the electron interacts with a fixed

number of excitons at a given instant of time (Fig. 3).   Unless this

condition is fulfilled, however, what we obtain would be just a

smearing-out of the polaron peak.   In other words, we have a distribu-

tion in polaron mass.   The fact is the appearance of a definite peak

during the passage of time.

      Our original interpretation of the phenomenon through the idea of

polaron had nothing to do with the presence of electron-hole drops.

We find, however, some difficulty if we proceed by completely neglect-

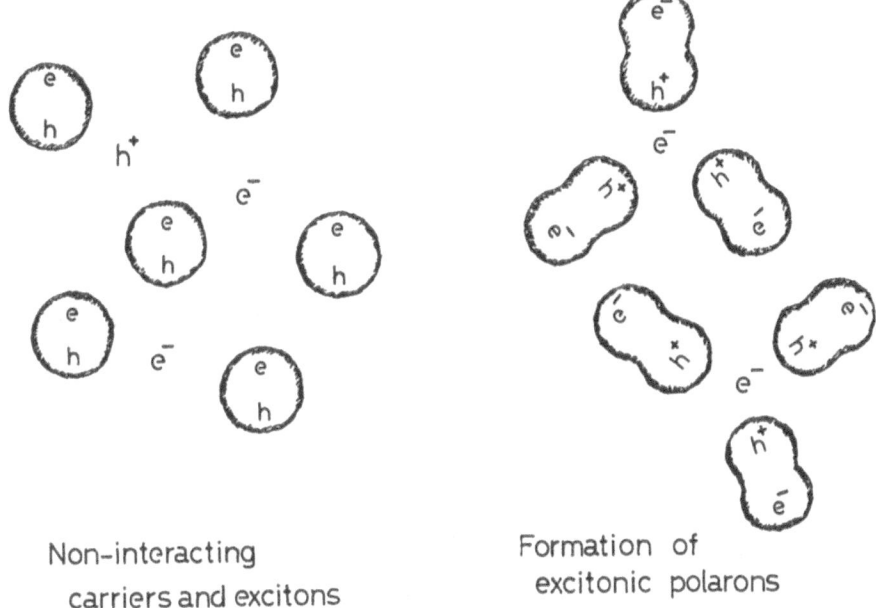

Non-interacting
carriers and excitons

Formation of
excitonic polarons

Fig.3. Difficulty of the polaron model: At a given delay-time, each
electron should be accompanied by an "equal" number of excitons.

ing the presence of drops. Figure 4 shows the time variation of the

relative intensities of ordinary and extraordinary peaks. The ordinary

resonance intensity falls exponentially with time. The extraordinary

peak, or the polaron peak, grows with time for some time-interval at

the beginning, saturates and then decays at last. Both the ordinary

and extraordinary peaks correspond to electrons. The only difference

is whether the electrons are coupled with excitons or not. The total

intensity remains practically constant for a certain interval of time

and then rapidly decays. Why does the total number of electrons not

decrease for this initial period of time? There should be a source of

producing electrons. When that source is exhausted, the observed

number of electrons, coupled or not coupled with excitons, starts to

decay. What is that source then? That must be electron-hole drops.

Excitons, evaporated from the drop, will subsequently yield free

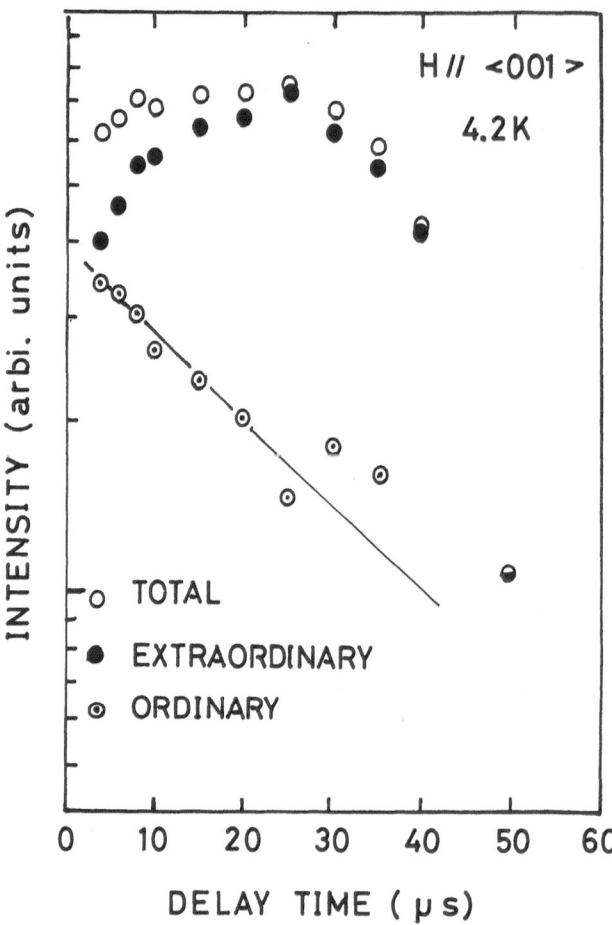

Fig.4. Intensities as a function of delay-time for ordinary, extraordinary and total electron resonances.

carriers on dissociation. One may expect a zone of uniform distribution of free carriers and excitons around an electron-hole drop. The uniformness is ensured by the uniform evaporation processes. If the distribution is uniform, there is no reason to denounce the uniform coupling between electrons and excitons. In other words, one can expect an average number of excitons with small fluctuation around an electron at a given instant of time. Even if, of course, the exciton evaporation is uniform around a drop, one has to expect a possible gradient in density along the direction perpendicular to the drop-surface. But the Auger processes due to exciton-exciton collisions would be more frequent near the surface and tend to make the density of excitons more uniform along that direction. In addition, if free carriers are produced dominantly through the Auger processes, one can forget the existence of the free carriers and hence the polarons in the region of less dense excitons. Thus we observe the cyclotron resonance signal arising only from the region of a high and "uniform" exciton density.

The sudden overshoot of the photoconductance signal at a proper delay-time as seen in Fig. 2 may correspond to the final exhaustion of drops or a single huge drop. Owing to the finite surface free energy, the unit volume free energy for the drop itself increases with decreasing drop size, eventually resulting in the small burst of the vapor pressure of the exciton gas.

REFERENCES

1) E. Otsuka, T. Ohyama and T. Sanada: Phys. Rev. Letters $\underline{31}$ (1973) 157.

2) E. Otsuka, T. Ohyama and T. Sanada: J. Phys. Soc. Japan $\underline{37}$ (1974) 114.

3) H. Numata: J. Phys. Soc. Japan $\underline{36}$ (1974) 309. See also Thesis, University of Tokyo, 1975.

Note Added in Proof.

Recently the authors have arrived at a conclusion that the strange peaks are not due to the polaron effect. They appear as a consequence of geometrical resonance, depending on the sample shape and the microwave frequency. The feature, however, is fairly complicated owing to the varying dielectric constant, which depends on the

degree of excitation. More details will shortly be published elsewhere.
The authors are indebted to Prof. C. D. Jeffries for suggesting the
possbility in this direction.

ELECTRON-HOLE DROPS IN SILICON

M. Kobayashi and S. Narita

Department of Material Physics,
Faculty of Engineering Science,
Osaka University,
Toyonaka, Osaka 560, Japan

ABSTRACT

Two problems concerning the emission from Si under high-excitation
have been investigated: One of them is the near-infrared emission from
highly excited Si-surface by a Q-switched ruby laser. Time-resolved
emission spectra make clear the electron-hole liquid phase extending
uniformly over the crystal surface. The second problem is the far-
infrared emission from Si under high-excitation. We observe the emission due to the 2p-1s transitions of the excitons in Si.

Up to present, the existence of electron-hole drops (EHD) has been

believed only for the indirect energy gap semiconductors, Ge and Si,

though there have been a fewer studies on EHD in Si comparing with Ge.

The cause of a fewer studies for Si is that the carrier concentration

inside EHD is one order of magnitude larger in Si than in Ge, and the

life-times of excitons and EHD in Si are also one order of magnitude

shorter comparing to Ge, so that for Si an appreciably stronger excita-

tion is necessary for making the same state of EHD as for Ge. However,

for making clear the whole aspect of EHD, the comparison between the

condensed phase in Si and that in Ge is indispensably important.

Here we treat two problems concerned with the highly excited state

in Si. (i) One of them is an experimental study of the highly excited

state of Si produced by a Q-switched ruby laser.

The experiment was performed by using phosphorus-doped 1200 Ωcm Si

samples. The resistivity corresponds to the doping level of 4×10^{12}

cm^{-3} impurity concentration. In the measurements, the samples were

cooled by being directly immersed in liquid helium or liquid hydrogen.

The excitation by a GaAs laser of 1 watt/pulse produced the excited

carriers of about 10^{14} cm^{-3}, while in the case of a Q-switched laser,
the excitation power was 200 kilowatt/pulse and the number of the
excited carrier amounted to more than 5×10^{18} cm^{-3}.

Near infrared emission line from Si in the case of GaAs laser
excitation is shown by the broken curve in Fig.1. This line was dis-
covered by Haynes[1] and ascribed to EHD by Pokrovskii *et al.*[2]   The
emission produced by a Q-switched ruby laser excitation is indicated by

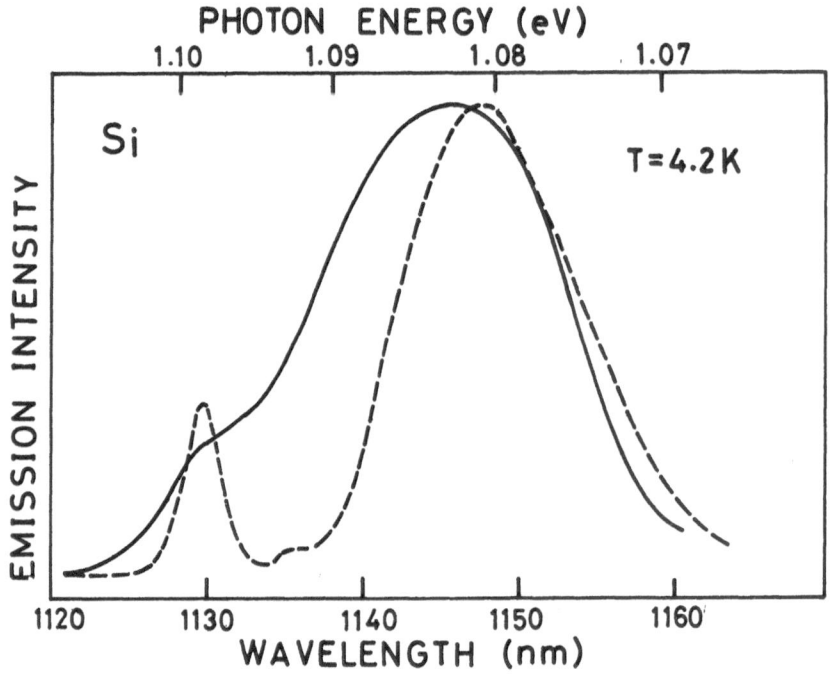

Fig.1. Comparison of near infrared emission spectra between two different
excitation. The broken curve corresponds to a GaAs laser (relatively
weak) excitation, while the solid curve to a Q-switched ruby laser
(extremely strong) excitation.

the solid curve in the figure. By the extremely high excitation the
spectral shape of the Haynes line changes and a new component grows
between the free exciton (FE) line and the Haynes line (N-line).
According to the conception of EHD, the spectral shape of EHD remains
unchanged with the change of the excitation power; that is, the carrier
density inside EHD is uniquely determined from the condition of the
minimum free energy, so that higher excitation increases only the number

of EHD's or their radii, which yields the same spectral shape. There-
fore, the new component produced at the higher energy side of the
Haynes line cannot be interpreted as due to the emission from normal
EHD.

On the other hand, as FE-component peaked at 1.098 eV changes with
the excitation level, we must separate the contribution of the FE-line
from the whole spectrum. Thus we study the time-resolved emission
spectra of highly excited Si.

Figure 2(a) shows the time-resolved emission spectra at 4.2 K. It

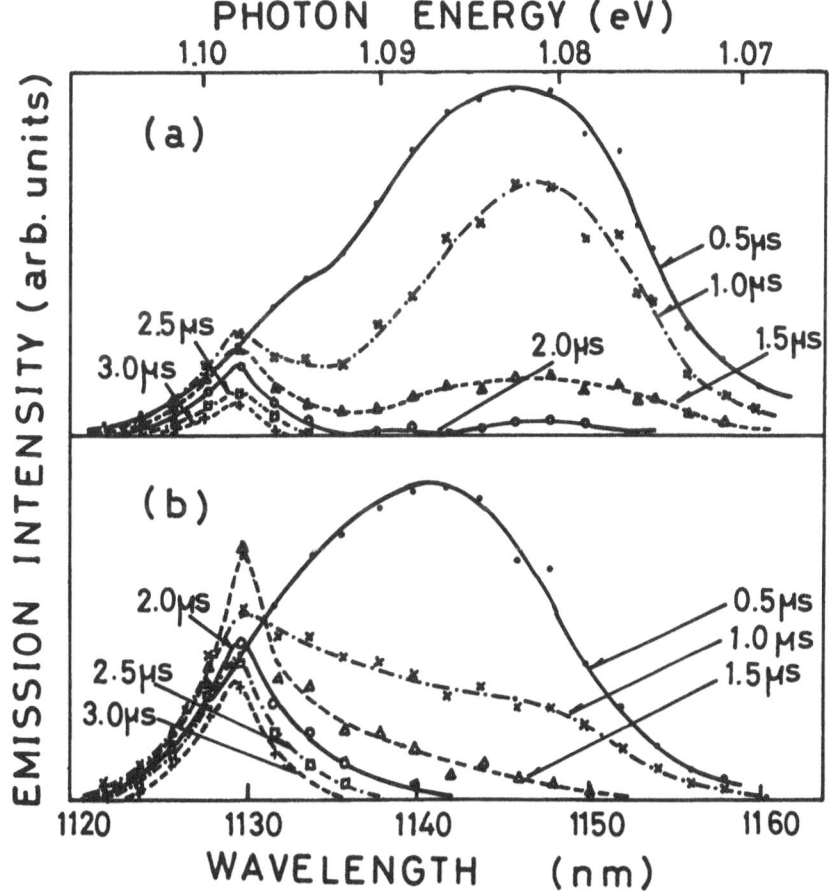

Fig.2. Time-resolved emission spectra at 4.2 K (a) and at 20 K (b). The
time delays between excitations and observations are indicated in
μsec units.

is seen in the figure that the new component (N'-line) decays faster than the N-line and the whole spectral shape tends to approach to the shape for the weaker excitation case (the Haynes line) as time passes. The decay time of the N'-line is within 1.5 μsec, while the FE-line grows after the decay of the N'-line and remains longer than the N'-line.

Figure 2(b) shows the time-resolved spectra at 20 K, where we hardly observe the N-line. As the spectral shape and the peak position of the line at 20 K peaked at 1.088 eV agree with those of the N'-line at 4.2 K, we regard the line as the N'-line. The line decays quickly within 1.5 μsec, whioh is close to the decay time of the N'-line at 4.2 K. A remarkable feature of the spectra at 20 K is that the FE-line requires about 1.5 μsec to reach its maximum after the excitation when the N'-line at 20 K disappears. This behavior of the FE-line is characteristic for high excitations and was not found in weaker excitation by Cuthbert[3] whose injection level was about 10^{17} cm^{-3}. The coincidence of the decay time of the N'-line with the growing time of the FE-line suggests that a source of the N'-line is formed initially in the sample and then it dissociates to the free excitons.

We have assumed so far that the new component at 4.2 K and the line peaked at about 1.088 eV are the same one and they have been designated commonly as the N'-line. This assumption is reasonable because the peak energies, the line widths and the decay times are in agreement between them. The line shape analysis for the N'-line was performed according to the assumption of band-to-band transition by Pokrovskii *et al.*[2] The density of state masses used are[4] $m_c = 1.08\ m_0$, and $m_h = 0.55\ m_0$. As the result, we obtained the carrier density of $n_0 = 8.3 \times 10^{18}$ cm^{-3}, which is appreciably larger compared with the concentration in equilibrium EHD, $n_0 = 3.7 \times 10^{18}$ cm^{-3}.

We suppose from the above experimental results that the crystal surface is filled with the electron-hole plasma just after the extremely high excitation, which is the origin of the N'-line. Free excitons cannot be generated at the initial stage because of the strong screening effect of the free carriers in the liquid.

From a phase-diagrammatical consideration,[5] the initial electron-hole liquid phase at 4.2 K by high-excitation changes to the coexistence phase of EHD and FE, which explains the time-resolved emission spectra of Fig.2(a), while at 20 K, the initial high pressure liquid phase passes through the coexistence phase of EHD and FE in a short time and runs into the single FE phase immediately, which can interpret the spectral figures of the time-resolved emission spectra of Fig.2(b).

The N'-line at 20 K might be also ascribed to EHD. However, this explanation cannot clarify the reason for the appearance of the FE-line after the decay of the N'-line as shown in Fig.2. If the N'-line is originated from EHD, it should coexist with FE, so that the post-excitation peak of FE cannot be observed. Besides, the carrier concentration inside EHD must decrease at higher temperatures, and therefore, the N'-line at 20 K must have a narrower line width compared with the N-line at 4.2 K. This is not the case in the present experiment.

(ii) The second problem is on the far-infrared emission from Si under high excitation. We expected the emission due to the plasma oscillation of EHD in Si as observed in Ge by Vavilov.[6]

Figure 3 shows the block diagram of the experimental set-up. The excitation power of the GaAs laser was 25 watt/pulse and supposed to make enough EHD in Si. We also used a halogen lamp of 150 watt. The light from the source was guided through a glass optical fiber to the Si sample which was set at the bottom of a metal cryostat. As the window of the cryostat we also used a pure Si plate which is transparent in the far-infrared region. The far-infrared luminescence from Si was introduced into a Michelson interferometer, Beckman model FS-720, then detected by the photoconductive semiconductor detectors of Ge-Sb or Ge-As. The interferometer was considered to be advantageous to measure the weak emission power less than 10^{-9} W. The Si samples

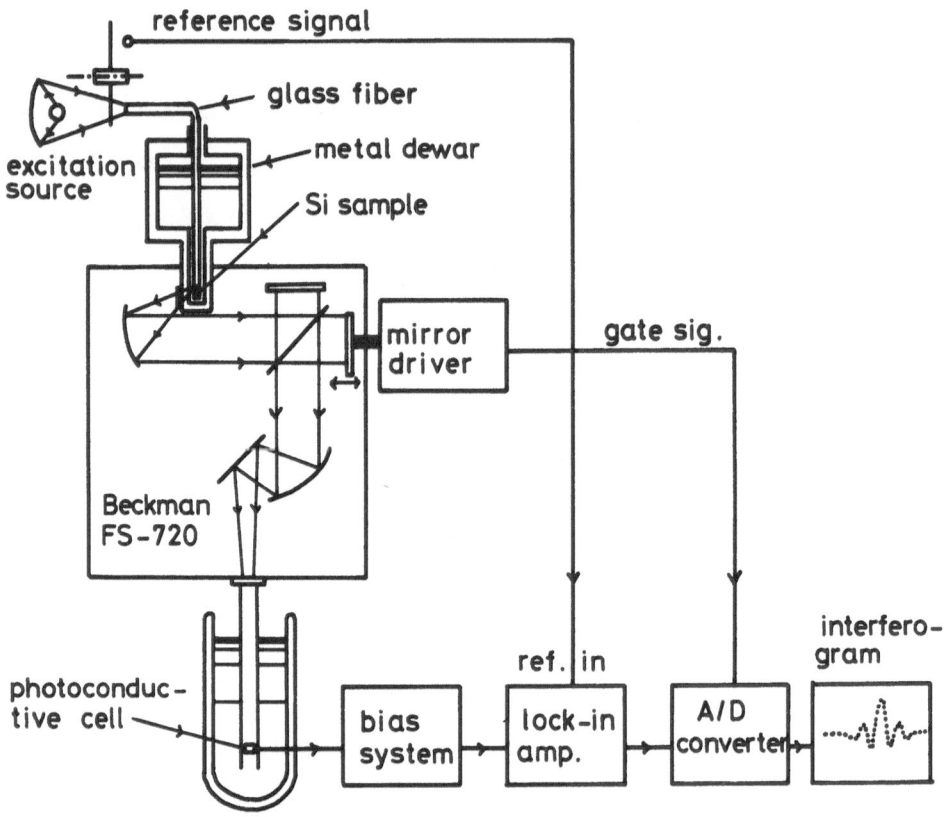

Fig. 3. Block diagram of experimental system for measuring far-infrared emission spectra of Si under high-excitation using Michelson interferometer.

used were cut from the same ingot as that used in the experiment (i).

Figure 4 shows the far-infrared emission spectrum from the Si sample at 4.2 K, which is obtained by using the Ge-Sb detector. The spectrum has a sharp peak at 93 cm^{-1} (11.5 meV) and tails off toward the high energy side.

As the origin of this far-infrared emission, three possibilities can be considered; the emission from plasma in EHD, the transitions between the discrete levels of the impurity state, and those of the exciton. The first possibility is excluded by the following reasons. If we calculated the plasma frequency, $\omega_p^2 = e^2 N/\varepsilon\mu$, by taking the carrier concentration in EHD of $N = 3 \times 10^{18}$ cm^{-3} and the reduced mass of $\mu =$

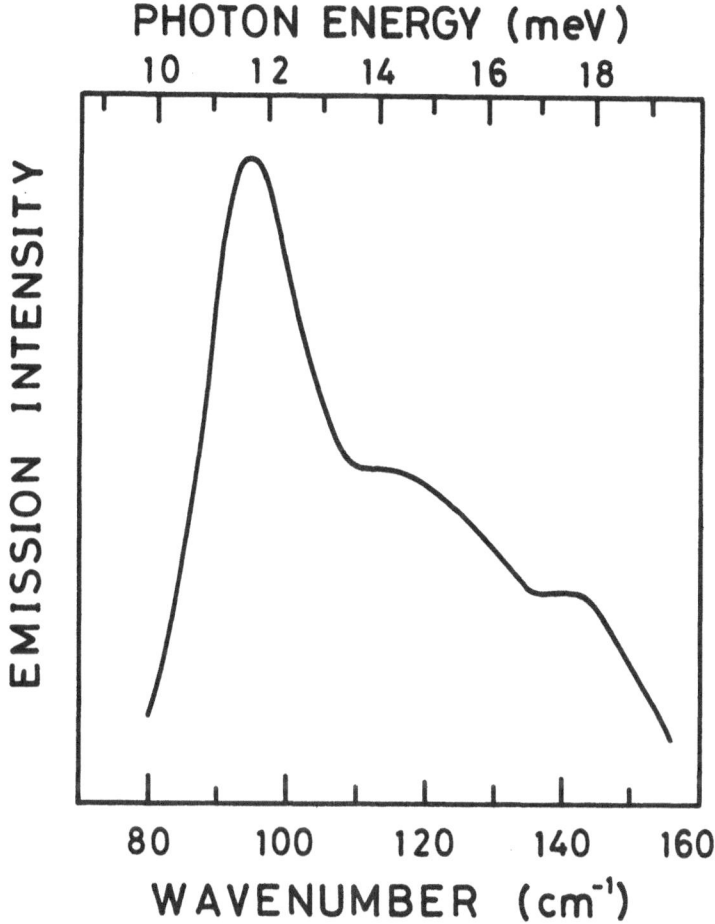

Fig.4. Far-infrared emission spectra of Si under high-excitation
using Michelson interferometer and Ge-Sb detector.

$0.15 \, m_0$, then we have the theoretical peak frequency $\omega_p/\sqrt{3}$ as $227 \, \mathrm{cm}^{-1}$,

which is too high compared with the experimental results. On the other

hand, the emission remains above the liquid nitrogen temperature up to

near the room temperature, though the intensity appreciably decreases.

These facts exclude the first possibility of the emission from EHD.

Secondary, the emission due to transitions between the discrete levels

of impurity is not probable, because the impurity concentration in the

present case is too small, $4 \times 10^{12} \, \mathrm{cm}^{-3}$ and also at $4.2 \, \mathrm{K}$ the ground

states of the impurities are occupied by electrons, so that such a transition cannot occur.

The experimental information regarding the excited states of the exciton in Si was given by Shaklee and Nahory.[7] They gave the 1s-2s energy separation for TA phonon assisted transition to be 10.7 meV and the separation for LO and TO assisted transitions to be 11.0 meV. Our experimental peak energy of 11.5 meV is very close to these values. We therefore attribute our emission peak in Fig.4 to the exciton transition from the 2p state to the 1s ground state. The energy difference of 0.5 ~ 0.8 meV between their results and ours may be explained as the energy difference in the 2s and 2p levels of the n = 2 state, whose degeneracy is removed on account of the mass anisotropy in the conduction band and the degeneracy of the valence band.[8] In the case of the interband absorption,[7] the 2s state was observed, while in the case of the far-infrared emission, the 2p state must be the initial state according to the selection rule.

The reason that the far-infrared emission from the plasma in EHD was not observed in the present experiment may be ascribed to the weakness of the emission intensity or the weak sensitivity of the present detector in the region. Therefore, further investigations are required.

ACKNOWLEDGEMENTS

The authors wish to express their sincere thanks to Messrs. Y. Tsuchihashi, T. Kawabata, and O. Kumagai for their helpful assistances, and they also thank to Mr. Saoyama of Asahi Glass Co., LTD for the Cerasolzer-Soldering of Si window.

REFERENCES

1) J. R. Haynes: Phys. Rev. Letters 17 (1966) 860.

2) Ya. Pokrovskii: Phys. Status solidi (a) 11 (1972) 385.

3) J. D. Cuthbert: Phys. Rev. B1 (1970) 1552.

4) A. S. Kaminskii: Ya. E. Pokrovskii and N. Y. Alkeev, Sov. Phys.

JETP <u>32</u> (1971) 1048.

5) M. Kobayashi, Y. Tsuchihashi and S. Narita: Solid State Commun. <u>15</u> (1974) 651.

6) V. S. Vavilov, V. A. Zayats, V. N. Murzin: *Proc. 10th Intern. Conf. Phys. Semiconductors,* Cambridge 1970 ed. S. P. Keller, J. C. Hensel and F. Stern (NBS, Springfield, Va. 1970) p.509.

7) K. L. Shaklee and R. E. Nahory: Phys. Rev. Letters <u>24</u> (1970) 942.

8) W. Kohn and J. M. Luttinger: Phys. Rev. <u>99</u> (1955) 915.

STOCHASTIC MODELS OF INTERMEDIATE STATE INTERACTION

IN SECOND ORDER OPTICAL PROCESSES

Ryogo Kubo, Toshihide Takagahara

Department of Physics,
University of Tokyo
Tokyo, Japan

and

Eiichi Hanamura

Institute for Solid State Physics
University of Tokyo
Tokyo, Japan

ABSTRACT

The stochastic theory of spectral line shape as developed by one
of the present authors is here extended to the problem of intermediate
state interaction in a second order optical process.   Depending on the
stochastic nature of the interaction, the process may appear as a pure
Raman process, or a luminescent process, or a mixed process.   Simple
but standard models are 1) the adiabatic random modulation and 2) the
off-diagonal random modulation of the intermediate states.   Limiting
cases of slow and fast modulation are discussed in a general way.   Some
model calculations are made in order to see how a spectrum changes its
nature from one limit to another.

I.   INTRODUCTION

In the course of an optical process, the system under observation
may interact with its environment.   An interaction present in the
initial or the final state of the process is called an initial or a
final state interaction, and affects the line shape of the absorption
or the emission spectrum of the system.   The nature of such an effect
is pretty well understood by now.   On the other hand, the role of an
intermediate state interaction, which will be abbreviated as IMSI in
the following, seems to be a matter of controversy, as is seen in a
number of recent papers on the relationship between a resonant Raman
process and a luminescent process.[1]

In order to look into the essential points of this sort of problem,

it is generally advantageous to simplify the problem somewhat by regarding the interaction as a random modulation acting on the system in
question.   This is the spirit of stochastic theories of line shape, as
developed some years ago by one of the present authors and by many
other authors.[2-5]   Such theories have been, however, so far limited
mostly to the initial or the final state interactions in the first
order processes.

The aim of the present work is to extend a stochastic theory to
the intermediate state interactions and to study how a second order
optical spectrum is affected by an IMSI.   It will be shown that the
spectrum bears simple characters in some ideal limits of the stochastic
IMSI.   If the model is simple enough, it is also possible to study in
more detail how the spectrum changes its character as the random IMSI
changes from the limit of slow modulation to the other limit of fast
modulation.   Such a treatment of stochastic models of IMSI is useful
for obtaining a deeper insight into the second order optical processes.
Since the work is still in progress, this is rather a sketchy report of
the results so far obtained.   A more complete paper will be published
elsewhere in the near future.

II.   GENERAL FORMULATION

We consider a system, denoted by S, in contact with its environment
R and the photon field Φ.   The quantum states and their energies of S
are denoted by A, B, and C, and the photon numbers by n_1, n_2, ...
The Hamiltonian of the total system is written as

$$\mathcal{H} = H + H_\Phi + V \quad , \tag{1}$$

where

$$H = H_S + H_R + H_{SR} \tag{2}$$

is the Hamiltonian for the system S + R, H_{SR} being the interaction between S and R, H_Φ the Hamiltonian of Φ, and V is the interaction of S with the photon field,

$$V = V_1^+ + V_1^- + V_2^+ + V_2^- + \cdots \quad . \tag{3}$$

The interaction $V_1^\pm$ creates or annihilates a photon ω_1 and combines the state A with B. Likewise $V_2^\pm$ creates or annihilates a photon ω_2 and combines C with B. Thus in a second order process starting from the initial state,

$$|a\rangle \ = \ | A, \ n_1+1, \ n_2, \ \ldots\rangle \quad \text{with the energy} \quad a = A + \omega_1 \ ,$$

and ending at the final state,

$$|c\rangle \ = \ | C, \ n_1, \ n_2+1, \ \ldots\rangle \quad \text{with the energy} \quad c = C + \omega_2,$$

a photon ω_1 is absorbed and a photon ω_2 is emitted. The energy of the photon field is referred to the state $(n_1, \ n_2, \ \ldots)$. In the inter- mediate states, S is in B and photon numbers are either $(n_1, \ n_2, \ \ldots)$ or $(n_1+1, \ n_2+1, \ \ldots)$. We ignore the latter state because we shall be mainly interested in the near resonance situation,

$$A + \omega_1 \sim B, \quad \text{or} \quad C + \omega_2 \sim B, \tag{4}$$

which may cause a real absorption process from A to B or a real emission process from B to C.

Since we focus our attention on the IMSI, the interaction H_{SR} is assumed to be present only in the state(s) B. The environment R is in an equilibrium at the initial time and is described by a density matrix ρ_R^e. It evolves in its own way throughout the whole process and causes a random modulation on S through the interaction H_{SR}.

The equation of motion of the density matrix of the total system,

$$i \frac{\partial \rho}{\partial t} = [\mathcal{H}, \rho], \tag{5}$$

is solved by iteration with the initial condition,

$$\rho(0) = |a\rangle \langle a| \times \rho_R^e \quad . \tag{6}$$

The solution gives the probability $P(t)$ of finding the final state c

at time t as

$$P(t) = \int_0^t dt_1 \int_0^{t_1} dt_2 \int_0^t dt_1' \int_0^{t_1'} dt_2'$$

$$\times Tr_R \langle c | e^{-i(H+\omega_2)(t-t_1)} V_2^+ e^{-iH(t_1-t_2)} V_1^- e^{-i(H+\omega_1)t_2} | a\rangle \rho_R^e$$

$$\times \langle a | e^{i(H+\omega_1)t_2'} V_1^+ e^{iH(t_1'-t_2')} V_2^- e^{i(H+\omega_2)(t-t_1')} | c\rangle \quad . \tag{7}$$

In this expression, propagation of the states is governed by the

Hamiltonian (2).   More rigorously, we could have included an appro-

priate part of the total Hamiltonian in the propagators, which should

give rise to the radiative damping or other sorts of damping effective

in bringing the system out of the states under consideration.   Instead

of doing such an explicit treatment of damping, we take here a somewhat

phenomenological approach in this respect by introducing an effective

damping γ_b for the intermediate state b.   Namely the propagators in

the intermediate states $\exp[-iH(t_1 - t_2)]$ and $\exp[iH(t_1' - t_2')]$ are

replaced by

$$e^{-(iH+\gamma_b)(t_1-t_2)} \quad \text{and} \quad e^{(iH-\gamma_b)(t_1'-t_2')} \quad ,$$

to take account of such damping.   It should be emphasized that this

damping is assumed to be caused by some mechanism other than the inter-

mediate state interaction on which we focus our attention.

    If the probability $P(t)$ is evaluated for a proper range of t, it

will be of the form,

$$P(t) = W_1 t + W_2 t^2 + \dots \quad . \tag{8}$$

Then the transition rate of the two-photon process is identified with W_1, which coincides in usual cases with the answer given by the golden rule. A question arises here whether or not a two-step process like a luminescence is ever contained in this calculation of the transition rate. In fact, a simple consideration shows that it is the case if the above-mentioned damping rate γ_b is large in comparison with the rate of real absorption or emission. If we write simple rate equations as

$$\dot{P}_a = - \lambda_{ab} P_a \, ,$$

$$\dot{P}_b = \lambda_{ab} P_a - \lambda_{bc} P_b - 2 \gamma_b P_b \, ,$$

$$\dot{P}_c = \lambda_{bc} P_b \, ,$$

and evaluate $P_c(t)$ for the initial condition,

$$P_a(0) = 1, \ P_b(0) = P_c(0) = 0 \, ,$$

we find that

$$P_c(t) \sim \frac{\lambda_{ab} \, \lambda_{bc}}{2 \, \gamma_b} t \qquad \text{for} \quad \gamma_b^{-1} \ll t \ll \lambda_{ab}^{-1}, \, \lambda_{bc}^{-1} \, , \tag{9}$$

if γ_b is much larger than other time rates. On the other hand, if $\gamma_b = 0$, we have only

$$P_c(t) \sim \frac{1}{2} \lambda_{ab} \, \lambda_{bc} \, t^2 \qquad \text{for} \quad t \ll \lambda_{ab}^{-1}, \, \lambda_{bc}^{-1} \, . \tag{10}$$

We now observe the following. The integrand of the expression (7), which we call the propagator of the second order process, is a function of the four time points $(t_2 < t_1, \ t_2' < t_1')$, independent of t, and furthermore is invariant with respect to translations along the time axis. This means that the Laplace transform of $P(t)$ has the form,

$$F(s) = \int_0^\infty dt \; e^{-st} \; P(t) = \frac{F_1(s)}{s^2} + \ldots \qquad ,$$

and the transition rate is obtained by

$$W_1 = \lim_{s \to +0} F_1(s)$$

$$= \lim_{s \to +0} \int_0^\infty d\tau \; e^{-s\tau} \int_0^\tau d\tau_1 \int_0^{\tau_1} d\tau_2 \; (I + II + III + c.c.). \qquad (11)$$

Here I, II, and III are decompositions of the original propagator of the second order process and are schematically represented by the graphs shown in Fig. 1.   The four time points appear in eq.(7) in different types of order.   The earliest of them can be chosen as the origin of time and the latest is called τ, so that now the four time points are renamed as $(0, \tau_2, \tau_1, \tau)$ according to their order on the time axis.   The upper arcs in each graph refer to propagation to the right and lower arcs to the left, in eq.(7), starting from the initial state and ending at the final state.   Each of arcs is associated with a particular state of the system.   The structure of these decomposed propagators will be seen in the examples to be discussed in the following.

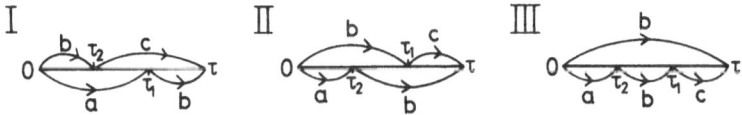

Fig.1. Decomposed propagators of a second order process.

III.   A THREE LEVEL SYSTEM WITHOUT IMSI

As a standard example, we now consider a three level system without IMSI.   From the graphs in Fig. 1, explicit forms of the propagators can be written down immediately by noticing that the phase factors for each of the time intervals $(0, \tau_2)$, (τ_2, τ_1) and (τ_1, τ) are determined by the difference of energies of propagation on the upper and lower arcs. Thus we have

$$I_0 = \exp[-(\gamma_b+i(b-c))(\tau-\tau_1)-i(a-c)(\tau_1-\tau_2)-(\gamma_b+i(a-b))\tau_2] \quad ,$$

$$II_0 = \exp[-(\gamma_b+i(b-c))(\tau-\tau_1)-2\gamma_b(\tau_1-\tau_2)-(\gamma_b+i(a-b))\tau_2] \quad ,$$

$$III_0 = \exp[-(\gamma_b+i(c-b))(\tau-\tau_1)-2\gamma_b(\tau_1-\tau_2)-(\gamma_b+i(a-b))\tau_2] \quad , \tag{12}$$

omitting the coefficient $|V_1|^2|V_2|^2$ hereafter. Equation (11) gives

$$W_1 = \lim_{s\to+0} [\frac{1}{\gamma_b+i(b-c)} \frac{1}{s+i(a-c)} \frac{1}{\gamma_b+i(a-b)} \tag{I}$$

$$+ \frac{1}{\gamma_b+i(b-c)} \frac{1}{2\gamma_b} \frac{1}{\gamma_b+i(a-b)} \tag{II}$$

$$+ \frac{1}{\gamma_b+i(c-b)} \frac{1}{2\gamma_b} \frac{1}{\gamma_b+i(a-b)} \tag{III}$$

$$+ \ c.c. \ ] \ . \tag{13}$$

The evaluation assumes $s \ll \gamma_b$ or $t \gg \gamma_b^{-1}$. By the relation,

$$\frac{1}{s+i(a-c)} = \pi\delta(a-c) - \frac{i}{a-c} \quad ,$$

the first term and its c.c. in eq.(13) yield a Raman term containing a delta-function and a negative term. As is expected from the conservation of energy, the latter term is exactly cancelled out by the contribution from the other terms, which amounts to

$$\gamma_b / \{ \gamma_b^2 + (b-a)^2 \} \{ \gamma_b^2 + (c-b)^2 \} \ . \tag{14}$$

So we are left with the pure Raman term,

$$W_1 = \pi\delta(a-c) \frac{\gamma_b^2 + (b-a)(b-c)}{\{ \gamma_b^2+(b-a)^2 \}\{ \gamma_b^2+(b-c)^2 \}} \tag{15}$$

$$= \frac{\pi \delta (A-C+\omega_1-\omega_2)}{\gamma_b^2 + (B-A-\omega_1)^2} \quad .$$

IV.   ADIABATIC RANDOM MODULATION IN IMSI

The environment R is now supposed to exert a random perturbation on S in its intermediate state.   The simplest model of this is a random adiabatic shift of B.   Thus we assume

$$b(t) = \bar{b} + b'(t), \quad \overline{b'(t)} = 0 \quad , \tag{16}$$

where $b(t)$ is a random process, $\bar{b}$ being the time-averaged intermediate level, and $b'(t)$ the fluctuation.   This causes random phase modulation of propagators.   From the graphs in Fig. 1, it is seen that the modulation works in the intervals $(0, \tau_2)$ and (τ_1, τ) but not in (τ_2, τ_1). It should be kept in mind, however, that the random modulation persists throughout the whole time interval $(0, \tau)$ even if it is inactive in the middle.

Now the propagators are modified by a common modulation factor,

$$F(\tau_2, \tau_1, \tau) = \langle \exp \left[i \int_0^{\tau_2} b'(t) dt - i \int_{\tau_1}^{\tau} b'(t) dt \right] \rangle \quad , \tag{17}$$

where the average is taken over all possible realizations of $b'(t)$. Without making any explicit calculations, we can see immediately what happens in the following two limits.

a) SLOW MODULATION LIMIT.   If the modulation is very slow, the adiabatic modulation is static.   The problem is essentially the same as that discussed in section III except that the intermediate level has a probability distribution, say $P(b)$.   Thus the transition rate W_1 should be

$$W_1 = \pi \delta (a-c) \int db \, P(b) \, \frac{\gamma_b^2 + (b-a)(b-c)}{\{ \gamma_b^2 + (b-a)^2 \} \{ \gamma_b^2 + (c-b)^2 \}} \quad ,$$

which is a pure Raman process.

b) FAST MODULATION LIMIT (NARROWING LIMIT).   If the modulation is
fast enough to satisfy the narrowing condition, the random phase modulation reduces to a phase memory relaxation familiar in spin resonance
phenomena.   In our case, the phase modulation factor given by eq.(17)
becomes

$$F(\tau_2, \tau_1, \tau) = \exp\{-\gamma'\tau_2 - \gamma'(\tau - \tau_1)\} \qquad , \qquad (18)$$

where

$$\gamma' \sim \langle (b')^2 \rangle \tau_m \qquad \left(\tau_m = \gamma_m^{-1}\right)$$

is the effective transverse relaxation rate of the phase memory.
This approximation is valid if the correlation time τ_m of the random
IMSI is so short as to satisfy the narrowing condition,

$$\langle (b')^2 \rangle^{\frac{1}{2}} \tau_m \ll 1 \quad ,$$

and if the resulting spectrum of W_1 is broad enough.   In this narrowing
limit the afore-mentioned cancellation of non-Raman term is no longer
complete and we are left with

$$W_1 = \pi\delta(a-c) \frac{\gamma^2 + (\bar{b}-a)(\bar{b}-c)}{\{\gamma^2 + (\bar{b}-a)^2\}\{\gamma^2 + (\bar{b}-c)^2\}}$$

$$+ \frac{\gamma'}{\gamma_b(\gamma_b + \gamma')} \frac{\gamma}{\{\gamma^2 + (\bar{b}-a)^2\}} \frac{\gamma}{\{\gamma^2 + (\bar{b}-c)^2\}} \qquad , \qquad (19)$$

where

$$\gamma = \gamma_b + \gamma' \quad .$$

Here we have, besides the Raman term, a luminescence-like continuous
spectrum inversely proportional to γ_b, in accordance with eq.(9).   Its
spectral function is the product of the resonance absorption factor and
the resonant emission factor, which seems quite natural.   For a vanish-

ing τ_m, everything is averaged out and only the Raman term remains.

In general cases of arbitrary rate of random modulation, it is necessary to define the problem more explicitly. As was discussed in a review article some years ago by one of the present authors,[2] it is possible to develop a general treatment if the random modulation is assumed to be basically Markoffian. Two standard examples are considered here.

1) TWO-STATE JUMP MODEL. The modulation $b'(t)$ in eq.(16) is assumed to take only two values, $b' = \pm \Delta$ corresponding to two possible states of R. This is of course a great idealization, but is sufficient to reveal some general features of the problem. The two states,

$$|+) = \begin{pmatrix} 1 \\ 0 \end{pmatrix} \quad \text{and} \quad |-) = \begin{pmatrix} 0 \\ 1 \end{pmatrix} \quad ,$$

alternate randomly as a Markoffian process characterized by the transition matrix,

$$\Gamma = \frac{\gamma_m}{2} \begin{pmatrix} -1 & 1 \\ 1 & -1 \end{pmatrix} , \tag{20}$$

and the random modulation is represented by the matrix,

$$\Omega = \begin{pmatrix} \Delta & 0 \\ 0 & -\Delta \end{pmatrix} . \tag{21}$$

Following the general formalism of such Markoffian random modulation, we easily find that the transition rate W_1 is given by the formula,

$$W_1 = \lim_{s \to +0} [(0| \frac{1}{\gamma_b + i(\bar{b}-c) + i\Omega - \Gamma} \frac{1}{s + i(a-c) - \Gamma} \frac{1}{\gamma_b + i(a-\bar{b}) - i\Omega - \Gamma} |0)$$

$$+ (0| \frac{1}{\gamma_b + i(\bar{b}-c) + i\Omega - \Gamma} \frac{1}{2\gamma_b - \Gamma} \frac{1}{\gamma_b + i(a-\bar{b}) - i\Omega - \Gamma} |0)$$

$$+ (0| \frac{1}{\gamma_b + i(c-\bar{b}) - i\Omega - \Gamma} \frac{1}{2\gamma_b - \Gamma} \frac{1}{\gamma_b + i(a-\bar{b}) - i\Omega - \Gamma} |0)$$

$$+ \text{c.c.}] \quad , \tag{22}$$

which is a generalization of eq.(13). Here each factor inside brackets is a two by two matrix and the initial state of R is the vector,

$$| 0) = \frac{1}{2} \begin{pmatrix} 1 \\ 1 \end{pmatrix} \quad ,$$

corresponding to the equilibrium eigenvector of Γ. Multiplication by $(0 | = (1, 1)$ from the left is summation over the final states of R. The right and left vectors $| 0)$ and $(0 |$ satisfy

$$\Gamma | 0) = 0 \quad , \quad (0 | \Gamma = 0 \ .$$

Evaluation of W_1 by eq.(22) is an elementary algebraic manipulation.

2) GAUSSIAN-MARKOFFIAN MODULATION. The modulation $b'(t)$ is assumed to be a Gaussian-Markoffian process, which is characterized by the correlation function,

$$\langle b'(t_1) b'(t_2) \rangle = \Delta^2 \exp [-\gamma_m | t_1 - t_2 |] \ . \tag{23}$$

The explicit form of the modulation factor eq.(17) is easily found to be

$$F_\pm (\tau_2, \tau_1, \tau) = \exp [-\frac{\Delta^2}{\gamma_m^2} \{ \gamma_m \tau_2 - 1 + e^{-\gamma_m \tau_2} + \gamma_m (\tau - \tau_1) - 1$$
$$+ e^{-\gamma_m (\tau - \tau_1)} \pm e^{-\gamma_m (\tau_1 - \tau_2)} (1 - e^{-\gamma_m \tau_2}) (1 - e^{-\gamma_m (\tau - \tau_1)}) \}] \ . \tag{24}$$

With the definitions of eqs.(12) and (24), the propagators I, II and III are expressed as $I_0 F_-$, $II_0 F_-$ and $III_0 F_+$, respectively. Equation (11) enables us to calculate W_1. It should be noted that the expression (22) is generally valid for a Markoffian modulation, not only for a two-state jump model. For a Gaussian-Markoffian modulation, Ω in this expression is considered as a real continuous variable and Γ is the diffusion operator,

$$\Gamma = \gamma_m \left[\frac{\partial}{\partial \Omega} \, \Omega + \Delta^2 \frac{\partial^2}{\partial \Omega^2} \right] \; . \tag{25}$$

The vectors $|0)$ and $(0|$ are eigenfunctions of Γ corresponding to the zero eigenvalue. Although the expression is compact, further analytical treatment is rather complicated, so that we have to work numerically.

Figures 2 and 3 are the results of such model calculations to show how the emission spectra vary as the rate of modulation changes from the slow limit to the fast limit. In each case, the levels A and C are taken at the same energy. The energies are normalized by the energy difference between these levels and the average energy $\bar{b}$ of the intermediate state B.

Figure 2 is for a two-state jump model. The state B randomly takes either of the two energies 0.8 and 1.2. The incident photon ω_1 is 1.5. The pure Raman peak with the delta-function $\delta(\omega_2 - 1.5)$ is omitted in the figure. When the modulation is slow with $\gamma_m = 0.01$, three resonance peaks are seen; one is a somewhat broadened Raman resonance which superposes on the delta-function peak and the other two peaks are in resonance with two possible excited levels and bear the character of luminescence. As the modulation becomes faster, say $\gamma_m = 0.1$ or 1.0, the narrowing takes place, which reduces the broad Raman-like peak and narrows the luminescence-like peaks into a sharp single peak at $\omega_2 = 1.0$ which finally disappears for $\gamma_m = \infty$.

Figure 3 is for a Gaussian model. The energy of the incident photon ω_1 is taken as 0.7. The general features are almost the same as the previous one. For slow modulation, a broadened Raman-like peak and a luminescence peak are seen. As the modulation becomes faster, the Raman-like peak disappears and the luminescence peak narrows. Both in Figs. 2 and 3, the zero-width Raman line, which is not shown in the figure, persists all the way. As is given in the figure caption, the intensity of this part does not so much vary with the rate of modulation through the slow and fast modulation limits.

Although the above results are obtained for some idealized models, the qualitative features should be common for more general cases of adiabatic modulation.

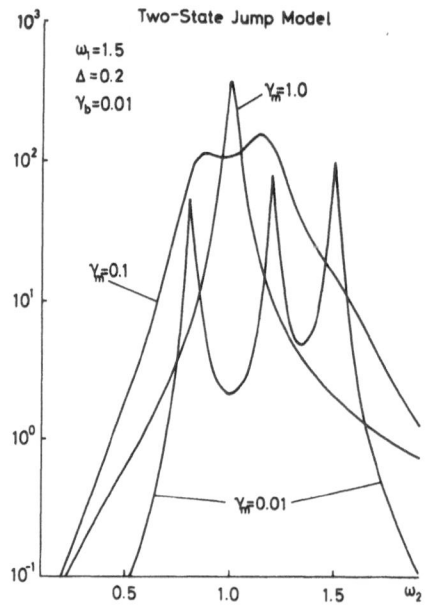

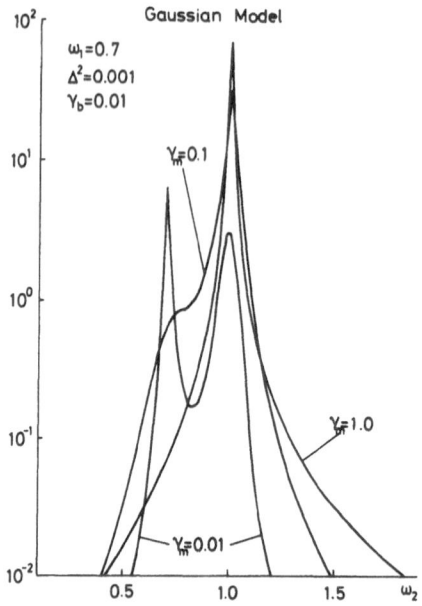

Fig.2. Emission spectra of a three-level system with adiabatic random modulation in a two-state jump model. The incident photon energy, the radiative damping rate and the rate and amplitude of modulation is denoted by ω_1, γ_b, γ_m and Δ, respectively. The emission intensity is plotted in an arbitrary scale. The intensity of the delta-function like Raman peak at $\omega_2 = 1.5$, which is omitted in the figure, is given as 3.997×10^1, 3.758×10^1 and 2.871×10^1 for $\gamma_m = 0.01$, 0.1 and 1.0, respectively.

Fig.3. Emission spectra of a three-level system with adiabatic random modulation in a Gaussian-Markoffian model. Notations are the same as in Fig.2. The intensity of the zero-width Raman line at ω_2 =0.7, which is not shown in the figure, is given as 3.567×10^1, 3.555×10^1 and 3.492×10^1 for $\gamma_m = 0.01$, 0.1 and 1.0, respectively.

V. OFF-DIAGONAL RANDOM MODULATION

The stochastic theory can be extended to more complex and more realistic models. The task is left for future studies. Here we

mention a simple model which may still simulate certain cases of
physical interest. Suppose that the state B consists of two levels
and the modulation from R acts as an off-diagonal random perturbation.
Namely, the Hamiltonian for B is assumed to be

$$H_B = \begin{pmatrix} b_1 & \Omega(t) \\ \Omega(t) & b_2 \end{pmatrix} \quad , \tag{26}$$

where $\Omega(t)$ is a random process. Furthermore we assume that $V_1^{\pm}$ connects
A with B_1 and $V_2^{\pm}$, C with B_2. Thus, in the absence of modulation,
there is no way of a second order process between A and C. When $\Omega(t)$
is very slow, the states B_1 and B_2 are mixed and A and C are connected,
giving rise to a Raman resonance in proportion to Ω^2. In the limit
of extremely fast modulation, $\Omega(t)$ is averaged out and there is no
second order process. For intermediate rates of modulation, there
occur the adiabatic and the nonadiabatic effects of modulation. If
the condition,

$$\gamma_m \ll E = |b_1 - b_2| \quad ,$$

is met, the effect is adiabatic and is similar to that discussed in
the previous section. The only difference is that the Raman process
is caused here by the off-diagonal modulation so that the resonance is
broadened; there is no zero-width line. If γ_m approaches E, the power
spectrum of modulation contains components which resonate with E to
induce nonadiabatic transitions between B_1 and B_2. This causes a
three-step transition, $A \rightarrow B_1$, $B_1 \rightarrow B_2$, $B_2 \rightarrow C$, a typical luminescence
process, emitting $\omega_2 \sim B_2 - C$ when photons $\omega_1 \sim B_1 - A$ are absorbed in
resonance. This model can be analysed in more detail if the process
$\Omega(t)$ is specified, for example, as a two-state jump model. Figure 4
shows a model calculation of this type. For the slow modulation,
excited states are shifted adiabatically to the levels at 1.672 and
0.628 for the set of parameters given in the figure. Since C = 0.1,

the Raman resonance peaks around $\omega_2 = 1.2$, and luminescent resonances are at 1.572 and 0.528. As the modulation becomes faster, the Raman peak decreases while the luminescent peaks dominate and coalesce into a single peak. These features are naturally understandable.

VI. CONCLUSION

Summarizing, we have shown that the effect of IMSI on a second order process can be studied by modeling it as a stochastic modulation. It is seen that the Raman-like and luminescence-like resonances may coexist in the emission spectra and the spectral shapes vary as the rate of modulation changes. This approach seems useful for understanding the nature of second order processes and the role of IMSI. Of course it is desired to clarify its relations to other theories. This is left for a future study.

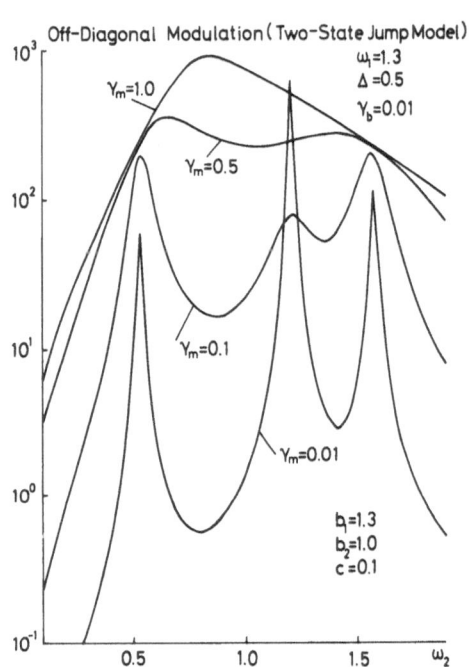

Fig.4. Emission spectra of a four-level system with off-diagonal modulation in a two-state jump model. The energy of the levels B_1, B_2 and C is denoted by b_1, b_2 and c, respectively. The level B_1 is optically connected only with the ground state A and the level B_2, only with C. The incident photon energy ω_1 is just resonant with the level B_1. There is no delta-function like Raman peak.

REFERENCES

1) V. Hizhnyakov and I. Tehver: Phys. Stat. sol. <u>21</u> (1967) 755; *ibid.* <u>39</u> (1970) 67. D. L. Huber: Phys. Rev. <u>170</u> (1968) 418; *ibid.* <u>178</u> (1969) 93.

M. V. Klein: Phys. Rev. B8 (1973) 919.

Y. R. Shen: Phys. Rev. B9 (1974) 622.

S. E. Schwartz: Phys. Rev. A11 (1975) 1121.

J. R. Solin and H. Merkelo: Phys. Rev. B12 (1975) 624.

D. L. Rousseau, G. D. Patterson and P. F. Williams: Phys. Rev. Letters 34 (1975) 1306.

S. Mukamel, A. Ben-Reuven and J. Jortner: Phys. Rev. A12 (1975) 947, and many other papers on related subjects.

2) R. Kubo: in *Stochastic Processes in Chemical Physics*, Adv. Chem. Phys. Vol. XV, ed. Kurt E. Shuler, (John Wiley & Sons, N. Y., 1969) p.101.

3) R. Kubo: in *Fluctuation, Relaxation and Resonance in Magnetic Systems*, ed. D. ter Haar, (Oliver and Boyd, Edinburgh, 1962) p.23.

4) P. W. Anderson: J. Phys. Soc. Japan 9 (1954) 316.

R. Kubo: *ibid.* 9 (1954) 1 63

5) R. Kubo: J. Phys. Soc. Japan, 26 Supplement (1969) 1.

PHOTOLUMINESCENCE IN HIGHLY EXCITED GaSe

A. Mercier, J. P. Voitchovsky, E. Mooser and A. Baldereschi
Laboratoire de Physique Appliquée, EPF-Lausanne,
Switzerland

ABSTRACT

The photoluminescence of GaSe at high excitation intensities is
reported. Four lines, whose intensity increases superlinearly with
excitation, are resolved and their behaviour is investigated as function
of excitation intensity, excitation frequency and temperature. According to their spectral position and their behaviour, two lines are attributed to different exciton-exciton scattering processes while the
remaining two which give rise to spontaneous and stimulated emission,
respectively, are tentatively attributed to exciton-carrier scattering.

The luminescence spectra of GaSe at high excitation intensity have
been studied in the past by various authors[1-7] who have proposed con-
tradictory interpretations.

An emission line which at 80° K lies 20 - 30 meV below the free-
exciton recombination line and whose intensity increases superlinearly
with excitation, has been observed by Ugumori et al.[1] and has been
attributed to exciton-exciton scattering. The same line has also been
observed by Leite et al.[2] who have interpreted it as originating from
exciton-carrier scattering. A new line which increases superlinearly
with excitation has been reported by Mercier and Voitchovsky.[3] At
4.2° K this line lies 4 meV below the free exciton line and has been
attributed to exciton-electron scattering. Even more confusing are the
various reports of a stimulated emission line which at 2° K lies approximately 45 meV below the free-exciton line. Nahory et al.[4] first reported this line and concluded that GaSe is a direct-gap semiconductor
in contradiction with other experimental data.[8] Stimulated emission
in GaSe has also been observed by Ugumori et al.[5] and has been attributed to exciton-exciton scattering. Recently, Kuroda et al.[6] have
interpreted the stimulated emission as the recombination of the direct

free exciton which is assisted by two-phonon emission.

In the present paper we briefly present the results of a detailed investigation of the photoluminescence of GaSe under high excitation intensity. A more extended version of the present work is given elsewhere.[7]

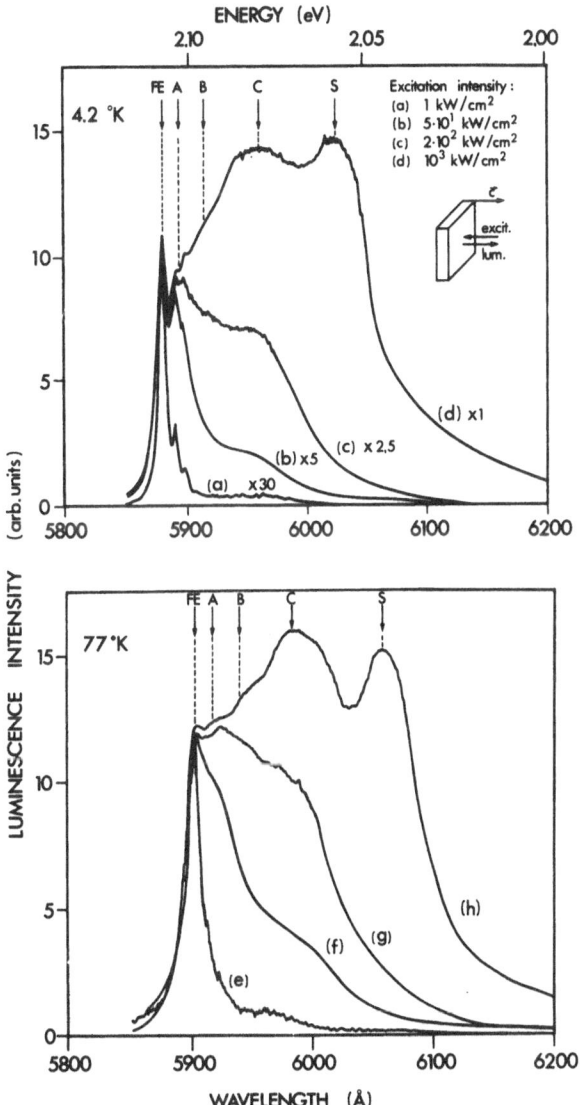

Fig.1. Photoluminescence spectra of GaSe at 4.2 K (upper portion) and 77 K (lower portion) for different excitation intensities. The geometry of the experiments is indicated.

The GaSe crystals used in the present experiments have low impurity concentration (about 10^{17} cm^{-3}) and therefore the luminescence spectra at low excitation intensity consist only of the direct free exciton emission, as is shown in Fig.1. As the excitation intensity increases, four new lines (labelled A, B, C, and S in Fig.1) appear in the low-energy tail of the spectra. At low temperature, these lines lie about 4, 15, 20, and 40 meV below the free exciton line.

The intensity of these lines varies superlinearly with excitation intensity as is shown in Fig. 2. In particular, the intensity of line S is strongly dependent on the excitation intensity as one expects in the case of stimulated emission. Using the intensity of the free-exciton recombination line as a measure of the exciton density D, the data given in Fig. 2 show that the intensity at line A is proportional

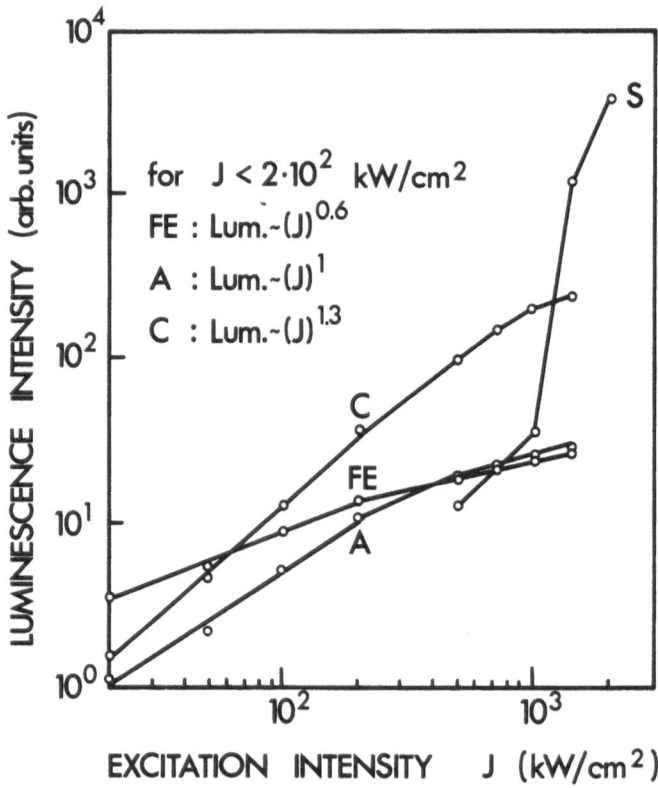

Fig.2. Intensity of the photoluminescence lines of GaSe as function of excitation intensity. The intensity variation of line B is similar to that of line C. The power laws, which are indicated, are valid at low excitation $(J < 2 \times 10^2$ kW/cm$^2)$.

to $D^{1.6}$ while that of lines B and C is proportional to D^2. With in-

creasing excitation intensity, the luminescence lines shift to lower

frequencies with the exception of line A which approximately keeps its

energy position. These results are given in Fig. 3 which shows that,

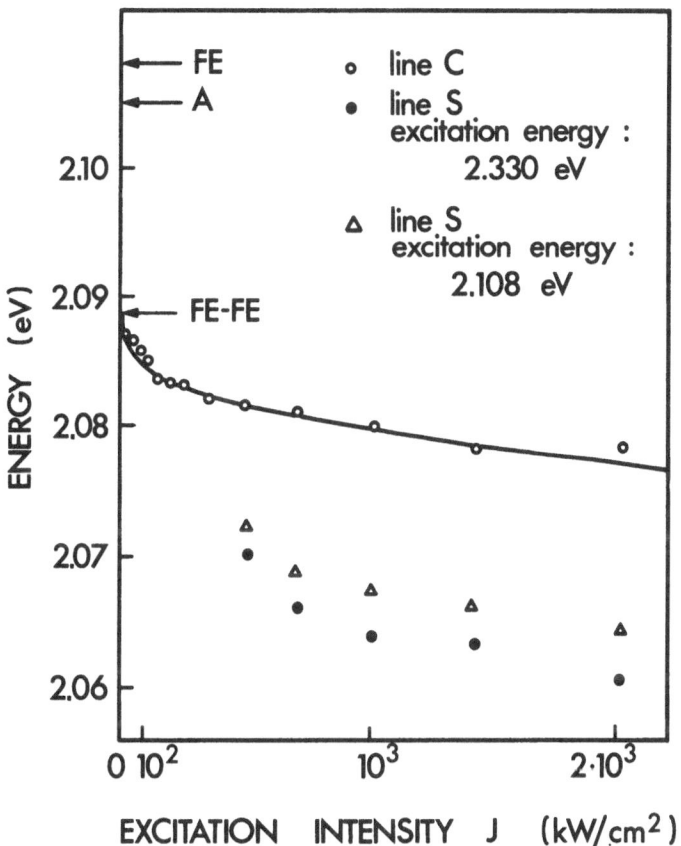

Fig.3. Spectral position at 4.2 K of the photoluminescence lines
of GaSe as function of excitation intensity. The free-
exciton (FE) and A lines do not shift with excitation
intensity. Notice the strong dependence on intensity of
the position of lines C and S. The arrow denoted FE –
FE shows the theoretical prediction for the position of
line C at J = 0.

for line S, the energy position is a function of the excitation frequency,

too. The temperature dependence of the energy position of the four

luminescence lines has also been studied and the results are shown in

Fig. 4 for temperatures between 4.2 K and 300 K.

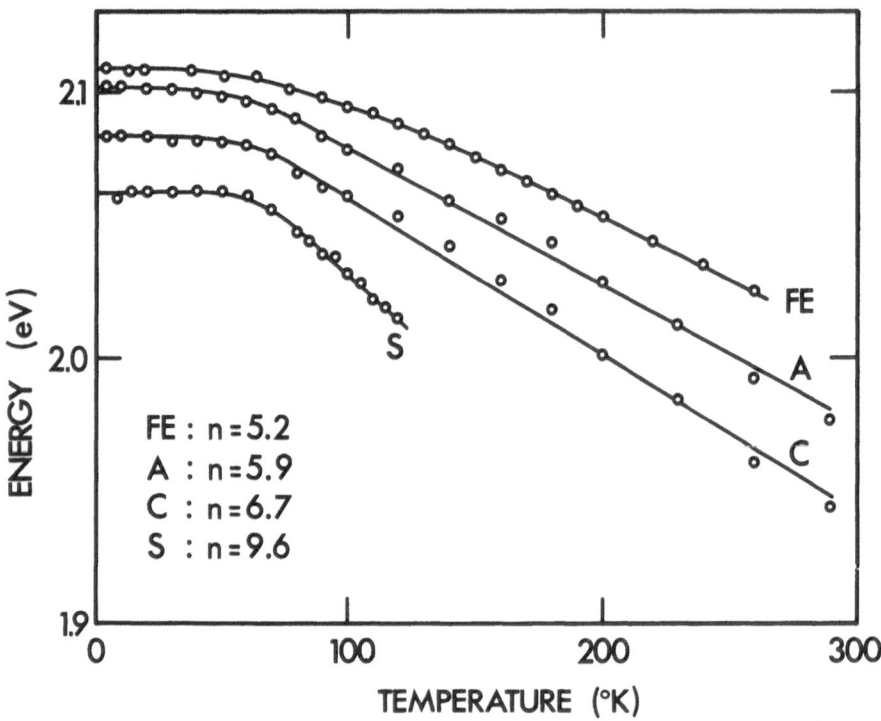

Fig.4. Spectral position of the photoluminescence lines of GaSe as function of temperature. The coefficients n of the linear shift (-nkT) of the various lines at higher temperatures are indicated.

The data presented above on the variation of the luminescence spectra as function of excitation intensity, excitation frequency and temperature, can be used to understand the mechanisms which are responsible for the various recombination processes.

The intensity of line A, which lies 4 meV below the free exciton line is proportional to the 1.6 power of the exciton density. This is approximately the expected behaviour for a line originating from exciton-carrier scattering.[9] The energy position as function of temperature of a line originating from exciton-carrier scattering depends on the effective-mass of the carrier, as shown by Bille.[10] The effective-mass values obtained by Ottaviani et al.[11] together with the experimental data shown in Fig. 4 suggest that the carriers are the electrons at the center of the Brillouin zone.

Lines B and C (15 and 20 meV below the free-exciton line) have an intensity which varies with the square of the exciton density. They are therefore interpreted as being due to inelastic exciton-exciton scattering processes in which one exciton recombines radiatively while the other is excited to the n = 2 state (line B) or is ionized into an electron-hole pair (line C). This interpretation is consistent with the observed value of the binding energy of the free exciton (19.8 ± 0.1 meV).[12]

Finally, line S (which lies about 40 meV below the free exciton) corresponds to stimulated emission as is evident from the data of Fig. 2. In order to understand the nature of this emission band, we give in Fig. 5 the excitation spectrum corresponding to the luminescence at λ =

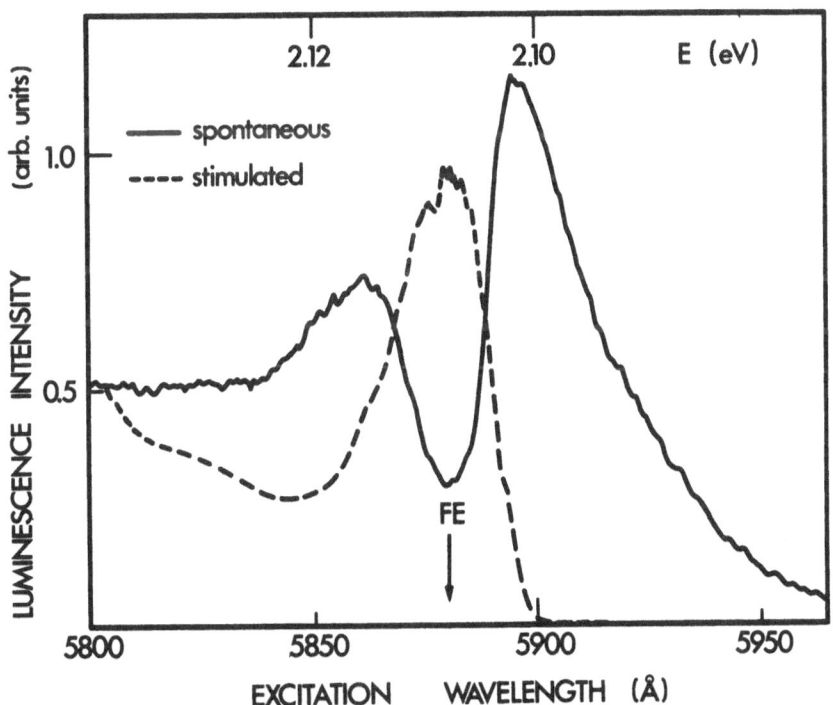

Fig. 5. Excitation spectra for the spontaneous (solid line) and stimulated (broken line) luminescence at λ = 6000 Å and at 4.2 K. The spectral position of the free exciton is indicated by the vertical arrow thus showing the relevance of the free exciton in the stimulated luminescence.

6000 Å and at 4.2 K. From these data, it is obvious that the free exciton plays an important role in the recombination process which gives
rise to this luminescence. This process, however, remains still unexplained. In particular, it is not clear at present why the stimulated
line is at much lower energy than the free-exciton line and why its
energy position is so strongly dependent on excitation intensity and on
excitation frequency.

REFERENCES

1) T. Ugumori, K. Masuda and S. Namba: Phys. Letters A 38 (1972) 117.

2) R. C. C. Leite, E. A. Meneses, N. Jannuzzi and J. G. P. Ramos:
   Solid State Commun. 11 (1972) 1741.

3) A. Mercier, E. Mooser and J. P. Voitchovsky: J. Luminescence 7
   (1973) 241.

4) R. E. Nahory, K. L. Shaklee, R. F. Leheny and J. C. DeWinter:
   Solid State Commun. 9 (1971) 1107.

5) T. Ugumori, K. Masuda and S. Namba: Solid State Commun. 12 (1973)
   389.

6) N. Kuroda, T. Nakanomyo and Y. Nishina: Jpn. J. Soc. appl. Phys.
   43 (1974) 63.

7) A. Mercier and J. P. Voitchovsky: Phys. Rev. B 11 (1975) 2243.

8) E. Aulich, J. L. Brebner and E. Mooser: Phys. Stat. sol. 31
   (1969) 129.

9) C. Benoit à la Guillaume, J. M. Debever, and F. Salvan: Phys. Rev.
   177 (1969) 567.

10) J. Bille, in *Festkörperprobleme XIII*   (Pergamon Press, New York, 1973)
    p.111.

11) G. Ottaviani, C. Canali, F. Nava, Ph. Schmid, E. Mooser, R. Minder
    and I. Zschokke: Solid State Commun. 14 (1974) 933.

12) E. Mooser and M. Schlüter: Nuovo Cimento B 18 (1973) 164.

STIMULATED EMISSION IN LAYER-TYPE SEMICONDUCTORS

Y. Nishina, N. Kuroda, M. Yashiro, K. Nakaoka and T. Goto

Research Institute for Iron, Steel and Other Metals
Tohoku University
Sendai, Japan 980

ABSTRACT

Stimulated photoluminescent spectra of several kinds of layer-type semiconductors, $GaSe_{1-x}S_x$, HgI_2 and PbI_2 have been measured at liq. N_2 and He temperatures. In GaSe, the nonlinear emission may be explained in terms of the Auger-type interactions between the direct exciton and the indirect one with their holes in common in the center of the Brillouin zone. The stimulated photoluminescence due to annihilation of excitonic molecule is found in HgI_2 at liq. He temperature. The gain spectrum for the nonlinear emission in 12R-type PbI_2 is compared with Haug's model of the exciton-LO-phonon interaction.

I.   INTRODUCTION

Electrical and optical properties of layer-type semiconductors under laser beam excitations have been investigated with the possibility in our mind that the strong anisotropy in the electronic structure might bear a stable metallic phase.[1] Also the technical ease in making the parallel cleavage planes out of the grown crystals would contribute to the stimulated emission with its high gain coefficient.[2] The mechanism of the stimulated emission, however, has various origins, depending upon the compound substances with their respective magnitude of the exciton-phonon, exciton-free carrier and exciton-exciton interactions. The present paper discusses the observed spectra of the stimulated emission in terms of the four different type of interactions played by the excitons at relatively high densities. The emission mechanisms are investigated with respect to changes in the time-resolved spectra for various values of chemical composition $(GaSe_{1-x}S_x)$, the excitation intensity and temperature.

II. $GaSe_{1-x}S_x$ $(0 \leq x \leq 0.2)$

ε-GaSe single crystal has been known for some time for the occurrence of high gain stimulated emission under the laser beam excitation with its photon energy above the direct exciton absorption edge of 2.109 eV (4.2 K).[2-5] At liq. He and N_2 temperatures the irradiation of the cleaved c-plane by the second harmonics (2.34 eV with pulse half-width = 12 nsec) of the Q-switched Nd-YAG laser gives rise to the photoluminescent spectra in reflection as shown in Fig. 1. Many of the

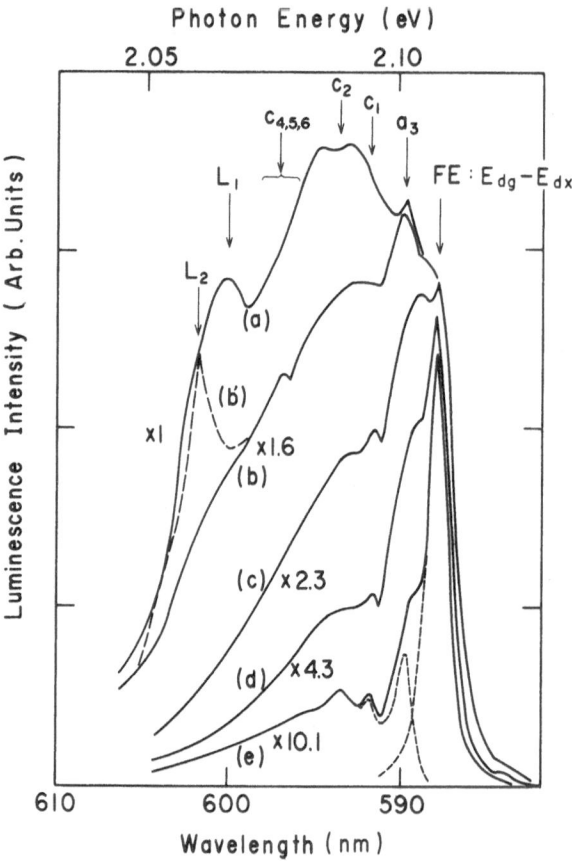

Fig.1. The luminescence spectra of GaSe measured in relection at 4.2 K.[14] The excitation intensities are (a) 4 MWcm^{-2}, (b) and (b)' 2 MWcm^{-2}, (c) 1 MWcm^{-2}, (d) 0.5 MWcm^{-2} and (e) 0.1 MWcm^{-2}, respectively.

spontaneous lines (a_3, c_1 through c_6 in the notation of Mercier[5]) have been identified in terms of the direct (a_3) or indirect (c_1 through c_6) bound exciton recombinations and their phonon replicas.[6] Above as-signment has been given on the basis of measurements of photon energy shifts of respective peaks with respect to chemical composition x. Namely, c_1 and c_6 follow the shift of the indirect edge with respect to x and extrapolates to the indirect edge of GaS ($x = 1$) minus the energy required for formation of the bound exciton states and its coupling with respective phonons. On the other hand, the a_3 peak follows the shift of the direct exciton with respect to x. Also the decay time of this peak is in the order of 10^{-8} sec (4.2 K) or less, whereas that of c_i's ($i = 1, \ldots\ldots 6$) lie in the range from 0.12 to 10 μsec, depending on x. These fine structures are resolved more clearly in our recent data of reflection measurement where many of the subsidiary peaks may be associated with stacking faults introduced by stress on cleaving the crystal.[7]

The stimulated emission line, L_1, at 2.067 eV becomes observable if the excitation on the cleaved c-plane exceeds the threshold intensity of approximately 2 MW/cm^2. For a particular configuration of the ex-citing beam along the c-axis, the threshold for the stimulated emission perpendicular to the c-axis may be as low as 0.7 MW/cm^2, depending upon the exciting beam width along the direction of emission. Various mech-anisms for the origin of this nonlinear emission have been proposed by many investigators in terms of,

i) direct exciton-direct exciton interaction.[4]

ii) recombination of the direct exciton associated with emission of two TO-phonons with small dispersion.[2,8]

iii) interaction between direct-exciton and conduction electron.[5]

The followings are the brief account of our arguments which indicate the difficulties in explaining our experimental results in any of the above-mentioned models: According to our magneto-optical absorption measurement[9] and the analysis of the photoluminescent spectra,[6] the

binding energy, E_{dx}, of the direct exciton is ~20 meV, for the direct

band gap, E_{dg} = 2.129 eV (4.2 K). If the Auger-type interaction between

the direct excitons is responsible for the stimulated emission (model

(i)), its main peak should be near the photon energy of $E_{dg} - 2E_{dx}$,

which is ~20 meV below the direct exciton edge, $E_{dg} - E_{dx}$, in contrast

to the observed difference of ~40 meV. Since the pure crystal is trans-

parent in the visible photon energy range below 2.10 eV (4.2 K) for the

polarization E⊥c, it is unlikely that the tail of the direct exciton

absorption induces such a large shift in the peak of the emission

spectrum. The reverse process of the stimulated emission,[10] if any,

would induce the peak shift in emission down to the energy near $E_{dg} -$

$2E_{dx}$, but never to $E_{dg} - 3E_{dx}$. As explained below, our observed spectra

do not show such a large shift in the photon energy.

The model for the stimulated emission through the interaction be-

tween the direct exciton and the two TO-phonons via indirect edge as

the intermediate state explains the characteristics of the emission

only on the semiquantitative basis. Although the TO-phonon has been

known to have very small dispersion,[11] this type of interaction would

produce the emission peak at ~50 meV below the energy $E_{dg} - E_{dx}$, corres-

ponding to the TO-phonon energy of 25 meV.[12] Also one may estimate

the temperature dependence of the threshold intensity for the stimulated

emission in terms of the statistics of exciton and phonon.[13] This

phenomenological treatment, however, cannot explain the fact that the

threshold intensity at 77 K is about 20 % lower than that at 4.2 K.

Mercier and Voitchovsky[5] have proposed that the Auger-type interaction

between the direct exciton and the conduction electron is the principal

mechanism of the stimulated emission on the basis of the large shift of

the free exciton peak toward higher level of excitation. Our data,

however, do not show such a large red shift as shown in Fig. 2. In

Mercier's measurement both pulse width (70 nsec at half maximum) of the

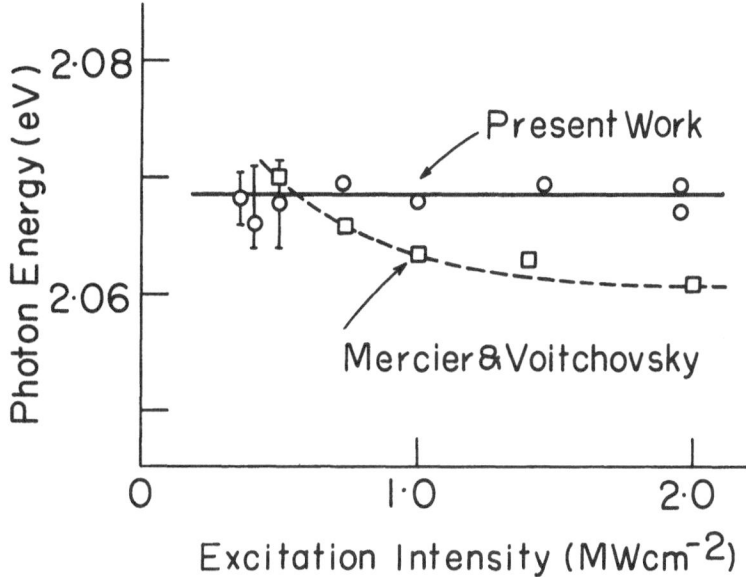

Fig.2. The energy shift of the stimulated emission line, L_1 in Fig. 1, with respect to the excitation intensity. The present data (o) are compared with those of Mercier and Voitchovsky (□).[5]

Q-switched Nd-YAG laser and the pulse repetition rate of 75 Hz are much larger (or higher) than our arrangement of 12 nsec and ~1 Hz, respectively, so that there is a good possibility that such a heavily irradiated sample may be damaged optically near the surface. If we irradiate our sample with the Q-switched pulse of $\cong$ 2 MW/cm^2 for 2×10^4 times or so, then the stimulated emission line shifts to the energy close to the results of Mercier. Such energy shift is shown by the dashed-line L_2 in Fig. 1.

So far, the gain coefficient spectrum near the threshold and the photon energy of the stimulated emission peak may be explained well in terms of the Auger type interaction between the direct exciton and the indirect exciton with their holes in common in the center of the Brillouin zone. Our analysis of the photoluminescent data[6] shows that the binding energy of the indirect exciton, E_{ix}, is 38 meV. Moreover, the indirect conduction band edge, E_{ig}, is found to lie

5 meV <u>above</u> the direct one, E_{dg}, so that the difference between the binding energy of the direct exciton, E_{dx}, and that of the indirect one, E_{ix}, makes the indirect exciton ground state <u>lower</u> than the direct by 13 meV. This relationship in excitonic energy levels makes the intensity of the stimulated emission as large as the direct transition accompanied by the Auger-type interaction between a pair of direct excitons. It also gives the comparable populations of both type of excitons at 77 K, where the threshold for the stimulated emission becomes the lowest. It should be noted that the Auger-type interaction between the direct exciton and the indirect one is enhanced in the present case by the exchange interaction between the holes of both types of excitons with their strong mass-anisotropy.[14] As shown in Fig. 3, the gain spectrum at 4.2 K near the threshold exhibits the line shape expressed by the phenomenological model by Benoît à la Guillaume[15] for the stimulated emission through the exciton-exciton collision, so that the gain coefficient, $g(\hbar\omega)$ for the emitted photon energy, $\hbar\omega$ is given by,

$$g(\hbar\omega) \propto (E_o - E_{ix} - \hbar\omega)^{1/2} (E_o - \hbar\omega)^{-6} , \tag{1}$$

where $E_o = E_{dg} - E_{dx} = 2.109$ eV (4.2 K) is the formation energy of the direct exciton and $E_{ix} \cong 40$ meV the binding of the indirect exciton. The experimental points show fine structures, corresponding to the Auger scattering of an indirect exciton to its excited states (n = 2 and 3) as indicated by the arrows in Fig. 3.

III. HgI$_2$

The photoluminescent spectra of red-HgI$_2$ with tetragonal structure have been measured at liq. He, N$_2$ and room temperatures for various levels of excitation by nitrogen laser (3.678 eV) and by Ar$^+$-ion laser (2.410 eV). Because of the spin-orbit coupling and the crystal field, the p-like valence band edge splits into three levels at the center of the Brillouin zone.[16] Consequently, the absorption spectrum shows

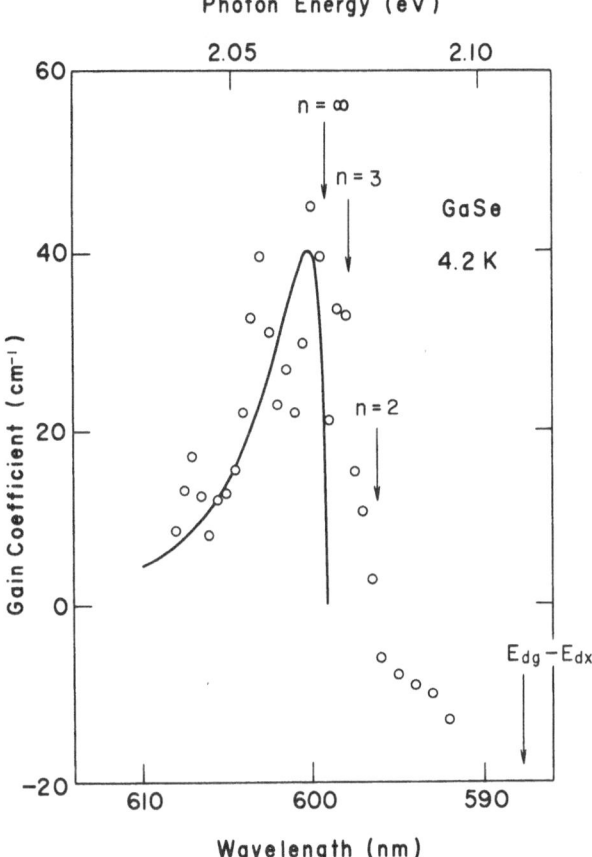

Fig.3. The gain spectrum of L_1 line for the
excitation intensity G = 1 MWcm^{-2} at
4.2 K.[14] The arrow denoted as E_{dg}
$- E_{dx}$ indicates the spectral position
of a direct free exciton. The solid
line is the calculated value according
to eq. (1).

three series of the exciton lines, namely A (electric vector, E⊥c), B

(E⊥ and ∥c) and C (E⊥ and ∥c) excitons for the ground state energy of

2.337 eV, 2.538 eV and 3.351 eV, respectively, at 4.2 K. The direct

transition to the s-like conduction band edge is allowed only for the

configurations of the electric vector shown in parentheses.

Figure 4 shows the photoluminescent spectrum just below the energy

for the A-exciton formation. For a weak excitation by Ar⁺-ion laser at

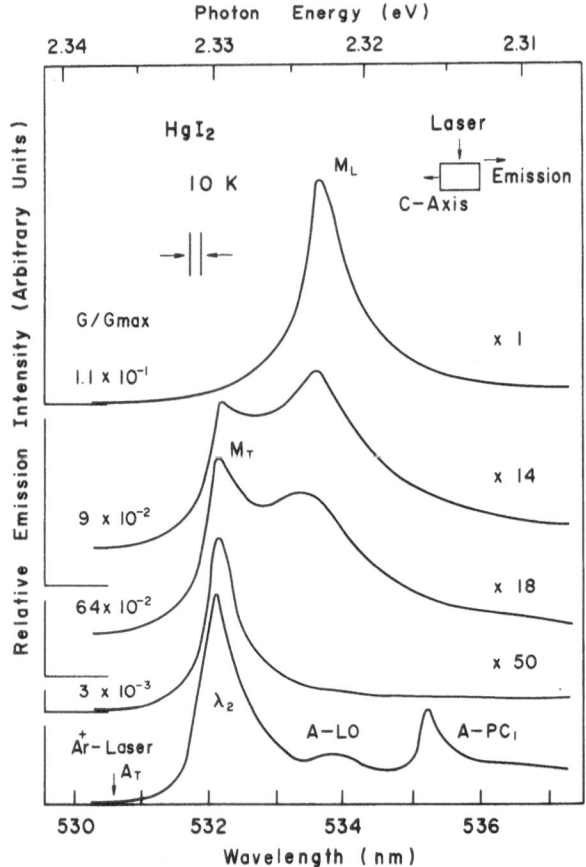

Fig.4. The photoluminescent spectrum of red HgI_2 at
10 K along the c-axis for the excitation
incident perpendicular to the c-axis by Ar^+-
laser or by N_2-laser pulse beams for various
excitation intensity, G, where the maximum of
G, G_{max} is 10 MWcm^{-2}.

10 K, the predominant emission peak, λ_2, comes from the recombination

of the A-exciton bound to a crystal imperfection as assigned by Novikov

et al.[17] The subsidiary peaks, A-LO and A-PC$_1$ are one LO-phonon- and

multiphonon-replica of the A-exciton line, respectively. For higher

levels of excitation, the λ_2-line merges into the M_T-line which is

due to the annihilation of the A-excitonic molecule, emitting one trans-

verse exciton.[18] Above the peak excitation intensity of ~500 kW/cm^2

by N_2-laser (pulse width: 10 nsec), another peak, M_L, begins to appear and shows the line narrowing with the sudden increase in intensity at this threshold of excitation, as shown in Fig. 5. The energy position

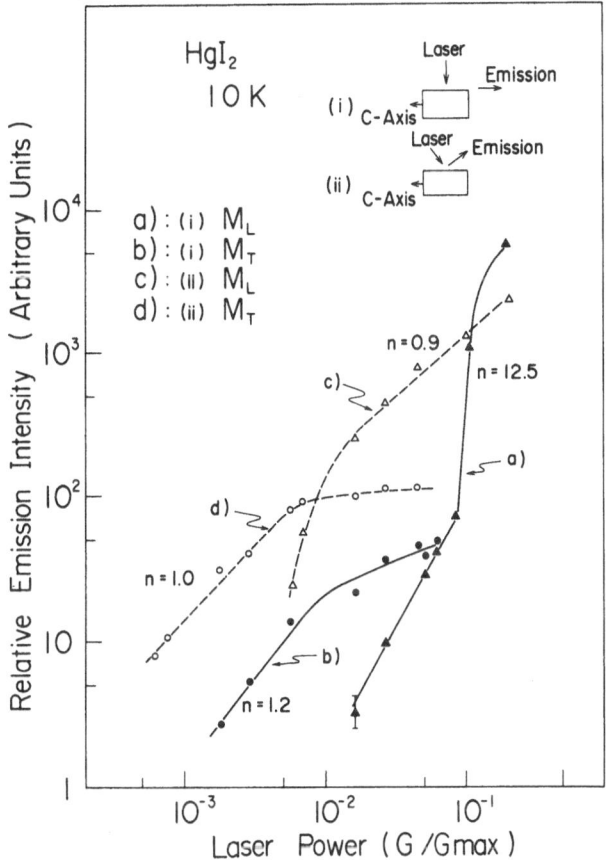

Fig.5. Plot of the relative emission intensities of M_L and M_T lines vs. N_2-laser beam intensity, G_{max} = 5 MWcm^{-2} at 10 K for both (i) normal and (ii) oblique incidence of the exciting beam on the crystal plane parallel to the c-axis. The nonlinear emission intensity is approximated by the form of $(G/G_{max})^n$, where the value of n is given in the figure.

of the M_L-peak and its temperature dependence up to 77 K indicate that this emission is caused by the annihilation of the A-excitonic molecule, emitting one longitudinal exciton. The following is the reason why the M_L-line shows the characteristic of the stimulated emission, whereas the M_T-line does not: The formation energy of the longitudinal exciton is generally higher than the transverse. Hence the transition for the less-populated final state (longitudinal exciton) becomes the predominant process of radiation.

At 77 K, the excitonic molecule is no longer stable, since its binding energy is found to be in the order of 3.8 meV from the analysis of the emission spectrum at 4.2 K.[18] For the N_2-laser beam excitation at this temperature, one may observe two principal lines. One is induced by the exciton-electron interaction (XE-line) and the other by the exciton-hole interaction (XH-line).[19] Only the former line shows the characteristic of the stimulated emission near the threshold excitation of ~4 MW/cm^2 for the beam incident perpendicular to the c-axis. The maximum gain coefficient is $(3.0 \pm 0.5) \times 10^3$cm^{-1} for the excitation intensity of 4 MW/cm^2. The photon energy of the XE-line just above the threshold shows a large shift from 2.29 to 2.15 eV as the temperature increases from 77 to 230 K, whereas the XH-line does from 2.31 to 2.24 eV for the same increase in temperature. The smallness in the energy shift of the XH-line is related to the fact that the effective hole mass, estimated from the peak shift with temperature is about ten times heavier than the electron,[19] so that the XH-line is located in the tail of the exciton absorption line. The presence of appreciable absorption explains why the XH-line does not become the stimulated emission.

IV. PbI_2

The stimulated emission in the single crystal of PbI_2 has been observed for the 2H-type or for the polytype of 2H and 12R under the

N_2-laser beam excitation at 4.2 K.[20] The exciting beam is focused by

a cylindrical lens on the c-plane. The emission spectrum is observed

in the direction perpendicular to the c-axis. For a weak excitation by

a Xe lamp the photoluminescent spectrum shows bound exciton lines for

the 2H-type, I(2H), at 2.4928 eV and for the 12R-type, I(12R), at

2.5029 eV, respectively. The spectrum also exhibits minor peaks at

2.4958 eV, A(12R) -LO, for the free exciton of the 12R-type coupled

with the LO-phonon of ~12 meV.[20]

For a higher level of excitation by N_2-laser, however, another

line A(2H)-LO for the free exciton of 2H-type coupled with the LO-

phonon begins to appear at 2.4868 eV and shows a sudden increase in

intensity and the narrowing of the line at a threshold excitation in-

tensity of ~6 MW/cm^2. The gain coefficient, $g(\hbar\omega)$, of this line near

the threshold agrees, except for the lower energy tail, with the

formula derived by Haug[21] for the stimulated emission induced by the

recombination of the free exciton assisted by emission of a LO-phonon,

so that, $g(\hbar\omega) \propto (\Delta/kT)^{3/2} \cdot \exp(-\Delta/kT) \cdot P$, where P = pumping rate, Δ

$\equiv \hbar\omega - (E_{1s} - \hbar\omega_{LO})$, with $\hbar\omega$, E_{1s} and $\hbar\omega_{LO}$ being the energies for the

emission peak, the ground state of the exciton and the LO-phonon, re-

spectively. The gain spectrum for the excitation intensity near 10^7

W/cm^2, however, disagrees appreciably with the above-mentioned model.

Namely, the peak in the gain coefficient spectrum shifts to the lower

energy by ~3 meV. Such discrepancy may be explained on qualitative

basis at least in terms of the polariton formation of the small wave

vector exciton. If the ratio of the probability of the zero phonon

emission to that of the phonon scattering becomes smaller with the

increase in the excitation intensity, the photon-like polariton would

be in more favorable position for the stimulated emission than the

exciton-like polaritons with higher energies. Hence the lower energy

range of the polariton would have higher gain coefficient than that of

the higher energy as the pumping rate increases. Also there is a possibility that the red shift may be present due to the increase in re-absorption in the neighborhood of exciton absorption band, which comes from the exciton-free carrier collision and gives rise to the quenching of stimulation in this energy range.

V.   CONCLUSIONS

Various mechanisms of stimulated emission have been identified in the layer-type semiconductors of (1) $GaSe_{1-x}S_x$, (2) red-HgI_2 and (3) 2H-PbI_2 in terms of (1) the direct-exciton-indirect-exciton interaction, (2) annihilation of excitonic molecule (~10 K) or direct exciton-electron (77 K) interaction and (3) the exciton-LO-phonon interaction, respectively. The efficiency of the stimulation is greatly assisted by the feasibility of the cleavage planes, so that a high value of the gain coefficient in the order of $10^3 cm^{-1}$ may be attained for the excitation intensity near several MW/cm^2 at low temperatures.

REFERENCES

1)   Y. Kuramoto and H. Kamimura: J. Phys. Soc. Japan 37 (1974) 716. Also see their paper in this volume.

2)   N. Kuroda, T. Nakanomyo and Y. Nishina: J. Japan Soc. appl. Phys. 43 suppl. (1974) 63.

3)   R. E. Nahory, K. L. Shaklee, R. F. Leheny and J. C. DeWinter: Solid State Commun. 9 (1971) 1107.

4)   T. Ugumori, K. Masuda and S. Namba: Solid State Commun. 12 (1973) 389.

5)   A. Mercier and J. P. Voitchovsky: Phys. Rev. 11B (1975) 2243.

6)   N. Kuroda and Y. Nishina: Phys. Status solidi (b) 72 (1975) 81.

7)   N. Kuroda and Y. Nishina: Il Nuovo Cimento 32 B (1976) 109.

8)   N. Kuroda and Y. Nishina: Suppl. Progr. theor. Phys. No.57 (1975) 51.

9) K. Aoyagi, A. Misu, G. Kuwabara, Y. Nishina, S. Kurita, T. Fukuroi, O. Akimoto, H. Hasegawa, M. Shinada and S. Sugano: *Proc. Int. Conf. Phys. Semicond., Kyoto 1966.* ed. G. M. Hatoyama (Phys. Soc. Japan Tokyo, 1966) p.174.

10) T. Kushida and T. Moriya: See this volume.

11) J. L. Brebner, S. Jandl, and B. M. Powell: Solid State Commun. 13 (1973) 1555.

12) N. Kuroda, Y. Nishina and T. Fukuroi: J. Phys. Soc. Japan 28 (1970) 981.

13) Y. Nishina: *Lecture Notes, Int. Conf. on Appl. High Magnetic Fields in Semicond, Würzburg,* 1974, Part 2, ed. G. Landwehr (Physikalisches Institut, Würzburg, 1975) p.64.

14) N. Kuroda and Y. Nishina: J. Luminescence 12/13 (1976) 623.

15) C. Benoît à la Guillaume, J. M. Debever and F. Salvan: *Proc. Int. Conf. Phys. Semicond., Moscow, 1968* ed. S. M. Ryvkin (Nauka, Leningrad, 1968) p.581.

16) K. Kanzaki and I. Imai: J. Phys. Soc. Japan 32 (1972) 1003.

17) B. V. Novikov and M. M. Pimonenko: Sov. Phys.-Semicond. 6 (1972) 671.

18) K. Nakaoka, T. Goto and Y. Nishina: to be published.

19) T. Goto, K. Nakaoka and Y. Nishina: J. Luminescence 12/13 (1976) 599.

20) M. Yashiro, T. Goto and Y. Nishina: Solid State Commun. 17 (1975) 765.

21) H. Haug: J. appl. Phys. 39 (1968) 4687.

EFFECTS OF EXCITATION-INDUCED OPTICAL ABSORPTION
IN HIGHLY EXCITED SEMICONDUCTORS

Takashi Kushida

The Institute for Solid State Physics
The University of Tokyo
Roppongi, Minato-ku
Tokyo 106, Japan

and

Tetsuo Moriya

Electrotechnical Laboratory, Tanashi Branch
Mukodai-machi, Tanashi,
Tokyo 188, Japan

ABSTRACT

The red shift of stimulated emission bands of various semiconductors with increasing pumping intensity or length of the excited region of the crystal is explained by the wavelength-dependent gain saturation due to relative increase in optical absorption. Several effects of excitation-induced negative gain are also discussed.

Stimulated emissions in undoped semiconductors have been studied by a number of investigators under intense optical or electron-beam excitations. It has often been observed that the stimulated emission band appearing on the lower energy side of the band edge shows a marked red shift with the increase of the excitation intensity or the length of the excited region of the crystal.[1] Although this phenomenon appears to be very common to wide-gap semiconductors, there exists no general theory of this red shift.

In the present paper, we show that this shift is explained well by the change of the gain spectrum caused by the relative increase in the optical absorption, especially that due to the reverse process of the stimulated emission which produces the gain. Several recent experiments on highly excited semiconductors are also interpreted in terms of excitation-induced optical absorption.

It is well known in the field of laser physics that the optical gain generally decreases when the input light level is increased. This gain saturation can be explained by the reduction of the population difference between the initial and the final states of the laser transition in the presence of intense resonant optical field. Since the optical gain is determined by the difference between the stimulated emission rate and the absorption rate, the population decrease in the initial state and the increase in the final state lead to the gain reduction, enhancing the relative contribution of the absorption process. In the case that the laser transition is related to some continuous energy band, as in semiconductors and organic dyes, this type of population change also causes variation of the gain spectrum, because the stimulated emission and reverse absorption spectra do not agree with each other in this case. Then, the spectrum of the emitted light changes with the intensity of the resonant radiation in the material. We attribute the red shift of the stimulated emission bands in various semiconductors mentioned above mainly to this effect. Some contribution may come from the population change due to other origin, which also gives a similar effect.

Figure 1 shows the emission spectra of a highly excited GaAs. The stimulated emission band denoted as $[P]$ emerges from the low energy side of the P band and shifts towards lower energies with the excitation intensity. Then, another mechanism, which we ascribe to the luminescence from metallic electron-hole liquid state,[2] becomes dominant and its stimulated emission band appears on the low-energy side. The $[P]$ band approaches asymptotically to an energy around 8 meV below the exciton energy $E_0^{1s} = (E_g - E_x^b)$ until it disappears. We attribute the spontaneous P emission band to a radiative exciton-exciton collision process and the $[P]$ band to its stimulated process.[3,4] The spontaneous emission and optical gain spectra due to this process were calculated using a simplified interaction potential between the two excitons.[4] The results agreed well with the experimentally obtained spontaneous emission and small-signal gain spectra. These two spectra were almost

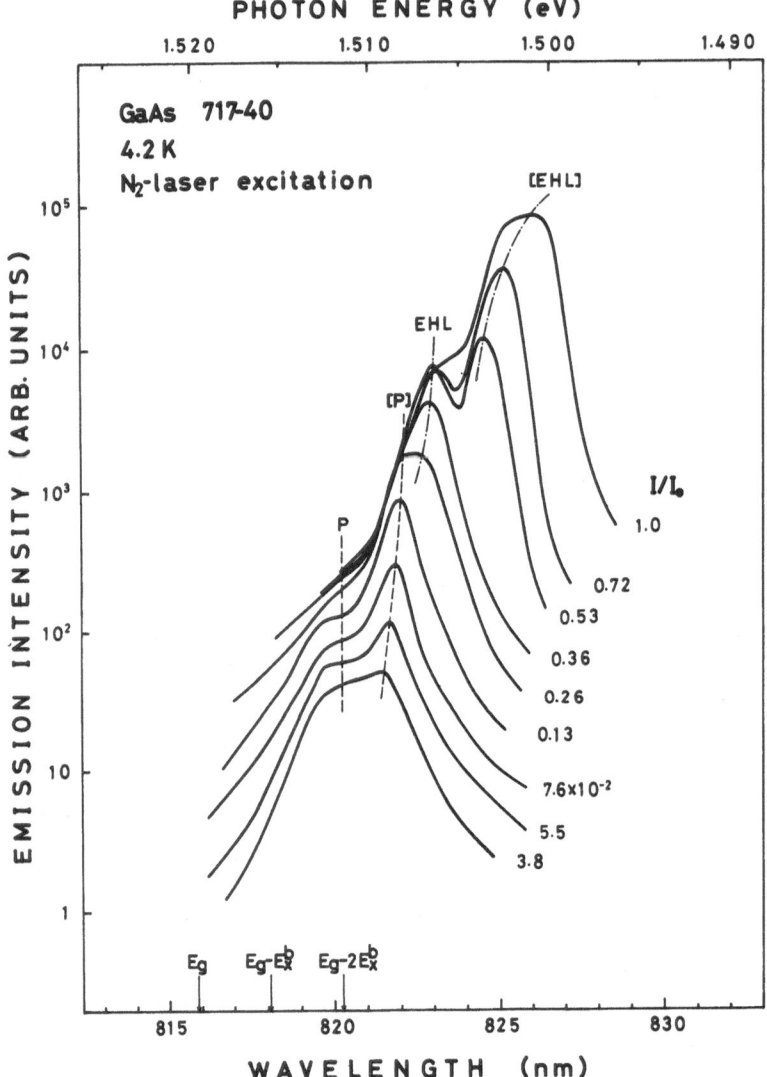

Fig.1. Emission spectra of nitrogen-laser-pumped GaAs observed in the
direction parallel to the excited surface of the crystal. The
excitation intensity is indicated by the ratio of the used
intensity I to the maximum intensity $I_0 \sim 5MW/cm^2$. The band
gap and exciton binding energies are denoted as E_g and E_x^b, respectively, and the stimulated emission bands are shown by [ ].

identical with each other.

Figure 2 shows the spectra calculated for the most important

collision process of two 1s excitons in which one exciton recombines to

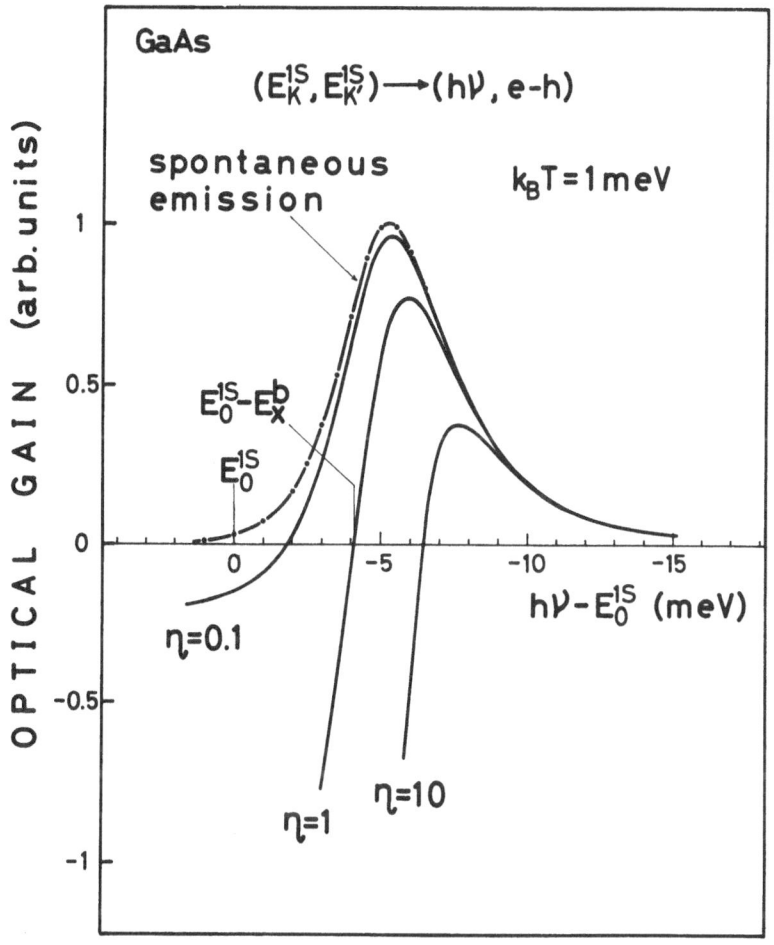

Fig.2. Calculated spontaneous emission (dashed curve) and optical gain spectra (solid curves) of GaAs due to radiative two 1s-exciton collision process. The parameter η is equal to $(N_e N_h / N_{1s}^2)$ $\times (M^2/m_e m_h)^{3/2}$, where N_{1s}, N_e and N_h are the densities and M, m_e and m_h are the masses of the 1s exciton, electron and hole, respectively. The energy of the 1s exciton with the momentum $\hbar K$ is denoted as E_K^{1s} .

emit a photon with energy $h\nu$, while the other dissociates into a free electron-hole pair.[4] Similarly evaluated absorption spectra due to the reverse process of the stimulated emission, $i.e.$, the optical absorption to create two 1s excitons with the annihilation of a free electron-hole pair, are plotted in Fig.3. The theoretical gain curve for η = 0 is identical with the spontaneous emission spectrum, which

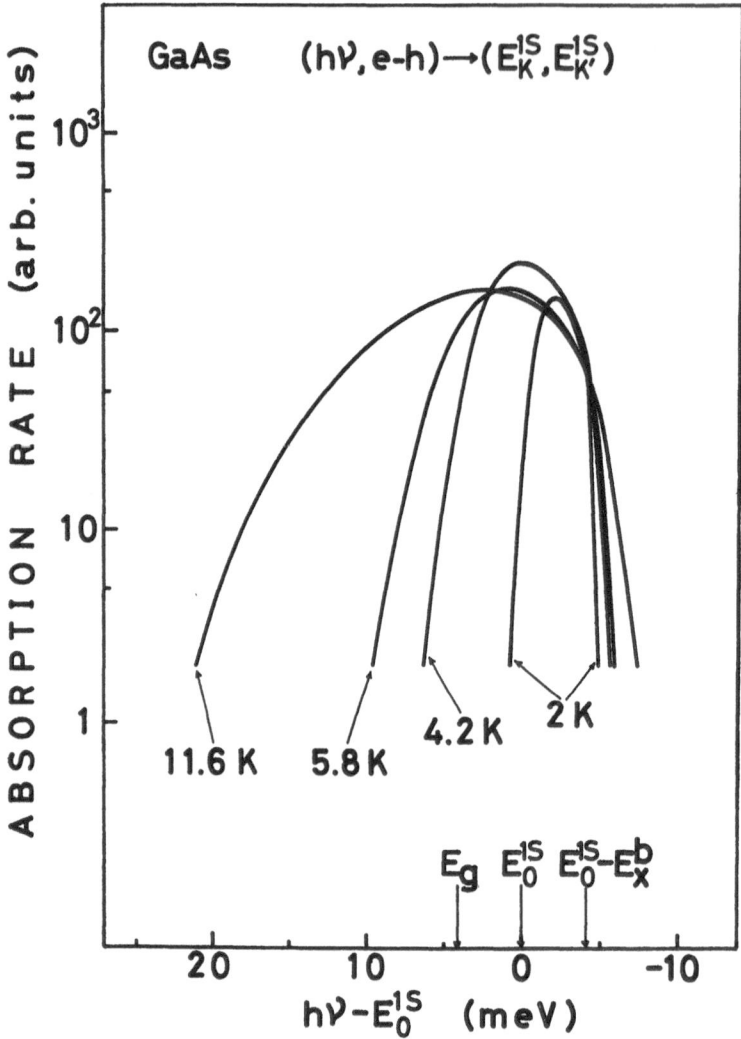

Fig.3. Optical absorption spectra of GaAs calculated for the process to create two 1s excitons with the annihilation of a free electron-hole pair.

agrees with the result on the small-signal gain spectrum in GaAs. As η is increased, the gain band is reduced and shifts towards lower energies. This is because the reverse absorption becomes significant with increasing η. As shown in Fig.3, this absorption band is located on the higher energy side of the emission band, so that the relative increase of the reverse absorption reduces the gain in the high-energy part of the spectrum and accordingly leads to the red shift of the gain band.

In highly excited GaAs, the spontaneously emitted radiation is amplified in the crystal. When the power of this resonant radiation becomes intense enough, an increase of the population ratio of free carriers to excitons occurs corresponding to the gain saturation for the large signal. Then, the gain spectrum changes, the gain becoming very small or even negative at the high-energy part, while the change being relatively small at the low-energy part. Thus the resonant radiation spectrum and accordingly the emitted output spectrum shift towards lower energies. Since η is dependent on the resonant radiation power inside the crystal, the magnitude of the red shift depends on the excitation intensity, the length of the excited region of the crystal, and so on. It may be reasonable to consider that the free carrier densities N_e and N_h become close to the exciton density N_{1s} at the high intensity limit of the resonant radiation. In the case of GaAs, this corresponds to η of ~ 10. Since the peak of the calculated gain curve for this value of η is shifted by ~ 8 meV from the exciton energy, the observed red shift of the $[P]$ band is explained very well by the above theory.

Benoit à la Guillaume *et al.*[5] observed a red shift of the gain peak as a function of excitation time in an electron-beam pumped CdS crystal. They ascribed this gain mechanism to the exciton-electron collision process and explained the time variation by the rise of the electron and exciton temperatures T_e and T_{ex}. The ratio of the reverse absorption rate to the stimulated emission rate due to this process increases very rapidly when T_e is increased or when N_{1s} is decreased. Since the location of the reverse absorption band is higher in energy than that of the emission band, the relative increase in the reverse absorption shifts the gain band to the red as in the case of two-exciton collision process. From the energy position of the small-signal gain peak and the spectral shape of the spontaneous emission band locating around the energy region of our interest, it seems to be more plausible to attribute the above gain mechanism to the exciton-exciton collision process. In this case also, we can explain the observed red shift by the relative increase of the excitation-induced optical absorption rates with time. The optical absorption due to the reverse processes of the radiative exciton-electron and exciton-exciton collisions will be enhanced by the rise of T_e and also by the increases of the free carrier density and the resonant radiation intensity (filling-up of trapping levels with time probably increases the lifetimes of the free carriers and excitons and also reduces the impurity-induced absorption below the band gap).

For the luminescence process of an excitonic molecule to leave a single exciton, similar red shift has been observed by several investigators.[6-8] In CuCl, for example, the M_L line increases in intensity much rapidly compared with that of the M_T line and shows an apparent red shift with the increase of the excitation intensity.[7] This result can be understood well again by considering the effect of the reverse absorption on the gain as follows. When the resonant radiation power is increased, the gain for the M_T line is saturated immediately because of the relatively long lifetime of the single transverse exciton. Gain saturation for the M_L line to leave the single longitudinal exciton with short lifetime becomes significant at higher radiation powers, and is stronger at the high-energy part than at the low-energy part for the same reason as mentioned above.

In Fig.2, the optical gain is negative in the high-energy part of the spectrum unless $\eta = 0$. This is because the stimulated emission is overcome by the reverse absorption in this spectral region. This negative gain explains the pumping-induced optical absorption below the band gap in highly excited CdS.[9,10] In GaSe, an intense stimulated emission band due to the exciton-exciton collision process was found to appear at 77 K under pulsed dye-laser excitation at 595 nm (~14 meV below the exciton energy) where the absorption coefficient in unexcited crystals is small.[11] This result is explained by our model as follows. The pumping light is first absorbed weakly to create free carriers, which then induce new absorption as discussed above. From the very high gain reported in the literature,[1] excitation-induced optical absorption around 595 nm is estimated to become fairly large. Thus, the efficient pumping below the exciton energy is understood well by our theory.

Hildebrand and Göbel[12] measured the excitation spectrum for the stimulated emission band due to the exciton-exciton collision process in GaAs, *i.e.*, the [P] band of Fig.1, at 2 K. The spectrum measured at the excitation density of $\sim 10^5$ W/cm^2 was found to be located between the energies E_0^{1s} and $(E_0^{1s} - E_x^b)$. Compared with the theoretical absorption spectrum of Fig.3, we find that this result can be explained by the pumping-induced negative gain exactly in the same way as in the case of GaSe mentioned above. Recently it has been found that the excitation spectrum for the excitonic molecule luminescence in CdS has a broad structure-less band peaking around the exciton energy.[13] We attribute this band again to the above-discussed excitation-induced optical absorption such as due to the reverse processes of the radiative exciton-exciton and exciton-electron collisions.

In conclusion, it has been shown that the excitation-induced optical absorption as well as the stimulated emission plays an important role in the optical spectra of highly excited semiconductors. Since the energy position of the stimulated emission band often depends on the pumping power, the length of the excited region and so on, the measurement of the unsaturated gain spectrum, especially under the condition of negligible absorption, is important for the identification of the dominant gain mechanism.[14]

REFERENCES

1) K. L. Shaklee, R. E. Nahory and R. F. Leheny: J. Luminescence 7 (1973) 284, and references therein.

2) T. Moriya and T. Kushida: Solid State Commun. 14 (1974) 245.

3) T. Moriya and T. Kushida: Solid State Commun. 12 (1973) 495.

4) T. Moriya and T. Kushida: *Proc. Intern. Conf. on Luminescence, Tokyo,* 1975, [J. Luminescence 12/13 (1976)] p.617: J. Phys. Soc. Japan 40 (1976) 1668, 1676.

5) C. Benoit à la Guillaume, J. M. Deveber, and F. Salvan: Phys. Rev. 177 (1969) 567.

6) H. Souma, T. Goto, T. Ohta and M. Ueta: J. Phys. Soc. Japan 29 (1970) 697.

7) M. Ueta and N. Nagasawa: *Proc. Oji Seminar on Physics of Highly Excited States in Solids, Tomakomai,* 1975 (This issue).

8) S. Shionoya, H. Saito, E. Hanamura, and O. Akimoto: Solid State Commun. 12 (1973) 223.

9) T. Goto and D. W. Langer: Phys. Rev. Letters 27 (1971) 1004.

10) Y. Oka and T. Kushida: J. Phys. Soc. Japan 33 (1972) 1372.

11) M. Higashida: Thesis, The University of Tokyo, 1975 (In Japanese).

12) O. Hildebrand and E. Göbel: *Proc. 12th Intern. Conf. Physics of Semiconductors, Stuttgart,* 1974, ed. M. H. Pilkuhn (B. G. Teubner, Stuttgart, 1974) p.147.

13) A. Mysyrowicz, A. J. Schmidt, Y. R. Shen, P. Robrish and H. Rosen: Solid State Commun. 17 (1975) 523.

14) T. Kushida and T. Moriya: Phys. Status solidi (b) 72 (1975) 385.

HIGH EXCITATION OF DIRECT SEMICONDUCTORS LIKE CdS

G. O. Müller, M. Rösler, H. H. Weber, and R. Zimmermann

Zentralinstitut für Elektronenphysik der Akademie
der Wissenschaften der DDR, GDR 108 Berlin

ABSTRACT

Electron-hole-plasmas in direct gap semiconductors show up in an
emission band the shape of which is always influenced by stimulation
or even gain saturation. Model calculations of rate equations in two
different simplified cases - non-thermal or non-uniform distributions -
are compared with qualitative experimental features. While saturation
of the maximum emission with excitation and red shift with spot length
can be understood from the competition of cool down of carriers with
pump-down by stimulation, other features need further refinement.

While it has become quite common within the last years[1] to think
of the break-down of excitons under high excitation in Ge and Si, this
idea remained constricted to a small number of papers considering
CdS,[2-4] GaAs,[5-7] and CdSe[8] amongst the direct gap materials. In the
case of CdS at very high excitation and especially in ultrapure crystals
the line Q assigned to the electron-hole plasma (EHP)[2,4] shows up with
a rather large displacement to lower energies relative to the exciton
region. The most direct confirmation of non-excitonic origin is the
fact that in highly Ga-doped crystals the same line is present, whereas
no excitonic emission is observed at normal excitation.[9] Until now -
contrary to the situation in GaAs - it has not been possible to obtain
a satisfactory explanation of its line shape. We summarize the most
obvious qualitative properties of the Q-line: i) broadening with increasing excitation,[2] ii) broadening almost entirely to lower
energies,[4] iii) saturation of maximum intensity, iv) increased spot
dimension of the excited region act qualitatively like increased excitation. Figure 1 illustrates the last property. The solid lines
represent spectra emitted along the largest dimension of the excitation
spot which is given as parameter. The dashed line is the emission from

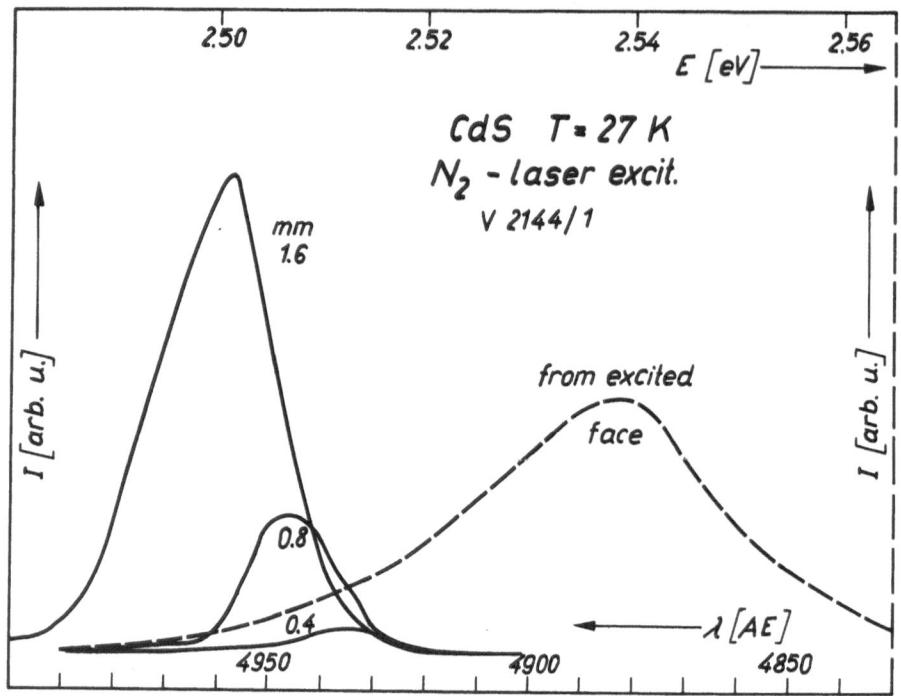

Fig.1. Emission spectra of CdS at 27 K (solid curves) along the largest
dimension of excitation spot (given as parameter) and (dashed) from
the excited surface of the 0.4 mm thick crystal.

the excited surface of the crystal (thickness about 0.4 µ). These

results as almost all other high excitation experiments evidence the

occurrence of stimulation and even its saturation. Therefore both

effects should be taken into account from the very beginning when EHP

in CdS (and other direct gap materials) is investigated.

Neglecting excitons we set up the following simple model. The

Hamiltonian consists of photons, LO-phonons, and electrons (holes) with

electron-photon and electron-phonon interaction. With the usual appro-

ximations[10] steady-state rate equations are derived for photon density

$v_{\vec{Q}}$ and electron (hole) distribution $n_{\vec{k}}$ ($p_{\vec{k}}$). The photon equation is

supplemented by a gradient term which allows for spatial inhomogeneity

$$0 = \dot{v}_{\vec{Q}} = c \cdot \alpha_0(\omega) \left[n_{\vec{k}} \, p_{\vec{k}} - (1 - n_{\vec{k}} - p_{\vec{k}}) \, v_{\vec{Q}} \right] - \vec{v}_{\vec{Q}} \cdot \mathrm{grad} \, v_{\vec{Q}} \quad . \tag{1}$$

$\vec{v_Q} = c \cdot \vec{e_Q}$ is the velocity of the $\vec{Q}$-photons. The energy of the emitted (absorbed) photon equals the energy of the electron-hole pair

$$\hbar\omega = hQc = E_G + E_e(\vec{k}) + E_h(\vec{k}) \quad , \tag{2}$$

E_G being the energy gap. $\alpha_0(\omega)$ is the absorption coefficient without excitation. Using known values for band masses and the momentum matrix element we obtain for CdS in the edge region

$$\alpha_0(\omega) = K \cdot (\hbar\omega - E_G)^{1/2} \quad , \quad K = 1.3 \times 10^3 \ cm^{-1} meV^{-1/2} \tag{3}$$

for the polarization perpendicular to the c-axis. For the sake of simplicity we neglect any polarization dependence in the rate equations. Let be L the large extension of the excited region. Then the x-axis is put into this direction, with x = 0 at the midpoint. Assuming for the time being uniform carrier distributions the rate eq.(1) is easily solved for $v_{\vec{Q}}$ with $\vec{Q}$ parallel to x

$$v_{\pm,\omega}(x) = \frac{n_{\vec{k}} \ P_{\vec{k}}}{1 - n_{\vec{k}} - P_{\vec{k}}} \left(1 - e^{-\alpha(\omega)\left(\frac{L}{2} \pm x\right)}\right) + v_{\pm,\omega}\left(\mp\frac{L}{2}\right) e^{-\alpha(\omega)\left(\frac{L}{2} \pm x\right)} \quad , \tag{4}$$

where $v_{+,\omega}(x)$ and $v_{-,\omega}(x)$ denote the right and left travelling photons. The total absorption coefficient of the excited system is

$$\alpha(\omega) = \alpha_0(\omega) \cdot (1 - n_{\vec{k}} - P_{\vec{k}}) = - \ gain \quad . \tag{5}$$

Assuming Fermi distributions for $n_{\vec{k}}$ and $p_{\vec{k}}$ and using (3,5) we have calculated gain values for CdS beyond $10^3 \ cm^{-1}$ at all temperatures below 25 K and densities above $3 \times 10^{17} \ cm^{-3}$ (Mott density). With realistic values for L enormously high photon densities would result which in turn influence the carrier distributions as clearly recognizable from the rate equation for electrons

$$0 = \dot{n}_{\vec{k}} = \frac{1}{\tau_{Sp}} \left[- n_{\vec{k}} \, p_{\vec{k}} + (1 - n_{\vec{k}} - p_{\vec{k}}) \frac{1}{4\pi} \int d\Omega_{\vec{Q}} \, v_{\vec{Q}} \right]$$

$$+ \sum_{\vec{k}'} W^{el}_{\vec{k}-\vec{k}'} \{ \delta(E_e(\vec{k}) - E_e(\vec{k}') - P) \left[(n_{\vec{k}'} - n_{\vec{k}}) N_p - n_{\vec{k}}(1 - n_{\vec{k}'}) \right] \qquad (6)$$

$$+ \delta(E_e(\vec{k}) - E_e(\vec{k}') + P) \left[(n_{\vec{k}} - n_{\vec{k}'}) N_p - n_{\vec{k}'}(1 - n_{\vec{k}}) \right] \}.$$

A similar equation determines $\dot{p}_{\vec{k}}$. The first line describes spontaneous
recombination (first part) and the net effect of stimulated recombination and generation due to reabsorption (second part), τ_{Sp} being the
carrier lifetime with respect to spontaneous recombination. The $\vec{k}'$-sum
comes from the interaction with LO-phonons having the constant energy
P. $W^{el}_{\vec{k}-\vec{k}'}$ is the squared Fröhlich interaction which gives rise to the
phonon scattering time τ_P. We have omitted the rate equation for the
phonon distribution N_P because we assume equilibrium of the phonons with
the bath temperature T. The phonon scattering forces the carriers to
approach equilibrium (Fermi function with T). At low excitations this
process is not disturbed by the recombination because of $\tau_P \ll \tau_{Sp}$.
However, as mentioned above, we expect enormous photon densities at
higher excitations, and stimulated recombination is able to compete
with all other processes ultimately. A decrease in the carrier distribution will result, especially in those regions where the photon
density is highest; i) at the borders of the excited region $x = \pm L/2$
and ii) at the energy of maximum gain. Thus we expect a spatially non-
uniform and energetically non-thermal distribution, which influences
also the spontaneous emission into other directions small compared with
L. Of course the gain decreases when $n_{\vec{k}}$ and $p_{\vec{k}}$ are reduced (5), and
finally saturation occurs. A full treatment of (1, 6) with energy and
spatial dependence was not possible. In a first attempt we have focused
our attention on the spatial non-uniformity. Only the two photon
densities $v_{\pm}(x)$ in the long dimension are important, they are confined

to the solid angle $4\pi\,\Omega_L$. In analogy to the Shaklee model[11] we assume

rapid hole relaxation, $p_{\vec{k}} = 1$. The laser input and the LO-phonon cool-

down of carriers is replaced by a formal excitation rate g of cold

carriers. Thus any energy dependence is relaxed, and we deal with the

electron density $n(x)$. (1, 6) are therefore simplified to

$$\pm \frac{\partial}{\partial x}\, \nu_\pm(x) = \alpha_0\, n(x)\, \big[1 + \nu_\pm(x)\big]\,, \tag{7}$$

$$g = \frac{n(x)}{\tau_{Sp}}\, \big[1 + \Omega_L(\nu_+(x) + \nu_-(x))\big]\,, \tag{8}$$

which are solved analytically[12] with the reflection R in the boundary

condition at $x = \pm L/2$. Results are shown in Figure 2 for normalized

density $n(x)L \cdot \alpha_0/2$ and excitation $gL\,\alpha_0\tau_{Sp}/2$. The relation between

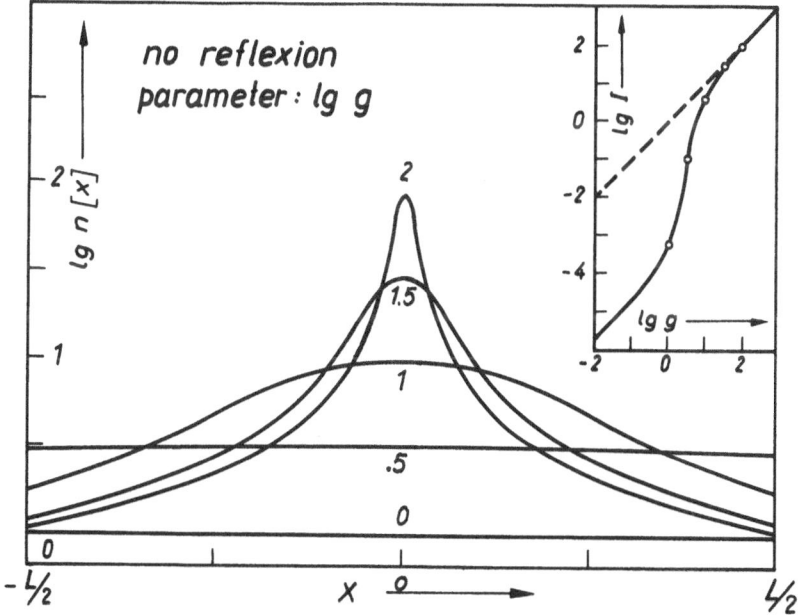

Fig. 2. Non-uniform electron density $n(x)$ at various excitations g.
Corresponding values are shown as circles in the inset, where
the emission I along the long dimension is plotted against
the excitation. In the saturation regime I approaches the
dashed line $I_{Sat} = g$. All quantities are given in normalized
units. $\Omega_L = 10^{-4}$ (see text).

emitted intensity I through both end faces at $x = \pm L/2$ and excitation is given by

$$(1 - 2\Omega_L) \cdot \text{arctanh} \left(\frac{1+R}{1-R} + \frac{2\Omega_L}{I} \right) + I = g \tag{9}$$

(inset in Figure 2). In the saturation region where I approaches g, the density becomes non-uniform. Only in the central part where $\nu_+ + \nu_-$ has its minimum we find an increase of n with g, ultimately as $n \sim g^{1/2}$ for R = 0. In the outer parts the density is slightly reduced and reaches a constant value independent of g. At finite R the emission saturates earlier, and the peak of $n(x)$ is flattened out.

The energetic pecularities of saturation have been investigated in a second step. In order to get rid of the space dependence we have averaged the photon densities (4) over the excited region. Together with corresponding expressions for the two perpendicular directions of Q and multiplied by their respective solid angles, this has been inserted into (6) as total photon density. The exciting laser radiation now comes in quite naturally as boundary conditions at the laser frequency (reflection is neglected throughout this part). For equal parabolic energy dispersions in (2) the rate equation (6) reduces to a difference equation coupling n_E with n_{E-P} and n_{E+P} which was solved by iteration.[12] The starting value n_{E_G} was varied until convergence was achieved ($n_E \rightarrow 0$ for energies well above the laser input). It was checked that the excitation g (number of absorbed laser photons per second and cm^2) equals the total emission from all surfaces. Results are given in Figure 3. The parameter values used are not very realistic but produce typical curves. Especially we had to choose kT > P in order to get smooth curves. At low excitation the carrier distribution is close to a Fermi function as expected (curves 0.1, 1, 10 in Figure 3b). At higher excitation, the distributions broaden and exhibit a dip which reflects carrier pump down at the maximum gain (or maximum emission). Note that $\hbar\omega = 2E + E_G$ (2). Obviously the distributions

Fig.3.
Non-thermal elec-
tron distribution
(b) and resulting
emission I (a) in
the long dimension
L of the excited
volume $V = L \cdot H^2$ at
various excita-
tions g. Dashed
line: limiting
distribution (see
text). Energy
unit is P, the
phonon energy.
Parameters used in
the calculation
are L = 0.5 mm, H
= 0.002 mm, ab-
sorption at the
laser frequency α_0
= 10^3 cm^{-1}, τ_p/τ_{Sp}
= 10^{-3}, kT = 5P.
Inset: emission
for different L at
fixed excitation.

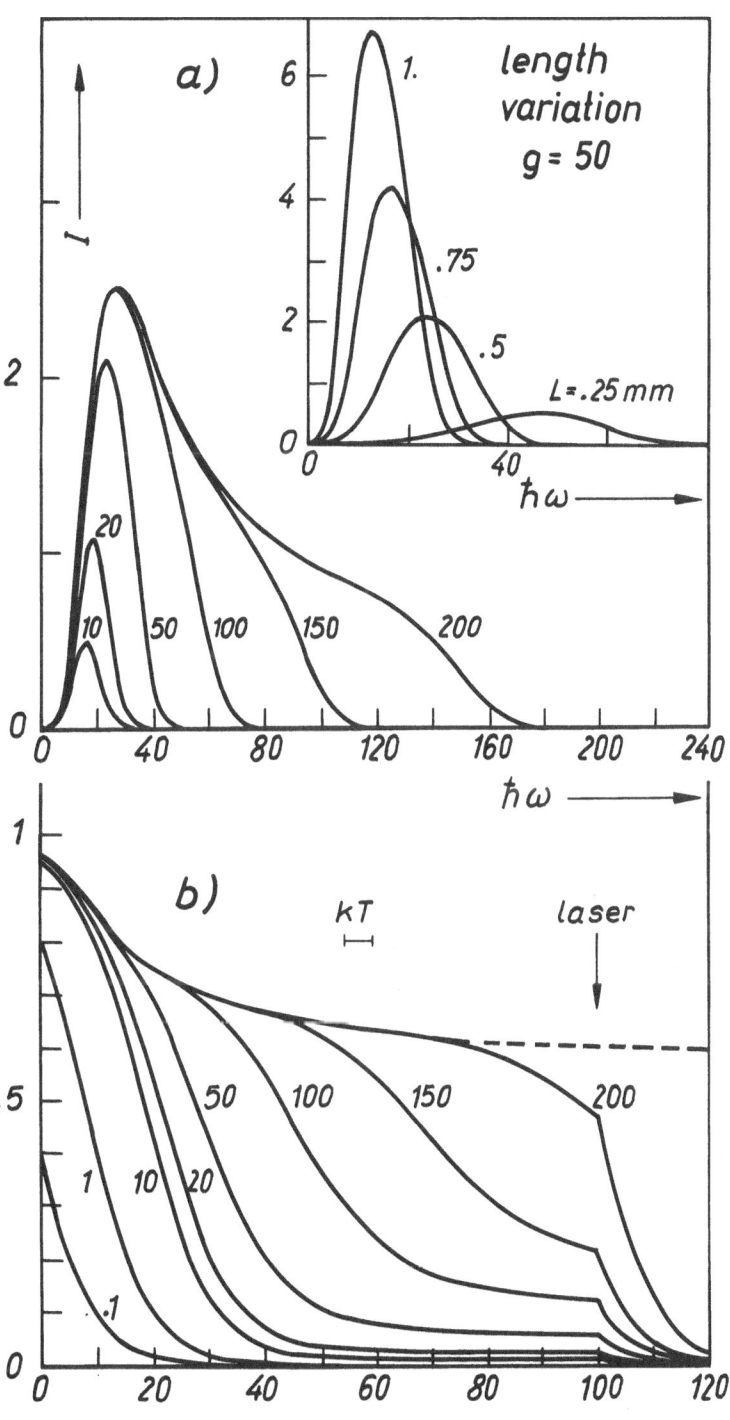

tend to approach a curve (dashed) which limits the attainable occupation n_E whatever the excitation might be. This is reflected in Figure 3a as a saturation of the emission line shape. We believe that this is the first theoretical derivation of line shape saturation. An additional effect described by our model is the absorption saturation at the laser frequency due to nonzero occupation in this region. With increasing excitation length L at fixed g the dip in the distributions becomes more pronounced. The emission shows a red shift together with a reduction of line width (inset in Figure 3a). The density is reduced by a factor 2 going from 0.25 to 1 mm.

Although being very crude the model could explain two features of the Q-line in CdS: i) saturation of maximum emission with excitation and ii) red shift of the emission peak with increasing L. The electron-electron interaction which was omitted in our model leads to a gap shrinkage with increasing carrier density. This explains the experimentally found shift of the red border of the emission.[9] On the other hand, the high energy cut-off may be understood by reabsorption in the low-density outer regions (Figure 2). However, difficulties arise for the length variation, where in both model calculations a reduction of the density was found. The above mentioned gap shrinkage would then result in a shift of the red border towards higher energies, in clear disagreement with the experiment.

Futher investigations should take into account the gap shrinkage explicitly and additionally lifetime broadening due to the very fast stimulated recombination.

Useful discussions with Dr. V. B. Timofeev and Dr. A. F. Dite are gratefully acknowledged.

REFERENCES

1) L. V. Keldysh: *Proc. IX Int. Conf. Phys. Semicond. Moscow 1968*. ed. S. M. Ryvkin, (Hauka, Leningrad, 1968) p.1303.

2) M. A. Jacobson, G. V. Michailov, B. S. Razbirin, I. N. Ural'tsev, G. O. Müller and H. H. Weber: *Proc. XII Int. Conf. Phys. Semicond. Stuttgart* 1974, ed. M. H. Pilkuhn (Teubner, Stuttgart, 1974) p.123.

3) V. B. Timofeev *et al.*: JETP **68** (1975) 335.

4) G. O. Müller *et al.*: *ECPS Conf. on Dielectrics and Phonons, Budapest* 1974.

5) W. F. Brinkmann, P. A. Lee: Phys. Rev. Letters **31** (1973) 237.

6) G. Göbel: Appl. Phys. Letters **24** (1974) 492.

7) T. Moriya, T. Kushida: Solid State Commun. **14** (1974) 245.

8) I. H. Akopjan, B. S. Razbirin: FTT **16** (1974) 189.

9) G. O. Müller, M. Rösler, H. H. Weber, R. Zimmermann, M. A. Jacobson, G. V. Michailov, B. S. Razbirin and I. N. Ural'tsev: J. Luminescence **12/13** (1975) 557.

10) H. Haken and S. Nikitine: Springer Tracts **73** (1975) 192.

11) K. L. Shaklee, R. E. Nahory and R. F. Leheny: *Proc. XI Int. Conf. Phys. Semicond. Warsaw* 1972, ed. M. Miasek (Elsevier, Amsterdam, 1972) p.853.

12) R. Zimmermann: to be published.

PICO-SECOND SPECTROSCOPY OF HIGHLY EXCITED SEMICONDUCTORS

Shigeo Shionoya

The Institute for Solid State Physics
The University of Tokyo
Roppongi, Minato-ku,
Tokyo 106, Japan

ABSTRACT

The importance of pico-second spectroscopy in the investigation of highly excited semiconductors is pointed out with a survey of dynamical aspects of phenomena to be studied. The present status of technical aspects of pico-second spectroscopy is reviewed. Some results of pico-second experiments performed recently for CdSe are described. Finally, it is emphasized that the study of coherent interactions of highly excited states, especially of the Bose-condensed state, with radiation fields is very important.

Pico-second spectroscopy is very important in investigating the physics of highly excited semiconductors. Most of phenomena due to high intensity excitation effects in semiconductors, especially in direct gap materials such as CdS, GaAs and CuCl, take place with time constants in the pico-second range, $i.e.$ $10^{-12} - 10^{-9}$s. By using pico-second spectroscopic technique one may obtain firsthand information on dynamical processes in such transient phenomena. Any other method of study, such as ordinary spectral analysis, gives only secondhand information, and may not substitute for the direct measurement in the time domain.[1] Some pioneering work recently made by the author's group and also by some others has already proven how pico-second spectroscopy is useful and important in the field under discussion.

First let us survey dynamical aspects of phenomena to be studied (Fig.1). When electrons and holes are created by the band-to-band transition in high densities, they will relax emitting longitudinal optical phonons with a time constant in an order of magnitude of $10^{-13} - 10^{-12}$ s, or they will spatially diffuse from the surface into the interior of a crystal with a time constant of probably $10^{-12} - 10^{-11}$ s.

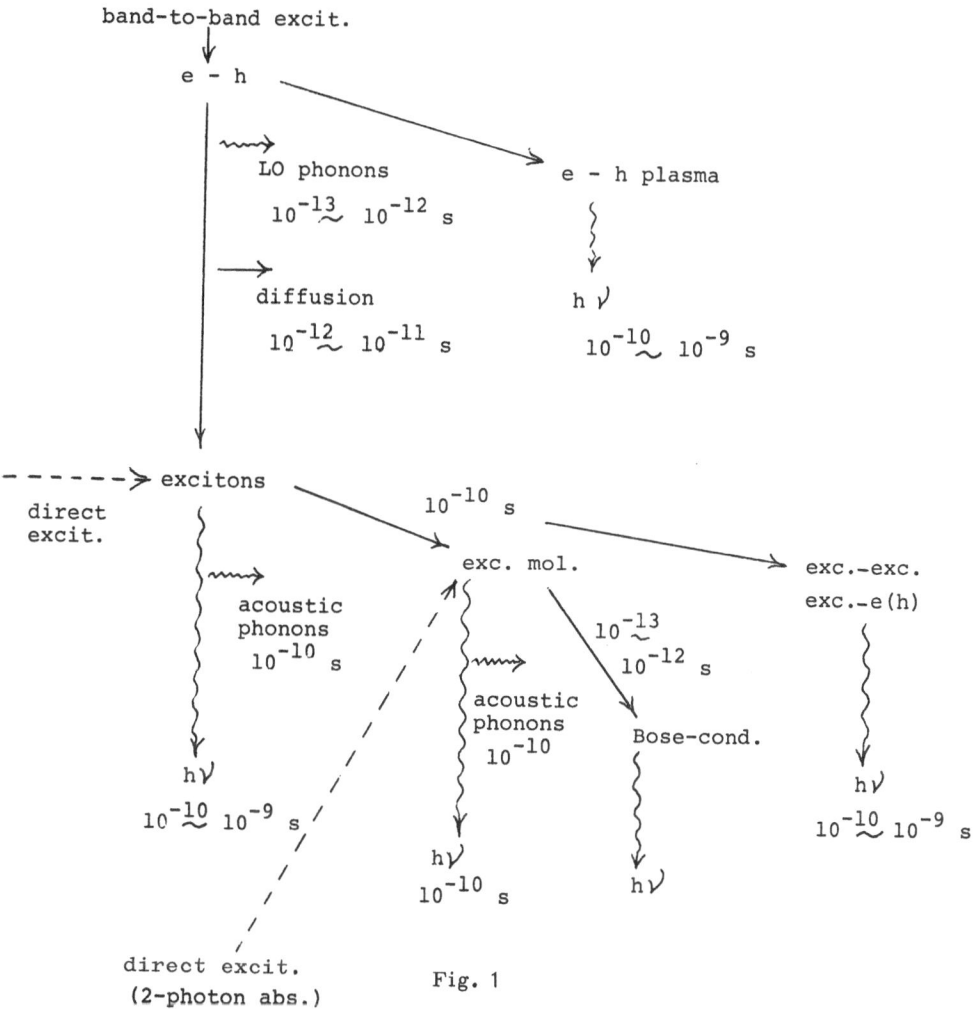

Fig. 1

In some conditions, electron-hole plasma will be generated, which will be annihilated radiatively in a lifetime of $10^{-10} - 10^{-9}$ s.

Excitons are formed from pairs of electrons and holes first in a hot state, and then the relaxation takes place with the emission of acoustic phonons, the time constant of which will be 10^{-10} s. Excitonic molecules will be formed from two single excitons, provided the electrons and holes have antiparallel spins. In some conditions, exciton-exciton and exciton-electron (or-hole) collision processes accompanied by the emission of a photon take place. The radiative dacay times of single excitons and excitonic molecules and those of the collision processes are in an order of magnitude of 10^{-9} - 10^{-10} s. If the necessary conditions are satisfied, excitonic molecules will exhibit the Bose condensation. Even if excitonic molecules are directly created by the two-photon absorption, some time will be required for the excitonic molecules to make rearrangement for the condensation as a result of repulsive interactions. The time required will be as short as 10^{-12} - 10^{-13} s.

Utilizing the technique of pico-second spectroscopy, it is possible to directly pursue these dynamical processes in the time domain. Further it should be noticed that the study of coherent interactions of highly excited states, especially of the Bose-condensed state, with radiation fields is very important. This will be mentioned later, but it is pointed out here that for this study pico-second technique is indispensable.

Next the present status of technical aspects of pico-second spectroscopy[1] will be briefly reviewed, and what we can do with this technique at present will be described. As pico-second pulse sources, mode-locked neodymium glass lasers producing 1.06 μm pulses or ruby lasers producing 694 nm pulses are usually used. The duration of pulses is 7 ps for glass lasers and 10 - 15 ps for ruby lasers. As to the peak power of output pulses, it is not difficult to obtain 1 GW, which gives excitation density of 1 - 10 GW/cm^2 corresponding to 10^{27} - 10^{28} photons/cm^2·s. By utilizing second (or higher) harmonic generation, higher frequency pulses are obtained. In ordinary semiconductors, if one gives band-to-band excitation using a 10 ps pulse at excitation density of 1 GW/cm^2, the number of electron-hole pairs generated is 10^{20} cm^{-3} in case of one-photon absorption, while 10^{17} cm^{-3} in case of two-photon absorption. These numbers are usually enough to observe high density excitation effects.

These mode-locked lasers are difficult to operate with high repetition rates. Therefore, one must measure the spectrum of a pico-second pulse to be measured with one-shot operation of laser. For this purpose a television camera system is very useful. The system very recently constructed by the author's group[2] has high sensitivity, which is 10^5 times that of high sensitive photographic films, and high spectral resolution of about 0.5 Å. Using this system it is possible to catch the spectrum of luminescence produced by excitation of a pico-second pulse as a one-frame picture of television and to reproduce it on the chart of a recorder.

Time-resolved spectroscopy in the pico-second region is possible to perform by using an ultrafast shutter, which is operated by utilizing the optical Kerr effect in CS_2. The time resolution is 10 ps. Combining this shutter and an optical delay system, one can pursue the time change of emission spectra in the pico-second region.

It is possible to produce a white pico-second pulse by making a strong pico-second laser pulse pass through glass, CCl_4 or some other substances. Using such a pulse, one can observe spectra of absorption induced by pico-second pulse excitation. In the pico-second spectro-scopy of highly excited semiconductors, it is very important to produce frequency-tunable pico-second pulses. There are at least two ways; one is to use the lasing of dyes and the other is to use the optical para-metric effect. In both cases, mode-locked neodymium lasers or ruby lasers are used for pumping. The parametric laser seems to be superior to the dye laser at least in two points. The first is that it covers a wider tuning range extending from visible to infrared. The second is that it can be operated in a single mode, so that the beam quality is better.

Kushida and coworkers[3] in our laboratory are attempting to gener-ate tunable pico-second pulses with high power by pumping a $LiNbO_3$ crystal as a parametric crystal with the second harmonic pulses (530 nm) of a neodymium glass laser. The tuning is made by changing the tempera-

ture of the crystal. The generation of pico-second pulses with wave-
lengths of 640 to 900 nm has been confirmed. Efforts are being made to
increase the output of the pulses and to make the spectral width narrower.   Attainment of a peak power of 100 MW in the above wavelength
range is expected in the near future. The second harmonics of these
tunable pulses could be used to directly create excitonic molecules
through two-photon absorption in some important materials such as CuCl
and CdS, if they are strong enough to give excitation density of 10 -
100 MW/cm^2.

Next some of the results of pico-second experiments recently performed in the author's laboratory will be described.[4-7] As to CdSe,
absorption spectra of single excitons and their time-change in the
pico-second region were measured with white pico-second pulses. It was
found that the absorption line of the n=1 A-exciton shifts to the high
energy side if the crystal is simultaneously excited intensely by pico-
second neodymium laser pulses. The blue shift is ascribed to the cut-
ting-down effect of the conduction and valence band edges caused by the
formation of high density excitons. A shift as large as 4 meV was observed, which corresponds to the number of excitons of 5×10^{16} cm^{-3}.
The time-change measurements of this blue shift showed that the shift
becomes the maximum at 20 ps after the pulse excitation and then disappears at 50 ps. The time resolution was 10 ps, so that the data were
not accurate enough, but these results are regarded as indicating that
it takes about 20 ps for the formation of n=1 A-excitons from electron-
hole pairs after band-to-band excitation. The disappearance of the
blue shift at 50 ps may be ascribed mainly to the formation of excitonic molecules.

Spectra of excitonic molecule luminescence (M-line) in CdSe were
observed under excitation with 1.06 μm pico-second pulses through two-
photon absorption. It was found that a sharp line appears under some
excitation conditions, which is not seen in the spectra under excitation

by nano-second pulses from a nitrogen laser. The position of the sharp line coincides with that expected for excitonic molecules at the k=0 state. These results have been interpreted as indicating the attainment of the Bose condensation of excitonic molecules. This sharp line is observed only by two-photon pico-second excitation, and not by one-photon pico-second excitation.

Further the time-dependence of the M, P_M, and P luminescence lines in the pico-second region was measured in CdSe. The P_M line is produced by the collision of two excitonic molecules, and the P line is due to the collision of two single excitons. It was observed that the M line reaches the maximum at 100 ps after the pulse excitation and decays with a time constant of 100 ps, and that the P_M and P lines show maxima at 200 and 300 - 350 ps with decay constants of 100 and 350 ps, respectively. Also for GaAs, the time dependence of the P line and the emission line due to electron-hole liquid was measured.[2] Both lines show nearly the same behavior, rising up at about 300 ps and decaying with time constants of about 300 ps. Very recently, the decay of excitonic molecule luminescence in CuCl was measured, and was found to have a time constant of 300 ps.

Recently Hanamura and Inoue[8] studied theoretically population dynamics of single excitons and excitonic molecules. They showed on the basis of some assumptions that the time-dependence of the concentration of single excitons and excitonic molecules is expressed by simple kinetic equations. In the case where single excitons are directly excited by a pico-second pulse, their calculation shows that the concentration of excitonic molecules has the maximum around at t= α^{-1}, α being the reciprocal time constant of the radiativel decay of excitonic molecules. The above-mentioned experimental results are understood by this calculation.

Finally, it is emphasized that the study of coherent interactions of highly excited states, especially of the Bose-condensed excitonic molecules, with radiation fields is very important. To obtain conclusive evidence for the attainment of the Bose condensation, it will be required to experimentally reveal the coherent nature of the Bose-condensed state. We must challenge this important target.

Two-kinds of coherent interactions of Bose-condensed excitonic molecules have been discussed theoretically by Hanamura.[9] One is super-radiance and the other is self-induced transparency. The Bose-condensed excitonic molecules should have a common phase, and the transverse relaxation time will be quite long, probably determined by the radiative decay time. Such excitonic molecules emit radiations cooperatively and spontaneously. The intensity of the radiation should be proportional to the square of the number of excitonic molecules. This corresponds to the case of super-radiance. Since the Bose-condensed excitonic molecules are expected to have a much longer relaxation time than uncondensed molecules, it will be possible to observe self-induced transparency.

From experimental points of view, the observation of self-induced transparency will be easier than that of super-radiance. From this observation, one may obtain some important information on the relaxation properties of the condensed excitonic molecules, which are closely connected with their coherent nature. To make this observation, one must directly create excitonic molecules by means of two-photon absorption and look at the propagation of an exciting light pulse through a crystal. In such experiments, we need tunable pico-second pulses with sufficient intensity. It is hoped that the use of such laser pulses is realized in the near future so that we could obtain fruitful outcomes on this important problem in the physics of highly excited semiconductors.

REFERENCES

1) R. R. Alfano and S. L. Shapiro: Phys. Today 28 (1975) No.7, p.30.

2) H. Kuroda, S. Shionoya, H. Saito, H. Masuko and K. Mogi: Opt.
    Commun. _12_ (1974) 107; H. Saito, S. Kuribayashi and S. Shionoya:
    Japan. J. appl. Phys. _15_ (1976) 947.

3) T. Kushida, Y. Tanaka, M. Ojima and Y. Nakazaki: Japan. J. appl.
    Phys. _14_ (1975) 1097.

4) H. Kuroda, S. Shionoya, H. Saito and E. Hanamura: J. Phys. Soc.
    Japan _35_ (1973) 534.

5) H. Kuroda and S. Shionoya: J. Phys. Soc. Japan _36_ (1974) 476.

6) S. Shionoya: *Proc. 12th Intern. Conf. Phys. Semiconductors, Stuttgart,* 1974,
    ed. M. H. Pilkuhn (Teubner, Stuttgart, 1974) p.113.

7) H. Saito, A. Kuroiwa, S. Kuribayashi, Y. Aogaki and S. Shionoya: J.
    Luminescence _12/13_ *(Proc. 1975 Intern. Conf. Luminescence, Tokyo)* (1976)
    575.

8) E. Hanamura and M. Inoue: Prog. theor. Phys. Suppl. No.57 (1975)
    p.35.

9) E. Hanamura: *Proc. 12th Intern. Conf. Phys. Semiconductors, Stuttgart,* 1974,
    ed. M. H. Pilkuhn (Teubner, Stuttgart, 1974) p.137.

COMMENT ON "SELF-INDUCED TRANSPARENCY
IN THE EXCITON SYSTEM"

Takeshi Watanabe, Toshiro Wada, and Chuji Horie

Department of Applied Physics
Tohoku University,
Sendai, Japan

ABSTRACT

Starting with the Hamiltonian for periodically arranged spin
operators with a spin 1/2 we derive the equations of motion of the spin
components under influence of an external field.   In order to find the
possibility of the self-induced transparency in this system a real
solution of a pulse-like propagating field is sought.   Following the
previous work by Haken and Schenzle we obtain a solution which contains
the influence of the dispersion and polariton effects of the system.
In the absence of these effects the solution corresponds to a well-known
hyperbolic-secant solution derived by McCall and Hahn.   Discussion is
made on the condition for SIT to be present.

I.   INTRODUCTION

Self-induced transparency (SIT) first studied by McCall and Hahn[1]
is a phenomenon that a short pulse of coherent light above a critical
power threshold propagates through a resonantly absorbing medium with
anomalously low energy loss.   In practice the critical power threshold
is characterized by field strengths sufficiently intense enough for 2π
pulse, and also the initially applied pulse width should be shorter
than any characteristic damping times of the excited state of the medium.

Recently this problem has been studied in connection with the
soliton problem.[2]   From this point of view the SIT in the exciton
system, if exists, is understood as the propagation of distortionless
light-pulse through the dispersive material interacting with light
fields nonlinearly.   In general, in a linear and dispersionless system
a pulse-like traveling wave can always exist.   However, if the dispersion is introduced to this system, then the various Fourier components of the initial wave will propagate at different velocities, and

therefore the initial wave form is not sustained.   On the other hand,

when a nonlinear interaction is introduced in the dispersion-less

system, the pulse energy is continually dissipated into higher harmonic

modes by the harmonic generation.   Therefore, if SIT is present in the

system a delicate balance should exist between these effects.

The study of SIT in the exciton system has been attracting much

theoretical interest,[3~5] and what has been shown so far is that the

polariton effect would prevent and the saturation effect would increase

the possibility of SIT in the exciton system.   The main differences

from the case of two-level atoms are; 1) there exists the energy dis-

persion in the exciton system, and the macroscopic polarization which

appears in Maxwell's equations should be defined in some artificial way,

and 2) the polariton effect would play an essential role in the propa-

gation of light fields in the medium.   The presence of these effects

and the nonlinear polarization indicates that if SIT exists in this

system there should be some critical restrictions on the field intensity

and the forms of the exciton dispersion and interaction with light

fields.

In the present work, instead of considering the exciton explicitly,

we consider a system consisting of two-level atoms locating periodically

at lattice sites l's.   By assigning to each atom the spin angular

momentum operator σ_l with a spin 1/2 and introducing spin-spin interac-

tions between atoms at different lattice sites we derive the equations

of motion for σ_l^+, σ_l^- and σ_l^z.   We define the macroscopic polarization

from these spin operators averaged over a region small compared with the

light wavelength.   The interaction between light fields and the medium

is expressed in terms of σ_l^+, σ_l^- and the circularly polarized vector

potential $\vec{\tilde{A}}_\pm$ of propagating light fields.   After making usual trans-

formations we derive the equation of motion for the amplitude and phase

of $\vec{\tilde{A}}_\pm$ by eliminating all variables associated with the medium.   We,

then, seek for a distortionless propagating pulse solution of light field and derive formally the conditions for SIT to be present in the system.

II.   EQUATIONS OF MOTION

Letting a_1^+, a_1^- and b_1^+, b_1^- be creation and annihilation operators of the ground and excited states of the atom at the 1-th site, we define the spin operators σ_1^+ and σ_1^- by

$$\sigma_1^+ = b_1^+ a_1 \quad \text{and} \quad \sigma_1^- = a_1^+ b_1 \ . \tag{1}$$

With the usual commutation relations σ_1^z is then given by

$$2\sigma_1^z = [\sigma_1^+, \sigma_1^-] = b_1^+ b_1 - a_1^+ a_1 \ . \tag{2}$$

In terms of these spin operators the Hamiltonian of our system is written as[5]

$$H = \Sigma_1 \omega_0 \sigma_1^z + \frac{1}{2} \Sigma'_{11'} J(1-1')(\sigma_1^+ + \sigma_1^-)(\sigma_{1'}^+ + \sigma_{1'}^-)$$

$$- \frac{i}{c} \vec{\mu}\omega_0 \cdot \Sigma_1 \vec{A}(1,t)(\sigma_1^+ - \sigma_1^-), \tag{3}$$

where $J(1-1')$ is the interaction potential between two atoms at the 1-th and 1'-th sites. ω_0 and μ are the energy difference and the atomic dipole moment between the two states, and $\vec{A}(1,t)$ represents the vector potential of the external field.

In order to write the equations of motion in the rotating frame we express the spin operators $\sigma_1^{\pm}$ and the vector potentials $\vec{A}$ as

$$\sigma_1^{\pm} = \tilde{\sigma}_1^{\pm} \exp \pm i(\omega t - k1) \quad ,$$

and $$\vec{A}_1^{\pm} = \vec{\tilde{A}}_{\pm}(1,t) \exp \mp i(\omega t - k1) ,$$

where $\tilde{\sigma}_1^{\pm}$ and $\vec{\tilde{A}}_{\pm}(1,t)$ are slowly varying functions of t and 1 in comparison to the exponential part.   Then the equations of motion of the spin operators are written as

$$\frac{\partial}{\partial t}\tilde{\sigma}_1{}^+ = +i\Delta\omega\tilde{\sigma}_1{}^+ + \frac{1}{c}\omega_0\vec{\mu}_+\vec{A}_-(1,t)\sigma_1{}^z - 2i\sigma_1{}^z\Sigma' J(1-1')e^{-ik(1'-1)}\tilde{\sigma}_{1'}{}^+ \quad, \tag{4}$$

$$\frac{\partial}{\partial t}\tilde{\sigma}_1{}^- = -i\Delta\omega\tilde{\sigma}_1{}^- + \frac{1}{c}\omega_0\vec{\mu}_-\vec{A}_+(1,t)\sigma_1{}^z + 2i\sigma_1{}^z\Sigma' J(1-1')e^{ik(1'-1)}\tilde{\sigma}_{1'}{}^- \quad, \tag{5}$$

$$\frac{\partial}{\partial t}\sigma_1{}^z = -\frac{\omega_0}{2c}\{\vec{\mu}_-\vec{A}_+(1,t)\tilde{\sigma}_1{}^+ + \vec{\mu}_+\vec{A}_-(1,t)\tilde{\sigma}_1{}^-\}$$

$$- i\{\tilde{\sigma}_1{}^+\Sigma' J(1-1')\tilde{\sigma}_{1'}{}^-e^{ik(1'-1)} - \tilde{\sigma}_1{}^-\Sigma' J(1-1')\tilde{\sigma}_{1'}{}^+e^{-ik(1'-1)}\} \quad, \tag{6}$$

where fast oscillating terms are neglected. Here, we have put $\Delta\omega=\omega_0-\omega$ and $\vec{\mu}_\pm=\vec{\mu}_x\pm i\vec{\mu}_y$. From these equations we can obtain the conservation law of the medium by multiplying $\tilde{\sigma}_1{}^-$ from the right of eq. (4) and $\tilde{\sigma}_1{}^+$ from the left of eq. (5) and making use of eq. (6) as follows;

$$\frac{\partial}{\partial t}(\tilde{\sigma}_1{}^+\tilde{\sigma}_1{}^-) + \frac{\partial}{\partial t}(\sigma_1{}^z\sigma_1{}^z - \sigma_1{}^z) = 0 \quad. \tag{7}$$

Concerning the last terms of eqs. (4)-(6) we treat $\tilde{\sigma}_{1'}{}^\pm$ as to be classical quantities and expand them formally as

$$\tilde{\sigma}_{1'}{}^\pm = \tilde{\sigma}_1{}^\pm + \frac{\partial\tilde{\sigma}_1{}^\pm}{\partial\vec{1}}(\vec{1}'-\vec{1}) + \cdots \cdots \quad.$$

Then, the summation over 1' can be approximated by

$$\Sigma'_{1'} J(1-1')\tilde{\sigma}_{1'}{}^\pm e^{\mp ik(\vec{1}'-\vec{1})} \simeq \tilde{J}(\pm\vec{k})\tilde{\sigma}_1{}^\pm - i\vec{v}(\pm\vec{k})\frac{\partial\sigma_1{}^\pm}{\partial\vec{1}} \quad, \tag{8}$$

where the Fourier transform $\tilde{J}(q)$ and the exciton velocity $\vec{v}(\vec{k})$ are defined by

$$J(1-1') = \frac{1}{N}\Sigma_q \tilde{J}(\vec{q}) e^{-i\vec{q}(\vec{1}-\vec{1}')} \quad, \tag{9}$$

and

$$\vec{v}(\pm\vec{k}) = \frac{\partial\tilde{J}(\vec{q})}{\partial\vec{q}}\bigg|_{q=\pm k} \quad.$$

It should be noticed that $\tilde{J}(\vec{k})=\tilde{J}(-\vec{k})$ since $J(\vec{1}-\vec{1}')$ depends on $|\vec{1}-\vec{1}'|$

only. The number of lattice sites N is defined in the next section.

III. TRANSFORMATION TO MACROSCOPIC EQUATION

In order to define the macroscopic polarization we introduce a transformation defined by

$$< \tilde{\sigma} >_{\vec{z}} = \sum_{1 < \Delta V} \Delta (\vec{z} - \vec{1}) \tilde{\sigma}_1 \qquad , \qquad (10)$$

where the summation over 1 is restricted within a volume ΔV. The dimension of ΔV is taken to be smaller than the wavelength of the light field and yet to be large enough so that the energy dispersion of ex- citon still has a meaningful effect. In the present case the trans- formation (10) can be interpreted as defining the average value of $\tilde{\sigma}_1^{\pm}$ or σ_1^z at a point z, since $\tilde{\sigma}_1^{\pm}$ as well as σ_1^z is a slowly varying function of 1. Now the Fourier transform of eq. (9) is to be defined in ΔV, and N in eq. (9) should be the number of lattice sites inside the volume ΔV.

With the above procedure and also the assumption that the average of product can be replaced by the product of the average divided by N the equations of motion (4)-(6) are written as

$$\frac{\partial}{\partial t} < \tilde{\sigma}^+ > = +i \Delta \omega < \tilde{\sigma}^+ > + \frac{1}{c} \omega_0 \vec{\mu}_+ \vec{A}_- < \sigma^z >$$

$$- 2i \frac{1}{N} < \sigma^z > < \tilde{\sigma}^+ > \tilde{J}(\vec{k}) - 2 \frac{1}{N} \vec{v}(\vec{k}) < \sigma^z > \frac{\partial}{\partial \vec{z}} < \tilde{\sigma}^+ > \qquad , \qquad (11)$$

$$\frac{\partial}{\partial t} < \tilde{\sigma}^- > = -i \Delta \omega < \tilde{\sigma}^- > + \frac{1}{c} \omega_0 \vec{\mu}_- \vec{A}_+ < \sigma^z >$$

$$+ 2i \frac{1}{N} < \sigma^z > < \tilde{\sigma}^- > \tilde{J}(-\vec{k}) + 2 \frac{1}{N} \vec{v}(-\vec{k}) \frac{\partial}{\partial \vec{z}} < \tilde{\sigma}^- > < \sigma^z > \qquad , \qquad (12)$$

$$\frac{\partial}{\partial t} < \sigma^z > = - \frac{\omega_0}{2c} \{ \vec{\mu}_- \vec{A}_+ < \tilde{\sigma}^+ > + \vec{\mu}_+ \vec{A}_- < \tilde{\sigma}^- > \} - \frac{1}{N} \vec{v}(\vec{k}) \frac{\partial}{\partial \vec{z}} (< \tilde{\sigma}^+ > < \tilde{\sigma}^- >). \qquad (13)$$

Also the conservation law given by eq. (7) is written as

$$\frac{1}{N}\frac{\partial}{\partial t}(<\tilde{\sigma}^+><\tilde{\sigma}^->) + \frac{1}{N}\frac{\partial}{\partial t}<\sigma^z>^2 - \frac{\partial}{\partial t}<\sigma^z> = 0 \quad . \tag{14}$$

Making use of eq. (10) we derive approximately the wave equation satisfied by $\vec{\tilde{A}}_\pm(z,t)$ as follows;

$$\frac{\partial}{\partial z}\vec{\tilde{A}}_\pm + \frac{\eta^2\omega}{kc^2}\frac{\partial}{\partial t}\vec{\tilde{A}}_\pm \pm \frac{i}{2}\frac{1}{k}(k^2 - \frac{\eta^2}{c^2}\omega^2)\vec{\tilde{A}}_\pm = \frac{2\pi\omega}{kc}\vec{\mu}_\pm <\tilde{\sigma}^\mp> \quad . \tag{15}$$

Here η is the index of refraction of the medium. The third term in the left hand side, which is neglected in the work of Haken and Schenzle, should be included self-consistently when the polariton effect is considered.

To obtain a solution of propagating fields we introduce a variable ζ by $\zeta = t - z/v$, where v is the velocity of the propagating fields and equals c/η in the case of nondispersive media. Then $\vec{\tilde{A}}_\pm$ are given as functions of ζ, and eq. (15) is transformed to

$$\frac{d}{d\zeta}\vec{\tilde{A}}_\pm(\zeta) \mp \frac{i}{2k\Delta}(k^2 - \frac{\eta^2}{c^2}\omega^2)\vec{\tilde{A}}_\pm(\zeta) = -\frac{2\pi\omega}{kc\Delta}\vec{\mu}_\pm <\tilde{\sigma}^\mp>_\zeta \quad , \tag{16}$$

where $\Delta = \frac{1}{v} - \frac{\eta^2\omega}{kc^2}$. For convenience we put

$$\beta = \frac{1}{2k\Delta}(k^2 - \frac{\eta^2\omega^2}{c^2}) \text{ and } \gamma = \frac{2\pi\omega}{kc\Delta} \text{, and transform } \vec{\tilde{A}}_\pm \text{ into}$$

$$\vec{\tilde{A}}_\pm(\zeta) = \vec{A}(\zeta) \exp \pm i(\beta\zeta + \phi(\zeta)) \quad . \tag{17}$$

When β is independent of ζ, it follows immediately from eq. (16) that

$$\frac{d}{d\zeta}(\vec{A} e^{\pm i\phi}) = (\dot{\vec{A}} \pm i\vec{A}\dot{\phi}) e^{\pm i\phi}$$

$$= -\gamma\vec{\mu}_\pm e^{\mp i\beta\zeta} <\tilde{\sigma}^\mp>_\zeta \quad , \tag{18}$$

where we put $d/d = $ $\dot{}$. We use eq. (18) to relate the macroscopic value of $<\tilde{\sigma}^\pm>_\zeta$ with the amplitude of the vector potential. It should be noticed that by introducing ζ the equations of motion (11) ~ (13) are expressed as ordinary differential equations.

IV. SEEK FOR SOLUTION OF SIT

We now seek for a real solution of $\vec{A}(\zeta)$ from eqs. (11)~(14) and (18). First, substitution of eq. (18) into eq. (13) gives

$$\frac{d}{d\zeta} < \sigma^z > = N\lambda^2 \frac{d}{d\zeta} \vec{A}^2 + \xi \frac{d}{d\zeta} Y \quad . \tag{19}$$

Here Y is defined by $Y = (<\tilde{\sigma}^+ > \cdot < \tilde{\sigma}^- >)/N$, which is equal to $N\lambda^4/g^2$ $\times(\dot{\vec{A}}^2 + \vec{A}^2\dot{\phi}^2)$ by means of eq. (18). For convenience we have put* λ^2 $= \omega_o/2N\gamma c$, $g = \omega_o |\vec{\mu}|/2c$ and $\xi = v(\vec{k})/v$. Equation (19) immediately gives

$$\frac{1}{N} < \sigma^z > = \lambda^2\vec{A}^2 + \xi\frac{1}{N} Y + C_1 \quad . \tag{20}$$

Also from eqs. (14) and (18) it follows that

$$\frac{d}{d\zeta} Y = \frac{d}{d\zeta} < \sigma^z > - \frac{1}{N} \frac{d}{d\zeta} <\sigma^z >^2 \quad , \tag{21}$$

and integration of eq. (21) gives

$$Y = < \sigma^z > - \frac{1}{N} (<\sigma^z >)^2 + C_2 \quad . \tag{22}$$

The constant C_2 is chosen so that $Y = A = 0$ at $\zeta = -\infty$. From eqs. (20) and (22), and the definition of Y it follows that

$$(\dot{\vec{A}})^2 = \vec{A}^2[-(\dot{\phi})^2 + \frac{g^2}{\lambda^2} \{ 1-\lambda^2\vec{A}^2-2C_1 \} \{ 1+\xi (1-2\lambda^2\vec{A}^2-2C_1) \}] \quad . \tag{23}$$

Here we retain the terms up to the first order in ξ.

By multiplying eq. (11) by $\gamma\vec{\mu}_-\tilde{A}_+$ and eq. (12) by $\gamma\vec{\mu}_+\tilde{A}_-$, and using eqs. (13) and (18) we obtain for a pulse-shape solution of $\vec{A}$

$$\vec{A}^2(\dot{\phi}+\frac{\beta}{2}) = - \frac{(1-\xi)}{2\lambda^2} \{ \Delta\omega (\frac{<\sigma^z >}{N} - C_1) - \tilde{J}(k) \{(\frac{<\sigma^z >}{N})^2 - c_1^2 \} \} \quad . \tag{24}$$

---

* These quantities have the same meaning as in reference 3), but differ by a factor 2.

Substituting eq. (20) we have for $\dot{\phi}$

$$\dot{\phi} = -\frac{\beta}{2} - \frac{\Delta\omega}{2} \{1-\xi \, (\lambda^2\vec{A}^2 + 2C_1)\} + \frac{1}{2} \, \tilde{J}(\vec{k}) \, \{2C_1 + (1+\xi)\lambda^2\vec{A}^2 \} \ . \tag{25}$$

Finally, by eliminating $\dot{\phi}$ from eqs. (23) and (25), the equation satisfied by $\vec{A}$ is derived as follows;

$$(\dot{\vec{A}})^2 = \vec{A}^2 \, [\, \frac{1}{\tau_0^2} - \frac{1}{\tau_1^2} \, \lambda^2\vec{A}^2 + \frac{1}{\tau_2^2} \, \lambda^4\vec{A}^4 \,] \ . \tag{26}$$

The constants, τ_0, τ_1 and τ_2, are defined as

$$\frac{1}{\tau_0^2} = \frac{g^2}{\lambda^2} \, (1-2C_1)(1+\xi \, (1-2C_1)) - \frac{1}{4} \, (\beta+\Delta\omega -2C_1\tilde{J}(\vec{k}))^2$$

$$-2C_1\xi\Delta\omega \, (\beta+\Delta\omega-2C_1\tilde{J}(\vec{k})) \quad , \tag{27}$$

$$\frac{1}{\tau_1^2} = \frac{g^2}{\lambda^2} \, (1+3\xi \, (1-2C_1)) + \frac{1}{2} \, (\beta+\Delta\omega -2C_1\tilde{J}(\vec{k}))\tilde{J}(\vec{k})$$

$$+ \frac{1}{2} \, \xi\{(\beta+\Delta\omega -2C_1\tilde{J}(\vec{k})) \, (\Delta\omega +\tilde{J}(\vec{k}))-C_1\Delta\omega \, \tilde{J}(\vec{k})\} \quad , \tag{28}$$

$$\frac{1}{\tau_2^2} = 2\xi \, \frac{g^2}{\lambda^2} - \frac{1}{4} \, \{(\tilde{J}(\vec{k}))^2 + 2\xi \, \tilde{J}(\vec{k}) \, (\Delta\omega +\tilde{J}(\vec{k})) \, \} \quad . \tag{29}$$

It is clear from eq. (26) that if τ_0^2 is negative there is no real solution of $\vec{A}$. In order to have a pulse-shape solution of $\vec{A}$ the following conditions are found to be required;

$$D = (\frac{1}{\tau_1^2})^2 - 4 \, (\frac{1}{\tau_0^2})(\frac{1}{\tau_2^2}) > 0, \text{ and } \frac{1}{\tau_1^2} +\sqrt{D} > 0 \ . \tag{30}$$

If these conditions are satisfied, then the solution is given by

$$\lambda^2\vec{A}^2 = \frac{2(1/\tau_0)^2}{(1/\tau_1)^2 +\sqrt{D} \, \cosh\frac{2}{\tau_0} \, (\zeta - \zeta_0)} \quad . \tag{31}$$

It is noticed that if $(1/\tau_2)^2 = 0$ the above solution becomes identical to that obtained for the two-level system.

V. INITIAL CONDITION AND POLARITON EFFECT

We discuss briefly the polariton effect in connection with choice of the integration constant C_1, which is to be determined by the initial condition of the medium. If we assume for simplicity the weak coupling between adjacent atoms as well as between exciton and light fields, we can put $\xi = 0$ and $\tilde{J}(\vec{k}) = 0$. Then, it follows from eq. (20) that

$$\frac{1}{N} < \sigma^z > = \lambda^2 \vec{A}^2 + C_1 \quad . \tag{32}$$

In general the first term of the right-hand-side is much smaller than unity. Also, the value of $<\sigma^z>/N$ is between $-1/2$ and $1/2$ in the present model. If the medium is in the ground state initially, then C_1 is taken to be $-1/2$, and this case with an additional condition $\beta = 0$ corresponds to the case treated by Haken and Schenzle.[3]

In the present formalism the polariton effect is reflected in β. As is usually done in the weak field polariton case, if we put $\dfrac{d}{d\zeta} = 0$ in eqs. (11), (12) and (16), then it follows that

$$\beta = \frac{1}{\Delta\omega} (-2C_1 - \lambda^2 \vec{A}^2) \frac{g^2}{\lambda^2} \quad . \tag{33}$$

It is clear from eq. (33) that the present case for $C_1 = -1/2$ and $\lambda^2 \vec{A}^2 \sim 0$ is essentially similar to the case treated by Hanamura.[4] Furthermore, eq. (33) shows that in the case $C_1 = -1/2$ and $\Delta\omega(=\omega_0-\omega) > 0$ the polariton effect is reduced as the field intensity increases. This corresponds to the giant polariton case, which is numerically analyzed by Inoue.[5] If this is the case, β becomes a function of ζ through $\vec{A}(\zeta)$ and eq. (18) will be modified by replacing $\dot{\phi}$ by $(\dot{\phi} + \dot{\beta}\zeta)$. However, taking $C_1 = -1/2$ in eq. (33) we obtain from eq. (27)

$$\frac{1}{\tau_0^2} = (\frac{3}{2} + \lambda^2 \vec{A}^2) \frac{g^2}{\lambda^2} - \frac{1}{4} \{ \frac{1}{(\Delta\omega)^2} \frac{g^4}{\lambda^4} (1-\lambda^2\vec{A}^2) + (\Delta\omega)^2 \} \quad , \tag{34}$$

which can be either negative or positive depending on g, λ, and $\Delta\omega$, and can find that if $1/\tau_o^2$ is positive for the weak field case, then it also holds for the giant polariton case.

VI. CONCLUSION

We derived formally the solution of a pulse-like propagating wave in the system with the dispersion and polariton effects, although the possibility of its occurrence in a real system would require more profound analyses of the parameters used. It was shown that SIT possibly takes place under some conditions.

REFERENCES

1) S. L. McCall and E. L. Hahn: Phys. Rev. 183 (1969) 457.

2) See, *e.g.*, A. C. Scott *et al.* : Proc. IEEE 61 (1973) 1443.

3) H. Haken and A. Schenzle: Z. Phys. 258 (1973) 231.

4) E. Hanamura: J. Phys. Soc. Japan 37 (1974) 1554.

5) M. Inoue: J. Phys. Soc. Japan 37 (1974) 1560.

COHERENT OPTICAL PULSES IN CRYSTALS

Okikazu Akimoto

The Institute for Solid State Physics,
The University of Tokyo,
Roppongi, Tokyo 106,
Japan

and

Kensuke Ikeda

Department of Physics,
Kyoto University
Kyoto 606,
Japan

ABSTRACT

Polariton effects on the steady propagation of an optical pulse
are studied. It is shown that pulse solutions exist even in the case
where the pulse width is sufficiently long compared with the reciprocal
of the polariton gap frequency. This suggests that the steady propagation may be realized also in crystals.

I.   INTRODUCTION

In the field of quantum optics, a phenomenon named self-induced
transparency (SIT) is well known.[1]   Consider a dilute absorbing medium
such as gaseous atoms or impurity ions in solids. If usual light comes
into the medium at resonant frequency, it is absorbed. However, if the
incoming light pulse is coherent and sufficiently intense, and if its
duration is shorter than the atomic relaxation times, the medium becomes
transparent and the pulse propagates without change of its shape. This
phenomenon has its origin in the nonlinearity of the medium: The
strongly excited atoms do not behave like a harmonic oscillator.

The purpose of this paper is to show theoretically that such a
propagation may be realized also in crystals and to propose a new concept *polariton-soliton* which is complementary to SIT.

The concept of polariton is well known. It is a mixed mode of the radiation field and the polarization wave in matter. In dilute media, this effect of mixing is very small and is smeared out by the inhomogeneous dipole-dipole interaction between atoms. In crystals, however, this effect becomes quite significant due to the high density of electric dipoles. The dispersion relation of the polariton differs markedly from that of the photon, giving rise to a gap in which the mixed mode cannot exist. Moreover, when the polarization wave has a dispersion, two modes having different wave numbers appear for one frequency. These two facts, the polariton formation and the spatial dispersion, must be reflected on the optical pulse progagation. In this paper, confining ourselves to the case of no spatial dispersion, we study the effect of polariton formation on the steady propagation of an optical pulse.

An optical pulse has a finite time width, so that it contains a spread of Fourier components of frequency. If this spread covers the polariton gap completely, the pulse cannot feel the existence of the gap. We call such a pulse *short*. In the opposite case, where the spread of Fourier components of frequency is completely inside or completely outside the polariton gap, we call the pulse *long*.

The orders of magnitude of the reciprocal of the polariton gap frequency are roughly microsecond in gases and picosecond in crystals. This tells us that a nanosecond pulse having nearly resonant frequency, for example, is short for gases, but long for crystals. The existing theories of SIT[1~4] applies only to the case of a short pulse, so it is desirable to construct a unified theory which covers both cases.

II. STEADY PULSE AND FUNDAMENTAL EQUATIONS

The starting point of our theory is the same as that of the existing theories. The crystal is represented by a continuous dielectric medium and the radiation field is treated classically. The steady pulse solution we wish to find out is of the form:

$$\vec{E}(t,z) = \hat{E}(t-z/V) \{\vec{1}\} , \tag{1}$$

$$\vec{P}(t,z) = \frac{1}{2} N\hbar\varkappa \left[u(t-z/V) \{\vec{1}\} + v(t-z/V)\{\vec{2}\} \right] , \tag{2}$$

where

$$\{\vec{1}\} = \vec{x} \cos\theta + \vec{y} \sin\theta , \quad \{\vec{2}\} = -\vec{x} \sin\theta + \vec{y} \cos\theta$$

$$\theta = \omega t - Kz + \phi (t-z/V).$$

The electric field $\vec{E}$ is factorized into two parts: the slowly varying pulse envelope $\hat{E}$ and the rapidly oscillating carrier wave $\{\vec{1}\}$. Steadiness of the propagation is expressed by letting the envelope be a function of only $t-z/V$, where V is the pulse velocity. In the carrier wave, ω is the frequency and K is the wave number at the pulse tail; the relation between them is to be determined in a self-consistent manner. Possible phase modulation ($\dot{\phi} \to 0$ for $t \to \pm\infty$) is also taken into account. The macroscopic polarization $\vec{P}$ is the sum of the in-phase component and the out-of-phase component; each component is also factorized in the same way. N is the dipolar density and $\varkappa$ is the dipole matrix element divided by $\hbar/2$.

The two components of the electric dipole, u and v, and the population inversion w, form a classical pseudo-spin vector, the motion of which is governed by a set of equations:

$$\left. \begin{array}{l} \dot{u} = (\omega - \omega_0 + \dot{\phi}) v , \\[2mm] \dot{v} = - (\omega - \omega_0 + \dot{\phi}) u + \varkappa\hat{E} w , \\[2mm] \dot{w} = - \varkappa\hat{E} v , \end{array} \right\} \tag{3}$$

which is called the optical Bloch equations. Here, ω_0 is the resonant frequency of the medium and all the relaxation times have been assumed to be infinity. Our task is to solve this set of equations together with the Maxwell equation for $\hat{E}$ and ϕ simultaneously. The method of solution is a power-series expansion in which all the quantities in the fundamental equations are expanded in terms of a small parameter related to the pulse width. The small parameter will be chosen in different ways depending on what kind of pulse we treat, a short pulse or a long pulse.

III. BEHAVIOR OF PULSE TAIL

Before solving our equations, let us determine the dispersion relation at the pulse tail. In order to see the behavior of the pulse tail where the excitation is very weak, it is sufficient to consider linearized version of our equations. If one solves the linearized equations, which are given by setting w equal to -1 corresponding to the atomic ground state, one can obtain exponential solutions $E \propto$ $\exp \pm(t-z/V)/\tau$ in general, besides the plane wave solution which corresponds to the usual polariton. Such a divergent solution is physically meaningless as itself in infinite media, and is usually left out of consideration. However, the tail of a steadily propagating pulse is not a plane wave but a growing (or decaying) wave, so that its behavior should be described at least locally by such an exponential solution. When the nonlinearity is taken into account, the divergence of this solution is suppressed and a pulse is formed; at the same time, the dispersion relation around the pulse peak is modulated through non-zero $\dot{\phi}$. The growth (or decay) constant τ, which is an integral constant, then gives a measure of the pulse width, so we call it the pulse width hereafter. The wave number K and the pulse velocity V determined as functions of ω and τ from the linearized equations are as follows:

$$\left(\frac{cK}{\omega}\right)^2 = \frac{1}{2} \left\{ 1 - \frac{\Delta}{\Delta^2 + \lambda^2} + \sqrt{\frac{(\Delta-1)^2 + \lambda^2}{\Delta^2 + \lambda^2}} \right\} , \tag{4}$$

$$\left(\frac{c}{v}\right)^2 = \frac{(\omega\tau)^2}{2}\left\{-1 + \frac{\Delta}{\Delta^2+\lambda^2} + \sqrt{\frac{(\Delta-1)^2+\lambda^2}{\Delta^2 + \lambda^2}}\right\}, \qquad (5)$$

where Δ and λ are the frequency difference and the reciprocal pulse width scaled by the polariton gap frequency, *i.e.*

$$\Delta = \frac{\omega-\omega_0}{2\pi N\hbar\varkappa^2} \quad, \qquad \lambda = \frac{\tau^{-1}}{2\pi N\hbar\varkappa^2} \quad. \qquad (6)$$

In eqs.(4) and (5), $\lambda \ll \omega\tau$ has been assumed.

The τ-dependent dispersion relation given by eq.(4) is plotted in Fig.1. When λ is large, namely in the case of a short pulse, the dispersion relation is close to that of the photon. When λ becomes small, *i.e.* when the pulse width becomes long, the dispersion relation approaches that of the polariton. An important fact seen in Fig.1 is

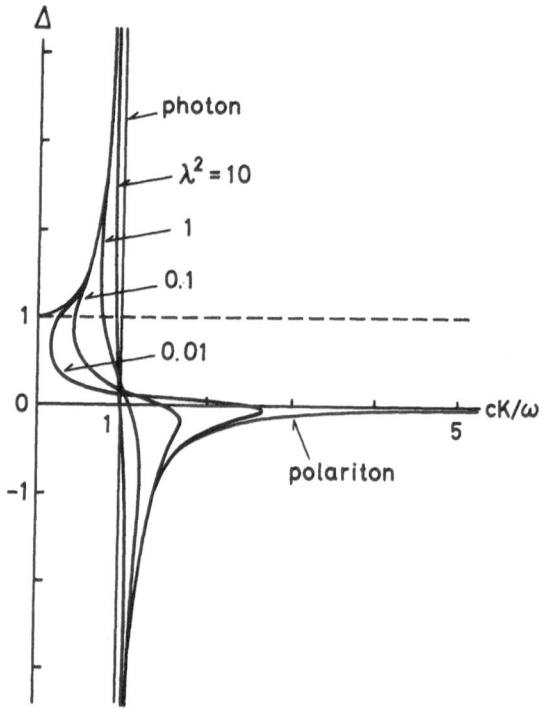

Fig.1. Dispersion relation at the pulse tail. Δ and λ are the frequency difference $\omega -\omega_0$ and the reciprocal of the pulse width τ^{-1}, respectively, scaled by the polariton gap frequency. The polariton gap corresponds to $0 < \Delta < 1$.

that K remains real even inside the polariton gap. This fact means that a growing (or decaying) wave can propagate inside the gap, although a plane wave cannot. It suggests that a long pulse having an exponential tail can also propagate inside the gap.

IV. PULSE SOLUTIONS

Now, we return to our nonlinear equations. By choosing a dimensionless time $(t-z/V)/\tau$ as the variable and using the inequality
$(\omega\tau)^{-1} \ll 1$, our equations are reduced to five ordinary differential
equations for $\hat{E}/2\pi N\hbar\varkappa$, ϕ, u, v and w, the coefficients of which are functions of λ and Δ. We expand all the coefficients and the functions
to be determined in power-series of λ/Δ or $1/\lambda$, according to the long
pulse $(\lambda \ll |\Delta|)$ or the short pulse $(\lambda \gg 1$ and $\lambda \gg |\Delta|)$, respectively.
By this expansion, the equations are reduced to simpler forms, which
can be solved easily.

The results are summarized as follows. Behaviors of the pulse
strongly depend on ω and τ. The short pulse is the same as the usual
SIT pulse, and is not affected by the existence of the polariton gap.
The pulse envelope has a hyperbolic secant shape, and the population is
completely inverted at the pulse peak; *i.e.*

$$\varkappa\hat{E} = 2\tau^{-1} \operatorname{sech}\left[(t-z/V)/\tau\right],$$

$$w = -1 + 2 \operatorname{sech}^2\left[(t-z/V)/\tau\right].$$

$$\tag{7}$$

The wave number K is close to that of the photon, and the group velocity
V is given by $c/V \sim 1+(\omega\tau/2\lambda)$ and is smaller than the light velocity.
A principal part of the polarization $\vec{P}$ is given by the out-of-phase
(absorptive) component v of the dipole moment; that is, the inequality
$u \ll v$ holds in this case.

The long pulse, on the other hand, shows different behaviors outside and inside the gap. Outside the gap $(\Delta > 1$ or $\Delta < 0)$, the pulse
has a hyperbolic secant shape, but as the pulse width becomes longer,
both the envelope and the population inversion tend to zero; their
explicit forms are

$$\varkappa\hat{E}= \sqrt{\frac{4\Delta-3}{\Delta-1}}\ \tau^{-1}\ \mathrm{sech}\left[\ (t-z/V)/\tau\ \right],$$

$$w =-1 + \frac{1}{2(\omega-\omega_0)^2}\ (\varkappa\hat{E})^2 \tag{8}$$

When $\tau\to\infty$, the wave number and the pulse velocity approach those of the polariton. Furthermore, in contrast with the case of a short pulse, a principal part of the polarization is given by the in-phase (dispersive) component u; that is $u\gg v$. This means that the electric dipoles adiabatically follow the electric field. Because of these facts, the pulse may be called a polariton-soliton.

The long pulse can propagate with real K and V also inside the gap. In this case, however, the envelope has no longer a hyperbolic secant shape, and does not vanish even in the limit of $\tau\to\infty$;

$$\varkappa\hat{E} = f(t-z/V;\ \omega-\omega_0)\cdot 2\pi N\hslash\varkappa^2 ,$$

$$w =-(\omega-\omega_0)\left[(\varkappa E)^2 + (\omega-\omega_0)^2\right]^{-1/2} . \tag{9}$$

Here, f is a dimensionless real function of the order of unity. Near the upper edge of the polariton gap, f is approximated by a hyperbolic secant, *i.e.* $f\sim 2\sqrt{1-\Delta}\ \mathrm{sech}\left[(t-z/V)/\tau\right]$, while near $\omega=\omega_0$, it is close to a period of cosine $\{1+\cos\left[\sqrt{\Delta}\ (t-z/V)/\tau\right]\}$ $(-\pi <\sqrt{\Delta}(t-z/V)/\tau <\pi)$ accompanied with an exponential tail. In the limit of $\tau\to\infty$, the pulse velocity and the wave number tend to zero, while the spatial width of the pulse V remains finite. This means that the pulse becomes a sort of standing wave without spatial oscillation in this limit. If we solve the linearized equations by assuming constant $\hat{E}$ inside the gap, we can obtain an exponentially decaying (or growing) solution without spatial oscillation, because the wave number is purely imaginary there. Such a solution can appear just inside the surface of the medium irradiated by light. The tail of the pulse solution we have obtained above continues to this type of linear solutions.

In the case of intermediate pulse width, our equations are not solved analytically, so they have to be treated numerically.

All the pulse solutions are classified in Fig.2.   In this figure,

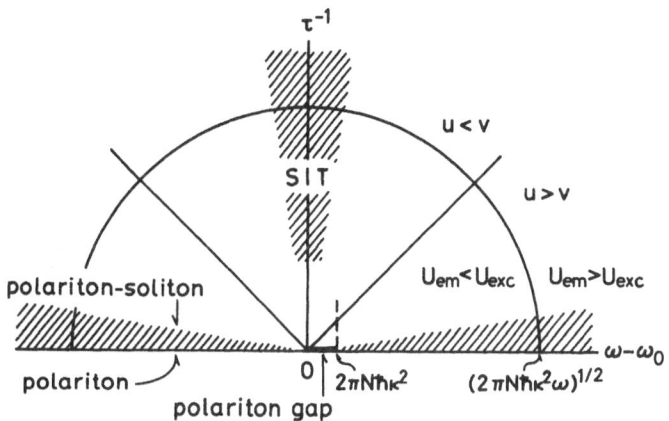

Fig.2. Classification of the pulse solutions.
See the text.

the abscissa is the optical frequency measured from the resonant frequency, and the ordinate the reciprocal of the pulse width.   The usual polariton corresponds to the line $\tau^{-1} = 0$.   Above the straight line $\tau^{-1} = |\omega - \omega_0|$, the out-of-phase component v of the dipole moment is larger than the in-phase component u; the pulses in this region may be called SIT-like.   Below this line, the in-phase component is larger; the pulses may be called polariton-like.   The usual SIT and the polariton-soliton we have studied above are indicated by the hatched regions. Outside the circle shown in Fig.2, the energy density of electromagnetic field $E^2/4\pi$, which is denoted by U_{em} in Fig.2, is larger than the energy density of excitation $N\hbar\omega(w+1)$, which is denoted by U_{exc} in the figure; inside the circle, it is reversed.

In conclusion, we have shown that, besides the usual SIT pulse, steady pulse solutions exist also in the case where the pulse width is sufficiently long compared with the reciprocal of the polariton gap frequency, and that such a long pulse having frequency outside the gap behaves as a polariton-soliton. Our conclusion suggests that the steady pulse propagation may be realized also in crystals, though it has been obtained by using a simplified model without taking into consideration realistic conditions such as spatial dispersion, relaxation times, surface effects *etc.* In view of recent remarkable development of picosecond pulse technique, it is hoped to observe such a propagation in near future.

REFERENCES

1) S. L. McCall and E. L. Hahn: Phys. Rev. 183 (1969) 457.

2) G. L. Lamb, Jr.: Rev. mod. Phys. 43 (1971) 99.

3) H. Haken and A. Schenzle: Z. Physik 258 (1973) 231.

4) R. A. Marth, D. A. Holmes and J. H. Eberly: Phys. Rev. A9 (1974) 2733; Z. Bialynicka-Birula, Phys. Rev. A10 (1974) 999.

AUTHOR AND PARTICIPANT INDEX

Name	Mailing Address	Page	Remarks
Goto, T.	Res. Inst. for Iron, Steel & Other Metals Tohoku University 1-1 Katahira 2-Chome Sendai Japan 980	327	P, OR . C
Grun, J.B.	Laboratoire de Spectroscopie et d'Optique du Corps Solide 5 rue de l'Université 67000 Strasbourg France	49	P A, S
Gurnee, M.N.		219	C
Haller, E.E.		270	C
Hanamura, E.	Institute for Solid State Physics University of Tokyo 7-22-1, Roppongi, Minato-ku, Tokyo Japan 106	25 304	P, OR A, S C
Hasegawa, H.	Dept. of Physics Kyoto University Kitashirakawa, Sakyo-ku, Kyoto Japan 606		P
Haug, H.	Institute für Theoretische Physik der Universität Frankfurt / Main 6 Frankfurt / M. 1 Robert-Mayer Straße 8-10 Fed. Rep. Germany	124	P A, S
Henneberger, F.		115	C
Hensel, J.C.	Bell Laboratories Murray Hill, New Jersey 07974 U.S.A.	166	P A, S
Horie, C.	Dept. of Applied Physics Tohoku University Aobayama, Sendai Japan 980	201 366	P, OR C, S C
Hvam, J.M.	Institute of Physics Odense University DK-5000 Odense Denmark	246	P A, S
Ikeda, K.		376	C
Inoue, M.	Institute for Solid State Physics University of Tokyo 7-22-1 Roppongi, Minato-ku, Tokyo Japan 106	25	P C
Ito, T.	Physics Dept. Tohoku University Aobayama, Sendai Japan 980		P
Jeffries, C.D.	Physics Dept. University of California Berkeley, Calif. 94720 U.S.A.	270	P C, S

Name	Mailing Address	Page	Remarks
Kamimura H.	Dept. of Physics University of Tokyo 7-3-1, Hongo, Bunkyo-ku, Tokyo Japan 113	237	P C
Kalia, R. K.		187	C
Kasuya, T.			OR
Kobayashi, K.	Institute for Solid State Physics University of Tokyo 7-22-1 Roppongi, Minato-ku, Tokyo Japan 106	80	P C, S
Kobayashi, M.	Dept. of Material Physics Osaka University 1-1 Machikaneyama-cho, Toyonaka Osaka Japan 560	295	P A
Kubo, R.	Dept. of Physics University of Tokyo 7-3-1, Hongo, Bunkyo-ku, Tokyo Japan 113	304	P A, S
Kuramoto, Y.	Dept. of Applied Physics Tohoku University Aobayama, Sendai Japan 980	237	P A, S
Kuroda, H.	Electrotechnical Laboratory 5-4-1 Mukodai-machi, Tanashi, Tokyo Japan 188		P
Kuroda, N.	Res. Inst. for Iron, Steel & Other Metals Tohoku University 1-1 Katahira 2-Chome, Sendai Japan 980	327	P C
Kushida, T.	Institute for Solid State Physics University of Tokyo 7-22-1 Roppongi, Minato-ku, Tokyo Japan 106	105 340	P C A, S
Lévy, R.		49	C
Martin, R.W.	Physikalisches Institut Teilinstitut 4 Universität Stuttgart 7 Stuttgart 80-Vaihingen Pfaffenwaldring 57 Fed. Rep. Germany		P
Meyer, J.R.		219	C
Mercier, A.		320	A
Miyamoto, S.	Broadcasting Science Research Lab. NHK 1-10-11 Kinuta, Setagaya-ku, Tokyo Japan 157		P
Mooser, E.		320	C

Name	Mailing Address	Page	Remarks
Morigaki, K.	Institute for Solid State Physics University of Tokyo 7-22-1 Roppongi, Minato-ku, Tokyo Japan 106	253	P C
Morimoto, M.	Physics Dept. Tohoku University Aobayama, Sendai Japan 980	230	P A,S
Morita, A.	Physics Dept. Tohoku University Aobayama, Sendai Japan 980	230	P,OR C
Moriya, T.	Electrotechnical Laboratory 5-4-1 Mukodai-machi, Tanashi Tokyo Japan 188	340	P C
Müller, G.O.	Akademie der Wissenschaften der DDR Zentralinstitut für Elektronenphysik 108 Berlin, Mohrenstraße 40/41 DDR	349 115	P A,S S
Mysyrowicz, A.	Groupe de Physique des Solides de l'Ecole Normale Supérieure Tour 23, 2, place Jussieu 75221, Paris, Cedex 05 France	57	P A,S
Nagaoka, Y.	Dept. of Physics Nagoya University Chikusa-ku, Nagoya Japan 464	137	P A,S
Nagasawa, N.	Physics Dept. Tohoku University Aobayama, Sendai Japan 980	1	P C
Nagashima, T.	Dept. of Applied Physics Tohoku University Aobayama, Sendai Japan 980	201	P A
Nakahara, J.		80	A
Nakajima, S.	Institute for Solid State Physics University of Tokyo 7-22-1 Roppongi, Minato-ku, Tokyo Japan 106	130	P A,S
Nakamura, A.	Institute for Solid State Physics University of Tokyo 7-22-1 Roppongi, Minato-ku, Tokyo Japan 106	253	P A,S
Nakaoka, K.		327	C
Namba, S.		98	C
Narita, S.	Dept. of Material Physics Osaka University 1-1 Machikaneyama-cho, Toyonaka, Osaka, Japan 560	295	P C,S

Name	Mailing Address	Page	Remarks
Nishina, Y.	Res. Inst. for Iron, Steel & Other Metals Tohoku University 1-1 Katahira 2-Chome, Sendai Japan 980	327	P,OR A,S
Ohyama, T.		262 288	C C
Oka, Y.	Institute for Solid State Physics University of Tokyo 7-22-1 Roppongi, Minato-ku, Tokyo Japan 106	105	P A,S
Ostertag, E.		49	C
Otsuka, E.	Dept. of Physics College of General Education Osaka University 1-1 Machikaneyama-cho, Toyonaka, Osaka, Japan 560	262 288	P,OR C A,S
Planel, R.		89	A
Port, H.	Physikalisches Institut Universität Stuttgart 7 Stuttgart 80, Pfaffenwaldring 57 Fed. Rep. Germany	49	P C
Rice, T.M.	Bell Laboratories Murray Hill, New Jersey 07974 U.S.A.	144	P A,S
Rösler, M.		349	C
Rößler, U.	Fachbereich Physik Universität Regensburg 84 Regensburg, Universitätsstraße 31 Fed. Rep. Germany		P
Runclman, W.A.	Dept. of Solid State Physics Australian National University P. O. Box 4, Canberra A.C.T. 2600 Australia		P
Saito, H.	Institute for Solid State Physics University of Tokyo 7-22-1 Roppongi, Minato-ku, Tokyo Japan 106		P
Sanada, T.	Physics Dept. College of General Education Osaka University 1-1 Machikaneyama-cho, Toyonaka, Osaka, Japan 560	262 288	P A,S C
Saner, R.	Physikalisches Institut Teilinstitut 4 Universität Stuttgart Stuttgart 80-Vaihingen Pfaffenwaldring 57 Fed. Rep. Germany		P S
Schenzle, A.	IBM Research Laboratory Tradewinds Drive 5 San Jose, California 95123 U.S.A.		P

Name	Mailing Address	Page	Remarks
Segawa, Y.	Institute for Physical and Chemical Research 2-1 Hirosawa, Wako, Saitama Japan 351	98	P A,S
Shindo, K.		230	C
Shionoya, S.	Institute for Solid State Physics University of Tokyo 7-22-1 Roppongi, Minato-ku, Tokyo Japan 106	358	P,OR A,S
Singwi, K.S.		187	C
Staehli, J.C.		270	C
Suga, S.	Institute for Solid State Physics University of Tokyo 7-22-1 Roppongi, Minato-ku, Tokyo Japan 106		P
Sugano, S.	Institute for Solid State Physics University of Tokyo 7-22-1 Roppongi, Minato-ku, Tokyo Japan 106		P
Takagahara, T.		304	C
Tandon. U.S.	Physics Dept. University of Allahabad Allahabad 211002 India		P
Toyozawa, Y.	Research Institute for Solid State Physics University of Tokyo 7-22-1 Roppongi, Minato-ku, Tokyo Japan 106		P S
Treusch, J.	Lehrstuhl für Theoretische Physik II Universität Dortmund 16 Dortmund-Hombruch Postfach 500 Fed. Rep. Germany		P
Ueta, M.	Physics Dept. Tohoku University Aobayama, Sendai Japan 980	1	P,OR A,S
Umeno, M.	Dept. of Electronics Nagoya University Chikusa-ku, Nagoya Japan 464		P
Vashishta, P.	Solid State Science Division Argonne National Laboratory Argonne, Illinois 60201 U.S.A.	187	P A,S
Voigt, J.	Sektion Physik der Humboldt-Universität zu Berlin Bereich 05, 102 Berlin Neue Schönhauser Str. 20 DDR	115	A

Name	Mailing Address	Page	Remarks
Voitchovsky, J.P.		320	C
Voos, M.		177	C
Vu Duy Phach, H.		49	C
Wada, T.		366	C
Watanabe, T.	Dept. of Applied Physics Tohoku University Aobayama, Sendai Japan 980	366 201	P,OR A,S C
Weber, H.H.		349	C
Westervelt, R.M.		270	A
Yashiro, M.		327	C
Zimmermann, R.		349	C

Selected Issues from

Lecture Notes in Mathematics

SPRINGER TRACTS IN MODERN PHYSICS

Ergebnisse der exakten
Naturwissenschaften

Editor: G. Höhler

Associate Editor:
E.A. Niekisch

Editorial Board:
S. Flügge, J. Hamilton,
F. Hund, H. Lehmann,
G. Leibfried, W. Paul

Springer-Verlag
Berlin
Heidelberg
New York

Volume 66

30 figures. III, 173 pages. 1973
ISBN 3-540-06189-4

Quantum Statistics

in Optics and Solid-State Physics

R. Graham: Statistical Theory of Instabilities in Stationary Nonequilibrium Systems with Applications to Lasers and Nonlinear Optics.
F. Haake: Statistical Treatment of Open Systems by Generalized Master Equations.

Volume 67

III, 69 pages. 1973
ISBN 3-540-06216-5

S. Ferrara, R. Gatto, A. F. Grillo:

Conformal Algebra in Space-Time

and Operator Product Expansion

Introduction to the Conformal Group in Space-Time. Broken Conformal Symmetry. Restrictions from Conformal Covariance on Equal-Time Commutators. Manifestly Conformal Covariant Structure of Space-Time. Conformal Invariant Vacuum Expectation Values. Operator Products and Conformal Invariance on the Light-Cone. Consequences of Exact Conformal Symmetry on Operator Product Expansions. Conclusions and Outlook.

Volume 68

77 figures. 48 tables. III, 205 pages. 1973
ISBN 3-540-06341-2

Solid-State Physics

D. Schmid: Nuclear Magnetic Double Resonance — Principles and Applications in Solid-State Physics.
D. Bäuerle: Vibrational Spectra of Electron and Hydrogen Centers in Ionic Crystals.
J. Behringer: Factor Group Analysis Revisited and Unified.

Volume 69

13 figures. III, 121 pages. 1973
ISBN 3-540-06376-5

Astrophysics

G. Börner: On the Properties of Matter in Neutron Stars.
J. Stewart, M. Walker: Black Holes: the Outside Story.

Volume 70

II, 135 pages. 1974
ISBN 3-540-06630-6

Quantum Optics

G. S. Agarwal: Quantum Statistical Theories of Spontaneous Emission and their Relation to Other Approaches.

Volume 71

116 figures. III, 245 pages. 1974
ISBN 3-540-06641-1

Nuclear Physics

H. Überall: Study of Nuclear Structure by Muon Capture.
P. Singer: Emission of Particles Following Muon Capture in Intermediate and Heavy Nuclei.
J. S. Levinger: The Two and Three Body Problem.

Volume 72

32 figures. II, 145 pages. 1974
ISBN 3-540-06742-6

D. Langbein:

Theory of Van der Waals Attraction

Introduction. Pair Interactions. Multiplet Interactions. Macroscopic Particles. Retardation. Retarded Dispersion Energy. Schrödinger Formalism. Electrons and Photons.

Volume 73

110 figures. VI, 303 pages. 1975
ISBN 3-540-06943-7

Excitons at High Density

Editors: H. Haken, S. Nikitine
Biexcitons. Electron-Hole Droplets. Biexcitons and Droplets. Special Optical Properties of Excitons at High Density. Laser Action of Excitons. Excitonic Polaritons at Higher Densities.

Volume 74

75 figures. III, 153 pages. 1974
ISBN 3-540-06946-1

Solid-State Physics

G. Bauer: Determination of Electron Temperatures and of Hot Electron Distribution Functions in Semiconductors.
G. Borstel, H. J. Falge, A. Otto: Surface and Bulk Phonon-Polaritons Observed by Attenuated Total Reflection.